CW00920811

# Geometric Modeling in Probability and Statistics

Ovidiu Calin • Constantin Udrişte

# Geometric Modeling in Probability and Statistics

Ovidiu Calin
Department of Mathematics
Eastern Michigan University
Ypsilanti, MI, USA

Constantin Udrişte
Faculty of Applied Sciences
Department of Mathematics-Informatics
University Politehnica of Bucharest
Bucharest, Romania

Additional material to this book can be downloaded from http://extras.springer.com

ISBN 978-3-319-07778-9 ISBN 978-3-319-07779-6 (eBook)
DOI 10.1007/978-3-319-07779-6
Springer Cham Heidelberg New York Dordrecht London

Library of Congress Control Number: 2014941797

Mathematics Subject Classification (2010): 53B20, 53C21, 53C22, 60D99, 62B10



Printed on acid-free paper

Springer is part of Springer Science+Business Media (www.springer.com)

# Preface

Statistical manifolds are geometric abstractions used to model information, their field of study belonging to Information Geometry, a relatively recent branch of mathematics, that uses tools of differential geometry to study statistical inference, information loss, and estimation.

This field started with the differential geometric study of the manifold of probability density functions. For instance, the set of normal distributions

$$p(x;\mu,\sigma) = \frac{1}{\sqrt{2\pi}\sigma} e^{-\frac{(x-\mu)^2}{2\sigma^2}}, \qquad x \in \mathbb{R},$$

with $(\mu, \sigma) \in \mathbb{R} \times (0, +\infty)$, can be considered as a two-dimensional surface. This can be endowed with a Riemannian metric, which measures the amount of information between the distributions. One of these possibilities is to consider the Fisher information metric. In this case, the distribution family $p(x;\mu,\sigma)$ becomes a space of constant negative curvature. Therefore, any normal distribution can be visualized as a point in the Poincaré upper-half plane.

In a similar way, we shall consider other parametric model families of probability densities that can be organized as a differentiable manifold embedded in the ambient space of all density functions. Every point on this manifold is a density function, and any curve corresponds to a one-parameter subfamily of density functions. The distance between two points (i.e., distributions), which is measured by the Fisher metric, was introduced almost simultaneously by C. R. Rao and H. Jeffreys in the mid-1940s. The role of differential geometry in statistics was first emphasized by Efron in 1975, when he introduced the concept of statistical curvature. Later Amari used the tools of differential geometry to develop this idea into an elegant representation of Fisher's theory of information loss.

A fundamental role in characterizing statistical manifolds is played by two geometric quantities, called dual connections, which describe the derivation with respect to vector fields and are interrelated in a duality relation involving the Fisher metric. The use of dual connections leads to several other dual elements, such as volume elements, Hessians, Laplacians, second fundamental forms, mean curvature vector fields, and Riemannian curvatures. The study of dual elements and the relations between them constitute the main direction of development in the study of statistical manifolds.

Even if sometimes we use computations in local coordinates, the relationships between these geometric quantities are invariant with respect to the selection of any particular coordinate system. Therefore, the study of statistical manifolds provides techniques to investigate the intrinsical properties of statistical models rather than their parametric representations. This invariance feature made statistical manifolds useful in the study of information geometry.

We shall discuss briefly the relation of this book with other previously published books on the same or closely related topic.

One of the well-known textbooks in the field is written by two of the information geometry founders, Amari and Nagaoka [8], which was published first time in Japan in 1993, and then translated into English in 2000. This book presents a concise introduction to the mathematical foundation of information geometry and contains an overview of other related areas of interest and applications. Our book intersects with Amari's book over its first three chapters, i.e. where it deals with geometric structures of statistical models and dual connections. However, the present text goes in much more differential geometric detail, studying also other new topics such as relative curvature tensors, generalized shape operators, dual mean curvature vectors, and entropy maximizing distributions. However, our textbook does not deal with any applications in the field of statistic inference, testing, or estimation. It contains only the analysis of statistical manifolds and statistical models. The question of how the new concepts introduced here apply to other fields of statistics is still under analysis.

Another book of great inspiration for us is the book of Kass and Vos [49], published in 1997. Even if this book deals mainly with the geometrical foundations of asymptotic inference and information loss, it does also contain important material regarding statistical manifolds and their geometry. This challenge is developed more geometrically in the present book than in the former.

# Overview

This book is devoted to a specialized area, including Informational Geometry. This is a field that is increasingly attracting the interest of researchers from many different areas of science, including mathematics, statistics, geometry, computer science, signal processing, physics, and neuroscience. It is the authors' hope that the present book will be a valuable reference for researchers and graduate students in one of the aforementioned fields.

The book is structured into two distinct parts. The first one is an accessible introduction to the theory of statistical models, while the second part is devoted to an abstract approach of statistical manifolds.

## Part I

The first part contains six chapters and relies on the understanding of the differential geometry of probability density functions viewed as surfaces.

The first two chapters present important density functions, which will offer tractable examples for later discussions in the book. The remaining four chapters devote to the geometry of entropy, which is a fundamental notion in informational geometry. The readers without a strong background in differential geometry can still follow. This part itself can be read alone as an introduction to information geometry.

Chapter 1 introduces the notion of statistical model, which is a space of density functions, and provides the exponential and mixture families as distinguished examples. The Fisher information is defined together with two dual connections of central importance to the theory. The skewness tensor is also defined and computed in certain particular cases.

Chapter 2 contains a few important examples of statistical models for which the Fisher metric and geodesics are worked out explicitly. This includes the case of normal and lognormal distributions, and also the gamma and beta distribution families.

Chapter 3 deals with an introduction to entropy on statistical manifolds and its basic properties. It contains definitions and examples, an analysis of maxima and minima of entropy, its upper and lower bounds, Boltzmann–Gibbs submanifolds, and the adiabatic flow.

Chapter 4 is dedicated to the Kullback–Leibler divergence (or relative entropy), which provides a way to measure the relative entropy between two distributions. The chapter contains explicit computations and fundamental properties regarding the first and second variations of the cross entropy, its relation with the Fisher information matrix and some variational properties involving Kullback–Leibler divergence.

Chapter 5 defines and studies the concept of informational energy on statistical models, which is a concept analogous to kinetic energy from physics. The first and second variations are studied and uncertainty relations and some thermodynamics laws are presented.

Chapter 6 discusses the significance of maximum entropy distributions in the case when the first $N$ moments are given. A distinguished role is played by the case when $N = 1$ and $N = 2$, cases when some explicit computations can be performed. A definition and brief discussion of Maxwell–Boltzmann distributions is also made.

## Part II

The second part is dedicated to a detailed study of statistical manifolds and contains seven chapters. This part is an abstractization of the results contained in Part I. Instead of statistical models, one considers here differentiable manifolds, and instead of the Fisher information metric, one takes a Riemannian metric. Thus, we are able to carry the ideas from the theory of statistical models over to Riemannian manifolds endowed with a dualistic structure defined by a pair of torsion-less dual connections.

Chapter 7 contains an introduction to the theory of differentiable manifolds, a central role being played by the Riemannian manifolds. The reader accustomed with the basics of differential geometry may skip to the next chapter. The role of this chapter is to accommodate the novice reader with the language and objects of differential geometry, which will be further developed throughout the later chapters.

A formulation of the dualistic structure is given in Chap. 8. This chapter defines and studies general properties of dual connections, relative torsion tensors and curvatures, $\alpha$-connections, the skewness and difference tensors. It also contains an equivalent construction of statistical manifolds starting from a skewness tensor.

Chapter 9 describes how to associate a volume element with a given connection and discusses the properties of dual volume elements, which are associated with a pair of dual connections. The properties of $\alpha$-volume elements are provided with the emphasis on the relation with the Lie derivative and vector field divergence. An explicit computation is done for the distinguished examples of exponential and mixture cases. A special section is devoted to the study of equiaffine connections, i.e. connections which admit a parallel $n$-volume form. The relation with the statistical manifolds of constant curvature is also emphasized.

Chapter 10 deals with a description of construction and properties of dual Laplacians, which are Laplacians defined with respect to a pair of dual connections. An $\alpha$-Laplacian is also defined and studied. The relation with the dual volume elements is also emphasized. The last part of the chapter deals with trace of the metric tensor and its relation to Laplacians.

The construction of statistical manifolds starting from contrast functions is described in Chap. 11. The construction of a dualistic structure (Riemannian metric and dual connections) starting from a contrast function is due to Eguchi [38, 39, 41]. Contrast functions are also known in the literature under the name of divergences, a denomination we have tried to avoid here as much as we could.[1]

Chapter 12 presents a few classical examples of contrast functions, such as Bregman, Chernoff, Jefferey, Kagan, Kullback–Leibler, Hellinger, and $f$-divergence, and their values on a couple of examples of statistical models.

The study of statistical submanifolds, which are subsets of statistical manifolds with a similar induced structure, is done in Chap. 13. Many classical notions, such as second fundamental forms, shape operator, mean curvature vector, and Gauss–Codazzi equations, are presented here from the dualistic point of view. We put our emphasis on the relation between dual objects; for instance, we find a relation between the divergences of dual mean curvature vector fields and the inner product of these vector fields.

[1] A divergence in differential geometry usually refers to an operator acting on vector fields.

The present book follows the line started by Fisher and Efron and continued by Eguchi, Amari, Kaas, and Vos. The novelty of this work, besides presentation, can be found in Chaps. 5, 6, 9, and 13. The book might be of interest not only to differential geometers but also to statisticians and probabilists.

Each chapter ends with a section of proposed problems. Even if many of the problems are left as exercises from the text, there are a number of problems, aiming to deepen the reader's knowledge and skills.

It was our endeavor to make the index as complete as possible, containing all important concepts introduced in definitions. We also provide a list of usual notations of this book. It is worthy noting that for the sake of simplicity and readability, we employed the Einstain summation convention, i.e., whenever an index appears in an expression once upstairs and once downstairs, the summation is implied.

The near flowchart will help the reader navigate through the book content more easily.

## Software

The book comes with a software companion, which is an Information Geometry calculator. The software is written in *C#* and runs on any PC computer (not a Mac) endowed with .NET Framework. It computes several useful information geometry measures for the most used probability distributions, including entropy, informational energy, cross entropy, Kullback–Leibler divergence, Hellinger distance, and Chernoff information of order $\alpha$. The user instructions are included in Appendix A. Please visit `http://extras.springer.com` to download the software.

## Bibliographical Remarks

Our presentation of differential geometry of manifolds, which forms the scene where the information geometry objects exist, is close in simplicity to the one of Millman and Parker [58]. However, a more advanced and exhaustive study of differential geometry can be found in Kobayashi and Nomizu [50], Spivak [78], Helgason [44], or Auslander and MacKenzie [9]. For the basics theory of probability distributions the reader can consult almost any textbook of probability and statistics, for instance Wackerly et al. [85].

The remaining parts of the book are based on fundamental ideas inspired from the expository books of Amari and Nagaoka [8] and Kass and Vos [49]. Another important source of information for the reader is the textbook of Murrey and Rice [60]. While the aforementioned references deal with a big deal of statistical inference, our book contains mainly a pure geometrical approach.

One of the notions playing a central role throughout the theory is the Fisher information, which forms a Riemannian metric on the space of probability distributions. This notion was first introduced almost simultaneously by Rao [70] and Jeffreys [46] in the mid-1940s, and continued to be studied by many researchers such as Akin [3], Atkinson and Mitchell [4], James [45], Oller [63], and Oller et al. [64, 65].

The role played by differential geometry in statistics was not fully acknowledged until 1975 when Efron [37] first introduced the concept of statistical curvature for one-parameter models and emphasized its importance in the theory of statistical estimation. Efron pointed out how any regular parametric family could be approximated locally by a curved exponential family and that the curvature of these models measures their departure from exponentiality. It turned out that this concept was intimately related to Fisher's theory of information loss. Efron's formal theory did not use all the bells and whistles of differential geometry. The first step to an elegant geometric theory was done by Dawid [33], who introduced a connection on the space of all positive probability distributions and showed that Efron's statistical curvature is induced by this connection.

The use of differential geometry in its elegant splendor for the elaboration of previous ideas was systematically achieved by Amari [6] and [7], who studied the informational geometric properties of a manifold with a Fisher metric on it. This is the reason why sometimes this is also called the Fisher–Efron–Amari theory.

The concept of dual connections and the theory of dually flat spaces as well as the $\alpha$-connections were first introduced in 1982 by Nagaoka and Amari [61] and developed later in a monograph by Amari [5]. These concepts were proved extremely useful in the study of informational geometry, which investigates the differential geometric structure of the manifold of probability density functions. It is worthy to note the independent work of Chentsov [26] on $\alpha$-connections done from a different perspective.

Entropy, from its probabilistic definition, is a measure of uncertainty of a random variable. The maximum-entropy approach was introduced by the work of Shannon [73]. However, the *entropy maximization principle* was properly introduced by Akaike [2] in the mid-1970s, as a theoretical basis for model selection. Since then it has been used in order to choose the least "biased" distribution with respect to the unknown information. The included chapter regarding maximum entropy with moments constraints is inspired by Mead [57].

The entropy of a continuous distribution is not always positive. In order to overcome this flaw one can use the relative entropy of two distributions $p$ and $q$. This concept was originally introduced by S. Kullback and R. Leibler in 1951, see [51, 52]. This is also referred in the literature under the names of divergence, information divergence, information gain, relative entropy, or Kullback–Leibler divergence. The Kullback–Leibler divergence models the information between a true distribution $p$ and a model distribution $q$; the reader can consult the book of Burnham and Anderson [21] for details.

In practice, the density function $p$ of a random variable is unknown. The problem is the one of drawing inferences about the density $p$ on the basis of $N$ concrete realizations of the random variable. Then we can look for the density $p$ as an element of a certain restricted class of distributions, instead of all possible distributions. One way in which this restricted class can be constructed is to consider the distributions having the same mean as the sample mean and the variance equal to the variance of the sample. Then, we need to choose the distribution that satisfies these constraints and is the most ignorant with respect to the other moments. This is realized for the distribution with the maximum entropy. The theorems regarding maximum entropy distributions subject to different constraints are inspired from Rao [71]. They treat the case of the normal distribution, as the distribution on $\mathbb{R}$ with the first two moments given, the exponential distribution, as the distribution on $[0, \infty)$ with the given mean, as well as the case of Maxwell–Boltzman distribution. The case of the maximum entropy distribution with the first $n$ given moments is inspired from Mead and Papanicolaou [57]. The novelty brought by this chapter is the existence of maximum entropy distributions in the case when the sample space is a finite interval. The book also introduces the curves of largest entropy, whose relevance in actual physical situations is worth examining.

The second part of the book deals with statistical manifolds, which are geometrical abstractions of statistical models. Lauritzen [54] defined statistical manifolds as Riemannian manifolds endowed with a pair of torsion-free dual connections. He also introduced an equivalent way of constructing statistical manifolds, starting from a skewness tensor. A presentation of statistical structure in the language of affine connections can be found in Simon [76].

The geometry of a statistical model can be also induced by contrast functions. The dualistic structure of contrast functions was developed by Eguchi [38, 39, 41], who has shown that a contrast function induces a Riemannian metric by its second order derivatives, and a pair of dual connections by its third order derivatives. Further information on contrast geometry can be found in Pfanzagl [69]. A generalization of the geometry induced by the contrast functions is the yoke geometry, introduced by Barndorff-Nilsen [11–13] and developed by Blæsid [18, 19].

## Acknowledgments

The monograph was partially supported by the NSF Grant # 0631541. The authors are indebted to Eastern Michigan University and University Politehnica of Bucharest for the excellent conditions of research. We wish to express many thanks to Professor Jun Zhang from University of Michigan for many fruitful discussions on the subject and for introducing the first author to the fascinating subject of this book. We express our gratitude to Professor Ionel Tevy from Politehnica University of Bucharest for clarifying certain mathematical problems in this book.

This work owes much of its clarity and quality to the numerous comments and observations made by several unknown reviewers whose patience and time spent on the manuscript are very much appreciated.

Finally, we would like to express our gratitude to Springer and its editors, especially Donna Chernyk for making this endeavor possible.

Ypsilanti, MI, USA — Ovidiu Calin
Bucharest, Romania — Constantin Udrişte
January, 2014

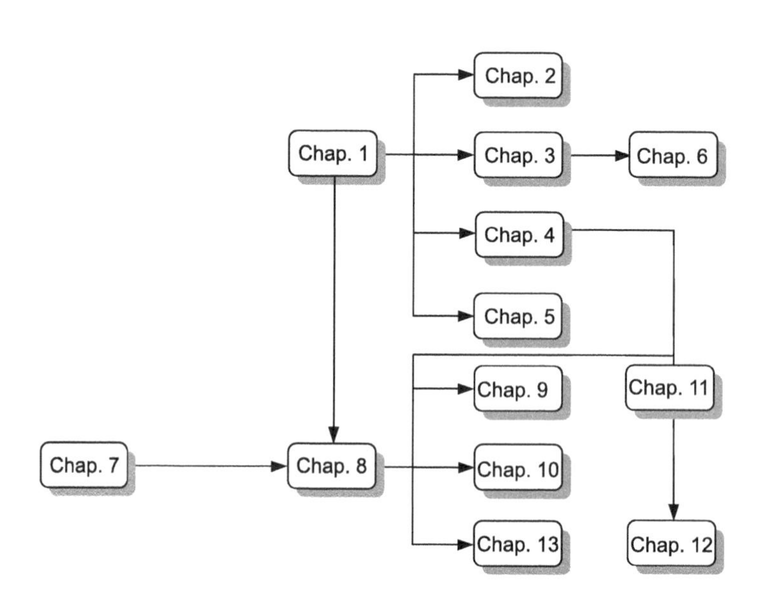
Chap. 2
Chap. 1
Chap. 3
Chap. 6
Chap. 4
Chap. 5
Chap. 9
Chap. 11
Chap. 7
Chap. 8
Chap. 10
Chap. 13
Chap. 12

# Contents

# List of Notations

The following notations are given in the order used in text.

| | |
|---|---|
| $(M, g)$ | Riemannian manifold |
| $\nabla$ | Linear connection |
| $Ric$ | Ricci tensor curvature |
| $R$ | Riemannian tensor of curvature |
| $\mathcal{X}$ | Sample space |
| $p(x)$ | Probability density function |
| $\mathcal{P}(\mathcal{X})$ | Statistical model of all probability densities on $\mathcal{X}$ |
| $\ell_x(\xi)$ | Log-likelihood function |
| $E_\xi$ | Expectation operator with respect to density $p_\xi$ |
| $\mathbb{E}$ | Parameter space |
| $g_{ij}(\xi)$ | Fisher–Riemann metric |
| $\nabla^{(0)}$ | Levi–Civita connection |
| $\nabla^{(\alpha)}$ | $\alpha$-connection |
| $\Gamma^{(\alpha)}_{ij,k}$ | Christoffel symbols of first kind for the $\alpha$-connection |
| $K_{ij}$ | Difference tensor |
| $C_{ijk}$ | Skewness tensor |
| $\psi$ | Digamma function |
| $\psi_1$ | Trigamma function |
| $\Gamma(\,\cdot\,)$ | Gamma function |
| $B(\,\cdot,\cdot\,)$ | Beta function |
| $d_H$ | Hellinger distance |
| $H(\xi)$ | Entropy function |

| | |
|---|---|
| $I(\xi)$ | Informational energy |
| $D_{KL}(\cdot\|\|\cdot)$ | Kullback–Leibler relative entropy |
| $\nabla, \nabla^*$ | Dual connections |
| $(M, g, \nabla, \nabla^*)$ | Statistical manifold |
| $R(X, Y)Z$ | Riemann tensor of curvature |
| $R^{(\alpha,\beta)}$ | Relative curvature tensors |
| $R^{(\alpha)}$ | Curvature tensor of $\nabla^{(\alpha)}$ |
| $dv$ | Volume form on $(M, g)$ |
| $\omega^{(\alpha)}$ | Volume form associated with connection $\nabla^{(\alpha)}$ |
| $Ric^{(\alpha)}$ | Ricci tensor associated with connection $\nabla^{(\alpha)}$ |
| $Ric, Ric^*$ | Ricci tensors associated with dual connections $\nabla$, $\nabla^*$ |
| $K = K^i\partial_i$ | Curvature vector field |
| $div, div^*$ | Divergence operators associated with dual connections $\nabla$, $\nabla^*$ |
| $div^{(\alpha)}$ | Divergence operator associated with $\nabla^{(\alpha)}$ |
| $L_X$ | Lie derivative with respect to vector field $X$ |
| $H^f$ | Hessian of $f$ associated with $\nabla$ |
| $H^{*f}$ | Hessian of $f$ associated with $\nabla^*$ |
| $\Delta^{(0)}$ | Laplacian associated with the Levi–Civita connection $\nabla^{(0)}$ |
| $\Delta^{(\alpha)}$ | Laplacian associated with the Levi–Civita connection $\nabla^{(\alpha)}$ |
| $D(\xi_1\|\|\xi_2)$ | Contrast function |
| $D^*(\xi_1\|\|\xi_2)$ | Dual contrast function |
| $D_{\mathcal{S}}(\xi_1\|\|\xi_2)$ | Contrast function on the manifold $\mathcal{S}$ |
| $\nabla^{(D)}$ | Linear connection induced by the contrast function $D$ |
| $\nabla^{(D^*)}$ | Linear connection induced by the dual contrast function $D^*$ |
| $\Gamma^{(D)}_{ij,k}$ | Coefficients of the connection induced by $D$ |

| | |
|---|---|
| $\Gamma_{ij,k}^{(D^*)}$ | Coefficients of the connection induced by $D^*$ |
| $D_f$ | $f$-divergence |
| $D^{(\alpha)}$ | Chernoff information of order $\alpha$ |
| $\mathcal{E}(p\|\|q)$ | Exponential contrast function |
| $D_X^{(\alpha)}Y$ | $\alpha$-connection on a submanifold |
| $L, L^*$ | Dual second fundamental forms |
| $R_{\mathcal{M}}, R_{\mathcal{M}}^*$ | Dual Riemann tensors on the (sub)manifold $\mathcal{M}$ |
| $H_p, H_p^*$ | Dual mean curvature vectors at point $p$ |
| $S^{(\alpha,\beta)}$ | Shape operator |
| $(M, h, D, D^*)$ | Statistical submanifold |

# Part I

# The Geometry of Statistical Models

# Chapter 1

# Statistical Models

This chapter presents the notion of statistical models, a structure associated with a family of probability distributions, which can be given a geometric structure. This chapter deals with statistical models given parametrically. By specifying the parameters of a distribution, we determine a unique element of the family. When the family of distributions can be described smoothly by a set of parameters, this can be considered as a multidimensional surface. We are interested in the study of the properties that do not depend on the choice of model coordinates.

## 1.1 Probability Spaces

Let $\mathbf{S}$ be a set (finite or infinite). A *$\sigma$-field* $\mathcal{F}$ over the set $\mathbf{S}$ is a collection of subsets of $\mathbf{S}$ that is closed under countable many intersections, unions, and complements. The pair $(\mathbf{S}, \mathcal{F})$ is called a *measurable space*. One may associate with each element of $\mathcal{F}$ a non-negative number which sets the "size" of each set. A function $\mu : \mathcal{F} \to \mathbb{R} \cup \{\pm\infty\}$ is called a *measure* on $\mathcal{F}$ if

(a) $\mu(E) \geq 0$, $\forall E \in \mathcal{F}$;

(b) $\mu(\emptyset) = 0$;

(c) $\mu\Big(\bigcup_{i\geq 1} E_i\Big) = \bigcup_{i\geq 1} \mu(E_i)$, for any pairwise disjoint sequence of sets $\{E_i\}_{i\geq 1}$ in $\mathcal{F}$.

The triple $(\mathbf{S}, \mathcal{F}, \mu)$ is called a *measure space*.

O. Calin and C. Udrişte, *Geometric Modeling in Probability and Statistics*,
DOI 10.1007/978-3-319-07779-6_1,

A function $f : \mathbf{S} \to \mathcal{Y}$ between two measurable spaces, $(\mathbf{S}, \mathcal{F})$ and $(\mathbf{Y}, \mathcal{G})$, is called *measurable* if $f^{-1}(G) \in \mathcal{F}$ for any $G \in \mathcal{G}$.

We shall deal next with a special type of measures useful for this book. A *probability measure* is a measure $\mu$ with $\mu(\mathbf{S}) = 1$. The measure $\mu$ in this case will be denoted customarily by $P$. A *probability space* is a measure space $(\mathbf{S}, \mathcal{F}, P)$, with a probability measure $P$. In this case the elements of $\mathcal{F}$ are called *events*. A probability measure assigns a non-negative number to each event, called *probability of the event*.

The set $\mathbf{S}$ is called *sample space*. It consists of all possible states that occur randomly as outcomes of a given physical experiment. The random outcomes (elements of sample space) can be "measured" by some variables which are random. A real-valued *random variable* $X$ on a probability space is a measurable function $X : \mathbf{S} \to \mathbb{R}$. This means that for any two values $a, b \in \mathbb{R} \cup \{\pm\infty\}$ we have

$$\{s \in \mathbf{S}; a < X(s) < b\} \in \mathcal{F},$$

i.e., all the sample states for which $X$ takes values between $a$ and $b$ is an event.

There are three distinguished classes of random variables: *discrete*, *continuous*, and *mixture* of the two. We shall discuss the first two types and leave the third as an exercise to the reader.

### 1.1.1 Discrete Random Variables

A *discrete random variable* $X$ takes finite or at most countably infinite values, $X : \mathbf{S} \to \mathcal{X} = \{x^1, x^2, x^3, \dots\}$. Its probability is described by a *probability mass function* $p : \mathcal{X} \to [0, 1]$

$$p_k = p(x^k) = P(X = x^k) = P(\{s \in \mathbf{S}; X(s) = x^k\}), \qquad \forall k \geq 1,$$

satisfying $\sum_{k \geq 1} p_k = 1$. The probability distribution characterized by a probability mass function is called a *discrete probability distribution*.

The associated *distribution function* can be written as the sum

$$F(x) = \sum_{k=1}^{N} p_k, \quad \forall x^N \leq x < x^{N+1}.$$

Among the most well-known discrete probability distributions used in this book are the Poisson distribution, the Bernoulli distribution, the binomial and geometric distributions.

### 1.1.2 Continuous Random Variables

A *continuous random variable* $X : \mathbf{S} \to \mathcal{X} \subset \mathbb{R}^n$ is a random variable taking a continuous range of values; its probability distribution has a *probability density function*, $p(x)$, which is a non-negative function, integrable with respect to the Lebesgue measure on $\mathcal{X}$, i.e. a function $p : \mathcal{X} \to \mathbb{R}$ satisfying

$$(i)\ p(x) \geq 0, \ \ \forall x \in \mathcal{X}; \quad (ii) \int_{\mathcal{X}} p(x)\, dx = 1.$$

For any $\mathcal{D}$ open set in $\mathcal{X}$, the relation between the probability measure $P$ and the density function $p(x)$ is given by

$$P(X \in \mathcal{D}) = \int_{\mathcal{D}} p(x)\, dx.$$

In the particular case when $\mathcal{X} = \mathbb{R}$, the foregoing relation becomes

$$P(a < X < b) = \int_a^b p(x)\, dx, \qquad \forall a, b \in \mathbb{R},\ a < b.$$

In fact, the integral formula can be written also in the differentiable form

$$p(x)dx = P(x < X < x + dx),$$

showing that the measure $p(x)dx$ represents the probability that the random variable $X$ takes values around the value $x$.

The function $p(x)$ defines in turn a probability distribution function $F : \mathbb{R} \to [0, 1]$

$$F(x) = \int_{-\infty}^{x} p(u)\, du.$$

This natural relation between probability distributions and probability densities will make us to use both denominations interchangeably.

We note that if $p_1(x)$ and $p_2(x)$ are two probability densities, then the convex combination

$$p(x) = \lambda p_1(x) + (1 - \lambda)p_2(x), \ \ \lambda \in [0, 1]$$

is also a probability density.

Among continuous probability distributions used in this book are the normal distribution, the lognormal distribution, the gamma and beta distributions.

It is worth noting that any discrete distribution can be seen as having a probability density function, which can be described in terms of Dirac measure as $p(x) = \sum_{k\geq 1} p_k \delta(x - x^k)$, where $(p_k)_{k\geq 1}$ are the probabilities associated with the discrete values $(x^k)_{k\geq 1}$ taken by the random variable $X$. For more details on Dirac distributions measures, see Example 3.2.7.

## 1.2 A Statistic Estimation

In practice a random variable is measured by a finite number of observations. Consider $N$ observations

$$u^1, u^2, \dots, u^N \in \mathcal{X} \tag{1.2.1}$$

of the continuous random variable $X$. We wish to estimate the underlying probability distribution, $p^*$, that produced the previous observations, which is in fact the probability density of $X$. This is also known under the name of *true distribution* of $X$. Using the observations data, as well as our experience, we can infer the shape of the distribution $p^*$. For instance, if the data (1.2.1) represents the weights of $N$ randomly chosen persons, using their symmetry and the bell shaped properties, a relatively simplistic inference is that $p^*$ is a normal distribution. In this case, the choice is unique up to two parameters, the mean and the variance. We need to choose the right parameters which fit data in the best possible way.[1] The general case is similar, in the sense that the shape of $p^*$ depends on several parameters, which makes the chosen distribution to belong to a parametrized space. We need to choose the point (i.e., a distribution) on this space that is the "closest" (with respect to a certain information measure) to the true distribution $p^*$.

## 1.3 Parametric Models

This section deals with a family of probability density functions which depends on several parameters, and hence it can be organized as a parameterized surface; each point of this surface represents a proba-

[1] There are several ways of fitting data to a distribution. For instance, minimizing the Kullback–Leibler divergence or the Hellinger distance are just a couple of ways of doing this

bility density. More precisely, let

$$\mathcal{S} = \{p_\xi = p(x;\xi) | \xi = (\xi^1, \dots, \xi^n) \in \mathbb{E}\}$$

be a family of probability distributions on $\mathcal{X}$, where each element $p_\xi$ can be parameterized by $n$ real-valued variables $\xi = (\xi^1, \dots, \xi^n)$.

### 1.3.1 Definition

The set $\mathbb{E} \subset \mathbb{R}^n$ is called the *parameters space*. The set $\mathcal{S}$ is a subset of the infinite dimensional space of functions

$$\mathcal{P}(\mathcal{X}) = \Big\{ f; f : \mathcal{X} \to \mathbb{R}, f \geq 0, \int_{\mathcal{X}} f\, dx = 1 \Big\}.$$

Being inspired and influenced by the definition of surfaces, we shall require the mapping

$$\iota : \mathbb{E} \to \mathcal{P}(\mathcal{X}), \qquad \iota(\xi) = p_\xi$$

to be an immersion, i.e.,

(i) $\iota$ is one-to-one;

(ii) $\iota$ has rank $n = \dim \mathbb{E}$.

For a suggestive diagram, see Fig. 1.1. The classical definition of rank, as the rank of the Jacobian of $\iota$, does not apply to infinite dimensional spaces. Therefore, here, by rank $n$ we understand that the functions

$$\frac{\partial p_\xi}{\partial \xi^1}, \dots, \frac{\partial p_\xi}{\partial \xi^n}$$

are linearly independent.

In the particular case when $\mathcal{X} \subset \mathbb{R}$, this condition is implied by the nonvanishing condition of the Wronskian at a single point

$$\begin{vmatrix} \varphi_1(x_0,\xi) & \dots & \varphi_n(x_0,\xi) \\ \varphi_1'(x_0,\xi) & \dots & \varphi_n'(x_0,\xi) \\ \dots & \dots & \dots \\ \varphi_1^{(n-1)}(x_0,\xi) & \dots & \varphi_n^{(n-1)}(x_0,\xi) \end{vmatrix} \neq 0, \qquad \forall \xi$$

for some $x_0 \in \mathcal{X}$, where we used the shorthand notation $\varphi_j(x,\xi) = \dfrac{\partial p_\xi(x)}{\partial \xi^j}$, and $\varphi_j^{(k)}(x_0,\xi) = \partial_x^k \varphi_j(x,\xi)_{|x=x_0}$. This can be shown by

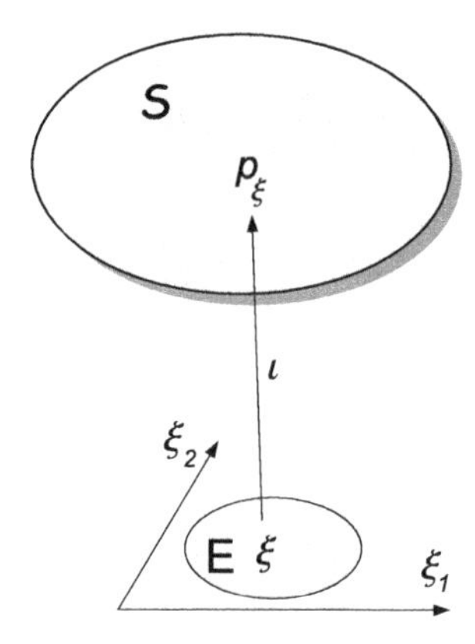


Figure 1.1: The immersion $\iota$ maps each parameter $\xi$ into a distribution $p_\xi \in \mathcal{S}$

assuming $\varphi_j$ dependent and differentiating $n-1$ times the linear combination $\sum_{j=1}^{n} c_j \varphi_j(x,\xi) = 0$ with respect to $x$ and then taking the value at $x_0$. We obtain a homogeneous system which has the trivial solution $c_j = 0$, $1 \leq j \leq n-1$. Hence, $\{\varphi_j\}$ are independent as functions of $x$, for all $\xi$.

The set $\mathcal{S}$, satisfying the aforementioned properties (i)–(ii), is called a *statistical model* or a *parametric model*[2] of dimension $n$. Sometimes, where there is no doubt about what is the parameters space or sample space, it is useful to use the abbreviate notations $\mathcal{S} = \{p_\xi\}$, or $\mathcal{S} = \{p(x;\xi)\}$. Other useful notations are the partial differentiation with respect to parameters $\partial_k = \frac{\partial}{\partial \xi^k}$. The condition *(ii)* states the *regularity* of the statistical model.

### 1.3.2 Basic Properties

The functions $\varphi_j(x,\xi) = \partial_j p_\xi(x)$ play the role of *basic vector fields* for the model surface $\mathcal{S} = \{p_\xi\}$. The vector field $\varphi_j$ acts on smooth mappings $f : \mathcal{S} \to \mathcal{F}(\mathcal{X}, \mathbb{R})$ as a differentiation[3]

---

[2]Some authors consider a more general definition for parametric models, without requiring properties (i) and (ii). All examples provided in this book satisfy relations (i) and (ii); we required them here for smoothness reasons.

[3]The reason why $f$ takes values in $\mathcal{F}(\mathcal{X}, \mathbb{R})$ is because we consider $f(p_\xi)$ as a real-valued function of $x$, $x \in \mathcal{X}$.

$$\varphi_j(f) = \frac{\partial(f(p_\xi))}{\partial \xi^j}.$$

As a consequence, Leibniz rule holds

$$\varphi_j(fg) = f\varphi_j(g) + \varphi_j(f)g,$$

for $f, g$ smooth mappings of $\mathcal{S}$ into the space of real-valued smooth functions on $\mathcal{X}$, denoted by $\mathcal{F}(\mathcal{X}, \mathbb{R})$.

A useful mapping is the *log-likelihood function* $\ell : \mathcal{S} \to \mathcal{F}(\mathcal{X}, \mathbb{R})$ defined by

$$\ell(p_\xi)(x) = \ln p_\xi(x).$$

Sometimes, for convenience reasons, this will be denoted by $\ell_x(\xi) = \ell(p_\xi(x))$. The derivatives of the log-likelihood function are

$$\partial_j \ell_x(\xi) = \frac{\partial \ln p_\xi(x)}{\partial \xi^j} = \varphi_j(\ell_x(\xi)), \quad 1 \leq j \leq n.$$

They are also found in literature under the name of *score functions*. Heuristically speaking, they describe how the information contained in $p_\xi$ changes in the direction of $\xi^j$. The functions $\partial_j \ell_x(\xi)$ will play a central role in defining the Fisher information matrix and the entropy on a statistical manifold.

Differentiating under the integral sign yields

$$\int_{\mathcal{X}} \varphi_j(x)\, dx = \partial_j \int_{\mathcal{X}} p_\xi(x)\, dx = \partial_j 1 = 0, \qquad \forall 1 \leq j \leq n. \quad (1.3.2)$$

It is worth noting that interchanging the derivative with the integral always holds if the sample space $\mathcal{X}$ is bounded. Otherwise, some fast vanishing conditions need to be required for the distribution $p_\xi(x)$, as $|x| \to \infty$.

In the case of discrete distributions we have

$$\sum_{k\geq 1} \varphi_j(x_k) = \partial_j \sum_{k\geq 1} p_\xi(x_k) = \partial_j 1 = 0, \qquad \forall 1 \leq j \leq n. \quad (1.3.3)$$

The derivative and the sum can be swapped if $\mathcal{X}$ is finite. Otherwise, the series should be supposed uniformly convergent in $\xi$.

Sometimes (see, for instance, the case of exponential families, Sect. 1.4), it is easier to check a condition of linear independence on the log-likelihood function rather than on the density functions. The following result is useful in practice.

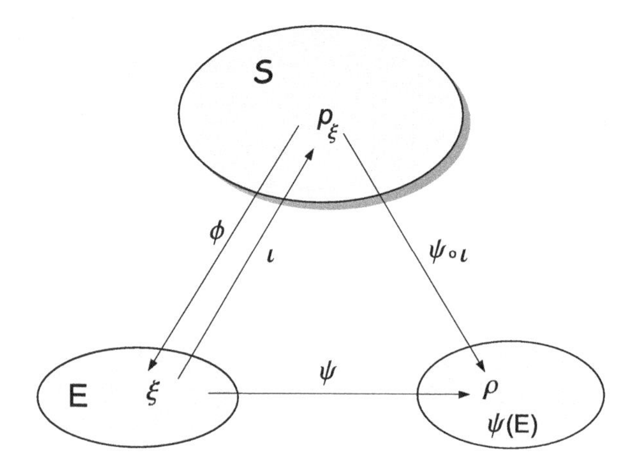


Figure 1.2: Two parameterizations of the statistical model $\mathcal{S}$

**Theorem 1.3.1** *The regularity condition (ii) of the statistical model $\mathcal{S} = \{p_\xi\}$ holds if and only if for any $\xi \in \mathbb{E}$ the set*

$$\{\partial_1 \ell_x(\xi), \partial_2 \ell_x(\xi), \dots, \partial_n \ell_x(\xi)\}$$

*is a system of $n$ linearly independent functions of $x$.*

*Proof:* Differentiating yields

$$\partial_j \ell_x(\xi) = \frac{\partial}{\partial \xi^j} \ln p(x;\xi) = \frac{1}{p(x;\xi)} \frac{\partial p(x;\xi)}{\partial \xi^j} = \frac{1}{p(x;\xi)} \varphi_j(x),$$

and hence the systems $\{\partial_j \ell_x(\xi)\}$ and $\{\varphi_j(x)\}$ are proportional, which ends the proof. ■

The following formula is useful in computations.

**Proposition 1.3.2** *Assume the conditions required for formulae (1.3.2) and (1.3.3) hold. If $E_\xi[\,.\,]$ denotes the expectation with respect to $p_\xi$, then*

$$E_\xi[\partial_j \ell_x(\xi)] = 0, \qquad 1 \le j \le n.$$

*Proof:* We have the following computation

$$\begin{aligned} E_\xi[\partial_j \ell_x(\xi)] &= E_\xi\Big[\frac{\partial_j p(x;\xi)}{p(x;\xi)}\Big] = \int_{\mathcal{X}} \partial_j p(x;\xi)\, dx \\ &= \partial_j\Big(\int_{\mathcal{X}} p(x;\xi)\, dx\Big) = \partial_j(1) = 0. \end{aligned}$$

Similarly, in the discrete case we have

$$\begin{aligned} E_\xi[\partial_j \ell_x(\xi)] &= \sum_{k\geq 1} p_\xi(x_k)\,\partial_j \ln p_\xi(x_k) \\ &= \sum_{k\geq 1} \partial_j p_\xi(x_k) = \partial_j \sum_{k\geq 1} p_\xi(x_k) = 0. \end{aligned}$$

■

In the language of Sect. 2.10, Proposition 1.3.2 states that the basis elements in the 1-representation have zero expectation.

### 1.3.3 Parameterizations

Since the statistical model $\mathcal{S} = \{p_\xi\}$ is the image of the one-to-one mapping $\iota : \mathbb{E} \to \mathcal{P}(\mathcal{X})$, $\iota(\xi) = p_\xi$, it makes sense to consider the inverse function $\phi : \mathcal{S} \to \mathbb{E} \subset \mathbb{R}^n$, $\phi(p_\xi) = \xi$. Since the mapping $\phi$ assigns a parameter $\xi$ with each distribution $p_\xi$, we can consider it as a *coordinate system* for our statistical model. The models of interest in this chapter can be covered by only one coordinate system, i.e., $(\mathbb{E}, \phi)$ forms a global coordinate system. This situation occurs most often in the case of statistical models.

The parameterization of a statistical model is not unique. We can change the coordinate system on $\mathcal{S}$ in the following way. Let $\psi : \mathbb{E} \to \psi(\mathbb{E}) \subset \mathbb{R}^n$ be a diffeomorphism (i.e., both $\psi$ and $\psi^{-1}$ are $C^\infty$-smooth, with $\mathbb{E}$ and $\psi(\mathbb{E})$ open sets in $\mathbb{R}^n$). Then $\psi \circ \iota : \mathcal{S} \to \psi(\mathbb{E}) \subset \mathbb{R}^n$ is another coordinate system, and if denote $\rho = \psi(\xi)$, then the statistical model can be also written as $\mathcal{S} = \{p_{\psi^{-1}(\rho)}; \rho \in \psi(\mathbb{E})\}$, see Fig. 1.2.

Even if the model may change its parameterization, its geometric properties are not affected. Since the geometric results proved in one parameterization are valid for all parameterizations, it is useful to choose a convenient parameterization to work with.

### 1.3.4 Usual Examples

**Example 1.3.1 (Exponential Distribution)** Let $\mathcal{X} = [0, \infty)$ and consider the one-dimensional parameter space $\mathbb{E} = (0, \infty)$, which is an open interval in $\mathbb{R}$. The exponential distribution with parameter $\xi$ is given by the formula

$$p(x; \xi) = \xi e^{-\xi x}.$$

The associated statistical model $\mathcal{S} = \{p(x;\xi)\}$ is one-dimensional, and hence can be considered as a curve in $\mathcal{P}([0,\infty))$. It is easy to verify that the mapping $\iota : \mathbb{E} \to \mathcal{P}([0,\infty))$ is one-to-one and that $\partial_\xi p_\xi(x) = (1-\xi x)e^{-\xi x} \neq 0$ for almost all $x \in [0,\infty)$.

The diffeomorphism $\psi : (0,\infty) \to (0,\infty)$, $\psi(\xi) = 1/\xi$ induces the new parameterization $p(x;\rho) = \frac{1}{\rho}e^{-x/\rho}$.

**Example 1.3.2 (Normal Distribution)** Let $\mathcal{X} = \mathbb{R}$ and $\mathbb{E} = \mathbb{R} \times (0,\infty)$. The normal distribution family is defined by the formula

$$p(x;\xi^1,\xi^2) = p(x;\mu,\sigma) = \frac{1}{\sigma\sqrt{2\pi}}\, e^{-\frac{(x-\mu)^2}{2\sigma^2}}, \quad x \in \mathcal{X},\ (\xi^1,\xi^2) \in \mathbb{E}. \tag{1.3.4}$$

Hence the statistical model $\{p(x;\mu,\sigma)\}$ can be seen as a two-dimensional surface parameterized by $\mathbb{R} \times (0,\infty)$.

Next we check the regularity condition (ii) of Sect. 1.3.1 using Theorem 1.3.1. Since the log-likelihood function is

$$\ell_x(\mu,\sigma) = \ln p(x;\mu,\sigma) = -\frac{(x-\mu)^2}{2\sigma^2} - \ln\sigma - \ln\sqrt{2\pi},$$

we find

$$\begin{aligned} \partial_\mu \ell_x(\mu,\sigma) &= \frac{x-\mu}{\sigma^2}, \\ \partial_\sigma \ell_x(\mu,\sigma) &= \frac{(x-\mu)^2}{\sigma^3} - \frac{1}{\sigma}. \end{aligned}$$

Making the variable change $y = x - \mu$, the condition reduces to show that the polynomials $y/\sigma^2$ and $y^2\sigma^3 - 1/\sigma$ are linearly independent functions of $y$, fact easy to check.

In order to show that the mapping $\iota : \mathbb{E} \to \mathcal{S}$ is one-to-one, assume that $p(x;\mu,\sigma) = p(x,\bar{\mu},\bar{\sigma})$. Equating the log-likelihood functions, $\ln p(x;\mu,\sigma) = \ln p(x,\bar{\mu},\bar{\sigma})$, we get an identity between two polynomials

$$\frac{(x-\mu)^2}{2\sigma^2} + \ln\sigma = \frac{(x-\bar{\mu})^2}{2\bar{\sigma}^2} + \ln\bar{\sigma}.$$

Equating the coefficients of similar powers yields $\mu = \bar{\mu}$ and $\sigma = \bar{\sigma}$.

We end up with an example of parameterization. Since the diffeomorphism $\psi : \mathbb{R} \times (0,\infty) \to \mathbb{R} \times (0,\infty)$, $\psi(\mu,\sigma) = (\mu+\sigma, 1/\sigma)$, has the inverse $\psi^{-1}(\rho,\eta) = (\rho - 1/\eta, 1/\eta)$, we arrive at the following parameterization

$$p(x;\rho;\eta) = \frac{\eta}{\sqrt{2\pi}} e^{-\frac{1}{2}(x\eta - \rho\eta + 1)^2}, \quad (\rho,\eta) \in \mathbb{E} \tag{1.3.5}$$

of the normal distribution. It is worth noting that the distributions (1.3.4) and (1.3.5) are geometrically identical, even if they look different in parameterizations.

**Example 1.3.3 (Gamma Distribution)** The sample space is given by $\mathcal{X} = [0, \infty)$ and the parameter space is $\mathbb{E} = (0, \infty) \times (0, \infty)$. We use the *Gamma function*

$$\Gamma(\alpha) = \int_0^\infty t^{\alpha-1} e^{-t} dt, \ \alpha > 0.$$

The Gamma family of distributions

$$p_\xi(x) = p_{\alpha,\beta}(x) = \frac{1}{\beta^\alpha \Gamma(\alpha)} x^{\alpha-1} e^{-x/\beta},$$

with parameters $(\xi^1, \xi^2) = (\alpha, \beta) \in \mathbb{E}$ forms a two-dimensional surface in $\mathcal{P}([0, \infty))$. We note that the exponential distribution $p_{1,\beta}(x)$ represents a curve family on this surface. The regularity condition can be checked out using Theorem 1.3.1.

**Example 1.3.4 (Beta Distribution)** We use the *Beta function*

$$B(a, b) = \int_0^1 t^{a-1}(1-t)^{b-1} dt, \qquad a, b > 0.$$

The Beta family of probability densities is

$$p_{a,b}(x) = \frac{1}{B(a, b)} x^{a-1}(1-x)^{b-1},$$

where $x \in \mathcal{X} = [0, 1]$ and $(a, b) \in \mathbb{E} = (0, \infty) \times (0, \infty)$, corresponds to a two-dimensional surface in the space $\mathcal{P}([0, 1])$.

**Example 1.3.5 ($\mathcal{P}(\mathcal{X})$, $\mathcal{X}$ finite)** Let $\mathcal{X} = \{x^1, x^2, \dots, x^n\}$ and

$$\mathbb{E} = \Big\{(\xi^1, \dots, \xi^{n-1}); \xi^i > 0, \ \sum_{i=1}^{n-1} \xi^i < 1\Big\},$$

$$p(x^i; \xi) = \begin{cases} \xi^i, & \text{if } 1 \le i \le n-1 \\ 1 - \sum_{i=1}^{n-1} \xi^i, & \text{if } i = n. \end{cases}$$

If $\{e_1, \dots, e_{n-1}\}$ denotes the natural basis of $\mathbb{R}^{n-1}$ (i.e., the vector $e_j$ has the entry 1 on the $j$th entry and 0 in rest), then $\partial_j p_\xi = (e_j, -1)^T$. This implies that $\{\partial_j p_\xi\}$, $j \in \{1, n-1\}$, are linearly independent, which implies the regularity of the model. The injectivity of $\iota$ is straightforward.

**Example 1.3.6 (Poisson Distribution)** The sample and parameter spaces are $\mathcal{X} = \{0, 1, 2, \dots\}$ and $\mathbb{E} = (0, \infty)$, and the distribution is given by

$$p(x; \lambda) = \frac{\lambda^x}{x!} e^{-\lambda}.$$

This forms a one-dimensional statistical model. There is only one score function in this case, $\partial_\lambda \ell_x(\lambda) = \dfrac{x}{\lambda} - 1$.

**Example 1.3.7 (Bivariate Normal Distribution)** Consider the distribution

$$p(x_1, x_2; \xi) = \frac{1}{2\pi\sigma_1\sigma_2\sqrt{1-\rho^2}} e^{-Q(x)/2}, \qquad x \in \mathcal{X} = \mathbb{R}^2,$$

where

$$Q(x) = \frac{1}{1-\rho^2}\Big[\frac{(x_1-\mu_1)^2}{\sigma_1^2} - 2\rho\frac{(x_1-\mu_1)(x_2-\mu_2)}{\sigma_1\sigma_2} + \frac{(x_2-\mu_2)^2}{\sigma_2^2}\Big],$$

with the parameter space

$$\xi = (\mu_1, \mu_2, \sigma_1, \sigma_2, \rho) \in \mathbb{E} = \mathbb{R}^2 \times (0, \infty) \times (0, \infty) \times [0, 1).$$

$\{p(x; \xi\}$ defines a five-dimensional statistical model.

**Example 1.3.8 (Multivariate Normal Distribution)** Let $\mathcal{X} = \mathbb{R}^k$, $n = k + \frac{1}{2}k(k+1)$, and $\xi = (\mu, A)$. The distribution is given by

$$p(x; \xi) = \frac{1}{(2\pi)^{k/2}(\det A)^{1/2}} e^{-\frac{1}{2}(x-\mu)^t A^{-1}(x-\mu)},$$

with the parameter space

$$\mathbb{E} = \{(\mu, A); \mu \in \mathbb{R}^k, \text{positive-definite } A \in \mathbb{R}^{k\times k}\}.$$

The log-likelihood function for the previous statistical model is

$$\ell_x(\mu, A) = \ln p(x; \xi) = -\frac{1}{2}(x-\mu)^t A^{-1}(x-\mu) - \frac{1}{2}\ln(\det A) - \frac{k}{2}\ln(2\pi),$$

Denote by $\mu = (\mu_1, \dots, \mu_k)$, $A = (A_{ij})$, $A^{-1} = (A^{ij})$, and $u_j = x_j - \mu_j$. Then a straightforward computation, see Problem 1.30, shows

$$
\begin{aligned}
\partial_{\mu_r} \ell_x(\mu, A) &= -\sum_{j=1}^{k} \frac{A^{rj} + A^{jr}}{2}(x_j - \mu_j) \\
&= -\sum_{j=1}^{k} \frac{A^{rj} + A^{jr}}{2} u_j, \\
\partial_{A^{\alpha\beta}} \ell_x(\mu, A) &= \frac{1}{2} A_{\alpha\beta} - \frac{1}{2}(x_\alpha - \mu_\alpha)(x_\beta - \mu_\beta) \\
&= \frac{1}{2} A_{\alpha\beta} - \frac{1}{2} u_\alpha u_\beta.
\end{aligned}
$$

The score functions $\partial_{\mu_r} \ell$ and $\partial_{A^{\alpha\beta}} \ell$ are linearly independent polynomials in $u_j$.

**Example 1.3.9 (Product of Statistical Models)** Consider two statistical models

$$
\mathcal{S} = \{p_\xi : \mathcal{X} \to \mathbb{R}; \xi \in \mathbb{E}\}, \qquad \mathcal{U} = \{q_\theta : \mathcal{Y} \to \mathbb{R}; \theta \in \mathbb{F}\}.
$$

Let

$$
\mathcal{S} \times \mathcal{U} = \{f_{\xi,\theta} : \mathcal{X} \times \mathcal{Y} \to \mathbb{R}; (\xi, \theta) \in \mathbb{E} \times \mathbb{F}\},
$$

where

$$
f_{\xi,\theta}(x, y) = p_\xi(x) q_\theta(y).
$$

Since $f_{\xi,\theta}(x, y) \geq 0$ and

$$
\begin{aligned}
\iint_{\mathcal{X}\times\mathcal{Y}} f_{\xi,\theta}(x, y)\, dxdy &= \iint_{\mathcal{X}\times\mathcal{Y}} p_\xi(x) q_\theta(y)\, dxdy \\
&= \int_{\mathcal{X}} p_\xi(x)\, dx \int_{\mathcal{Y}} q_\theta(y)\, dy = 1,
\end{aligned}
$$

it follows that $f_{\xi,\theta}$ is a probability density on $\mathcal{X} \times \mathcal{Y}$. In order to show that $\mathcal{S} \times \mathcal{U}$ is a statistical model, we still need to verify the regularity condition. Since the log-likelihood functions satisfy

$$
\ell_{(x,y)}(\xi, \theta) = \ln f_{\xi,\theta}(x, y) = \ln p_\xi(x) + \ln q_\theta(y) = \ell_x(\xi) + \ell_y(\theta),
$$

we obtain

$$
\begin{aligned}
\partial_{\xi^j} \ell_{(x,y)}(\xi, \theta) &= \partial_{\xi^j} \ell_x(\xi) \\
\partial_{\theta^i} \ell_{(x,y)}(\xi, \theta) &= \partial_{\theta^i} \ell_y(\theta).
\end{aligned}
$$

Since the systems of functions $\{\partial_{\xi^j}\ell_x(\xi)\}$, $\{\partial_{\theta^i}\ell_y(\theta)\}$ are linearly independent on $\mathcal{X}$ and on $\mathcal{Y}$, respectively, as they act on independent variables $x$ and $y$, it follows that the system of functions $\{\partial_{\xi^j}\ell_{(x,y)}(\xi,\theta)$, $\partial_{\theta^i}\ell_{(x,y)}(\xi,\theta)\}$ are linearly independent on $\mathcal{X}\times\mathcal{Y}$. Hence $\mathcal{S}\times\mathcal{U}$ becomes a statistical model, called the *product statistical manifold* of $\mathcal{S}$ and $\mathcal{U}$.

It is worth noting that if $p_\xi$ and $q_\theta$ are the probability densities of two independent random variables $X$ and $Y$ with sample spaces $\mathcal{X}$ and $\mathcal{Y}$, respectively, then their joint probability density $f_{\xi,\theta}(x,y)$ belongs to the product statistical manifold $\mathcal{S}\times\mathcal{U}$.

Some of the previous examples are special cases of some more general statistical models, which will be treated next. The following two families of probability densities occupy a central role in the study of informational geometry. For this reason we dedicate a separate section to each of them.

## 1.4 Exponential Family

Consider $n+1$ real-valued smooth functions $C(x)$, $F_i(x)$ on $\mathcal{X}\subset\mathbb{R}^k$ such that

$$1, F_1(x), F_2(x), \dots, F_n(x)$$

are linearly independent, where 1 denotes the constant function. Then define the normalization function

$$\psi(\xi) = \ln\Big(\int_{\mathcal{X}} e^{C(x)+\xi^i F_i(x)}\,dx\Big), \tag{1.4.6}$$

and consider the *exponential family* of probability densities

$$p(x;\xi) = e^{C(x)+\xi^i F_i(x)-\psi(\xi)}, \tag{1.4.7}$$

with $x\in\mathcal{X}$, and $\xi$ such that $\psi(\xi)<\infty$. Finally, choose the parameter space $\mathbb{E}$ to be a non-empty set of elements $\xi$ with $\psi(\xi)<\infty$. It is worth stating that such a non-empty set $\mathbb{E}$ does not exist for all choices of $F_i(x)$, see Problem 1.29.

Formula (1.4.7) can be written equivalently as

$$p(x;\xi) = h(x)e^{\xi^i F_i(x)-\psi(\xi)}, \tag{1.4.8}$$

with $h(x) = e^{C(x)}$, and $\psi(\xi)$ not depending on $x$. We note that the normalization condition, $\int p(x;\xi)\,dx = 1$, is equivalent with condition (1.4.6).

The statistical manifold $\mathcal{S} = \{p(x;\xi)\}$, with $p(x;\xi)$ given either by (1.4.6) or by (1.4.8), is called an exponential family and $\xi^j$ are its *natural parameters*. The parameter space $\mathbb{E}$ is an open subset of $\mathbb{R}^n$. The dimension of $\mathbb{E}$ is called the *order* of the exponential family. In the following we shall check the definition conditions of a statistical manifold.

*The injectivity of* $\iota : \mathbb{E} \to \mathcal{S}$. Assume $\iota(\xi) = \iota(\theta)$. Then $\ln p(x;\xi) = \ln p(x;\theta)$, and hence

$$\xi^i F_i(x) - \psi(\xi) = \theta^i F_i(x) - \psi(\theta).$$

Since $\{F_i(x)\} \cup \{1\}$ are linearly independent, then $\xi^i = \theta^i$.

*The regularity condition.* Differentiating in the log-likelihood function

$$\ell_x(\xi) = C(x) + \xi^i F_i(x) - \psi(\xi)$$

and obtain

$$\partial_j \ell_x(\xi) = F_j(x) - \partial_j \psi(\xi). \tag{1.4.9}$$

Since $\{F_j(x)\}$ are linearly independent, so will be $\{\partial_j \ell_x(\xi)\}$, and hence the regularity condition is satisfied.

The function $\psi(\xi)$ depends on $C(x)$ and $F_i(x)$ by formula (1.4.6). The next result shows a relation involving the expectation operator.

**Proposition 1.4.1** *Assume the conditions required for formula (1.3.2) hold. Suppose the functions* $\{F_i(x)\}$ *satisfy the integrability condition* $E_\xi[F_i] < \infty$, *where If* $E_\xi[\cdot]$ *denotes the expectation with respect to* $p_\xi$. *Then*

$$\partial_j \psi(\xi) = E_\xi[F_j], \quad 1 \le j \le n. \tag{1.4.10}$$

*Proof:* Differentiating in the normalization condition $\int_{\mathcal{X}} p(x;\xi)\,dx = 1$ yields

$$\begin{aligned}
\int_{\mathcal{X}} \partial_{\xi^j} p(x;\xi)\,dx &= 0 \Leftrightarrow \\
\int_{\mathcal{X}} p(x;\xi)\Big(F_j(x) - \partial_{\xi^j}\psi(\xi)\Big)\,dx &= 0 \Leftrightarrow \\
\int_{\mathcal{X}} p(x;\xi) F_j(x)\,dx &= \partial_{\xi^j}\psi(\xi) \int_{\mathcal{X}} p(x;\xi)\,dx \Leftrightarrow \\
E_\xi[F_j] &= \partial_{\xi^j}\psi(\xi) \\
&= \partial_j \psi(\xi).
\end{aligned}$$

■

As a consequence, relation (1.4.9) yields

$$\partial_j \ell_x(\xi) = F_j(x) - E_\xi[F_j].$$

This leads to the relation

$$\begin{aligned} E_\xi[\partial_i \ell_x(\xi)\partial_j \ell_x(\xi)] &= E_\xi[(F_i(x) - E_\xi[F_i])(F_j(x) - E_\xi[F_j])] \\ &= Cov_\xi(F_i, F_j), \end{aligned} \tag{1.4.11}$$

which will be used in later computations. Another important relation useful in future computations is obtained if differentiate in (1.4.9) with respect to $\xi^i$; we find

$$\partial_i \partial_j \ell_x(\xi) = -\partial_i \partial_j \psi(\xi). \tag{1.4.12}$$

It can be shown that the function $\psi(\xi)$ is convex, see Problem 1.1. Therefore, relation (1.4.12) implies the concavity of the log-likelihood function $\ell_x(\xi)$.

Many of the usual distributions belong to the exponential family. Here are some useful examples described in terms of functions $C(x)$ and $\{F_i(x)\}$. The explicit computations are left as an exercise to the reader.

**Example 1.4.1 (Exponential Distribution)** This refers to Example 1.3.1. Choose $n = 1$ and

$$C(x) = 0, \quad F_1(x) = -x, \quad \xi^1 = \xi, \quad \psi(\xi) = -\ln \xi.$$

**Example 1.4.2 (Normal Distribution)** Consider Example 1.3.2 and choose

$$C(x) = 0, \quad F_1(x) = x, \quad F_2(x) = x^2, \quad \xi^1 = \frac{\mu}{\sigma^2}, \quad \xi^2 = \frac{-1}{2\sigma^2},$$

$$\psi(\xi) = -\frac{(\xi^1)^2}{4\xi^2} + \frac{1}{2}\ln\Big(\frac{-\pi}{\xi^2}\Big).$$

We note that parameters $(\xi_1, \xi_2) \in \mathbb{E} = \mathbb{R} \times (-\infty, 0)$.

**Example 1.4.3 (Poisson Distribution)** In Example 1.3.6, choose $n = 1$, $C(x) = -\ln x!$, $F_1(x) = x$, $\xi = \ln \lambda$, $\psi(\xi) = \lambda = e^\xi$.

**Example 1.4.4 (Gamma Distribution)** In Example 1.3.3, let

$$C(x) = -\ln x, \quad F_1(x) = \ln x, \quad F_2(x) = x, \quad \xi^1 = \alpha, \quad \xi^2 = \frac{-1}{\beta},$$

$$\psi(\xi) = \ln(\beta^\alpha \Gamma(\alpha)) = \ln\Big(\Big(-\frac{1}{\xi^2}\Big)^\alpha \Gamma(\xi^1)\Big),$$

with parameters $(\xi^1, \xi^2) \in \mathbb{E} = \mathbb{R} \times (-\infty, 0)$.

**Example 1.4.5 (Beta Distribution)** Consider Example 1.3.4. In this case $n = 2$ and

$$C(x) = -\ln(x(1-x)), \quad F_1(x) = \ln x, \quad F_2(x) = \ln(1-x),$$

$$\xi^1 = a, \quad \xi^2 = b, \quad \psi(\xi) = \ln B(\xi^1, \xi^2),$$

with $\mathbb{E} = (-\infty, 0) \times (-\infty, 0)$.

## 1.5 Mixture Family

Let $F_i : \mathcal{X} \to \mathbb{R}$ be $n$ smooth functions, linearly independent on $\mathcal{X}$, satisfying the integral constraints

$$\int_{\mathcal{X}} F_j(x)\,dx = 0, \qquad 1 \le j \le n. \tag{1.5.13}$$

Consider another smooth function $C(x)$ on $\mathcal{X}$ with

$$\int_{\mathcal{X}} C(x) = 1. \tag{1.5.14}$$

Define the following family of probability densities

$$p(x;\xi) = C(x) + \xi^i F_i(x). \tag{1.5.15}$$

The statistical manifold $\mathcal{S} = \{p(x;\xi)\}$ is called a *mixture family* and $\xi^j$ are its *mixture parameters*. The dimension of the parameters space $\mathbb{E}$ is called the *order* of the mixture family. We note that $\mathcal{S}$ is an affine subspace of $\mathcal{P}(\mathcal{X})$.

Both the injectivity and the regularity conditions of the immersion $\iota : \mathbb{E} \to \mathcal{P}(\mathcal{X})$ result from the fact that $\{F_i(x)\}$ are linearly independent functions.

It is worth noting that the integral constraint (1.5.13) is compatible to the fact that $\partial_{\xi^j} p(x;\xi) = F_j(x)$ and $\int \partial_{\xi^j} p(x;\xi)\,dx = 0$, see (1.3.2). This constraints will be used in the next result, which deals with the properties of the log-likelihood function $\ell_x(\xi)$. In spite of its proof simplicity, given the importance of this result, a full proof is included.

**Proposition 1.5.1** *If $E_\xi[\cdot]$ denotes the expectation with respect to $p_\xi$, then*

*(i)* $\partial_j \ell_x(\xi) = \dfrac{F_j(x)}{p(x)}$;

*(ii)* $\partial_i \partial_j \ell_x(\xi) = -\dfrac{F_i(x)F_j(x)}{p(x;\xi)^2}$;

*(iii)* $\partial_i \partial_j \ell_x(\xi) = -\partial_i \ell_x(\xi)\, \partial_j \ell_x(\xi)$.

*Proof:*

(*i*) Differentiating and using that $\partial_{\xi^j} p(x;\xi) = F_j(x)$ yields

$$\partial_j \ell_x(\xi) = \partial_{\xi^j} \ln p(x;\xi) = \frac{\partial_{\xi^j} p(x;\xi)}{p(x;\xi)} = \frac{F_j(x)}{p(x;\xi)}.$$

(*ii*) Differentiating in (*i*) yields

$$\begin{aligned} \partial_i \partial_j \ell_x(\xi) &= \partial_{\xi^i} \frac{F_j(x)}{p(x;\xi)} = F_j(x) \frac{-\partial_{\xi^i} p(x;\xi)}{p(x;\xi)^2} \\ &= -\frac{F_i(x)F_j(x)}{p(x;\xi)^2}. \end{aligned}$$

(*iii*) From (*ii*) and (*i*) we get

$$\partial_i \partial_j \ell_x(\xi) = -\frac{F_i(x)}{p(x;\xi)} \frac{F_j(x)}{p(x;\xi)} = -\partial_i \ell_x(\xi)\, \partial_j \ell_x(\xi).$$

■

The next example provides the reason for the name "mixture family."

**Example 1.5.1 (Mixture of n + 1 Distributions)** Consider $n+1$ probability densities, $p_1, p_2, \ldots, p_n, p_{n+1}$, which are linearly independent on $\mathcal{X}$. Let $\mathbb{E} = \{\xi \in \mathbb{R}^n; \xi^j > 0; \sum_{j=1}^n \xi^j < 1\}$. The following weighted average is also a probability density

$$\begin{aligned} p(x;\xi) &= \xi^i p_i(x) + \Big(1 - \sum_{i=1}^n \xi^i\Big) p_{n+1}(x) \\ &= p_{n+1}(x) + \sum_{i=1}^n \xi^i (p_i(x) - p_{n+1}(x)), \end{aligned}$$

which becomes a mixture family with $C(x) = p_{n+1}(x)$ and $F_i(x) = p_i(x) - p_{n+1}(x)$. The integral constraints (1.5.13) and (1.5.14) follow easily from the properties of probability density functions.

**Proposition 1.5.2** *The statistical manifold $\mathcal{P}(\mathcal{X})$, with $\mathcal{X}$ finite (see Example 1.3.5), is a mixture family.*

*Proof:* Let $\mathcal{X} = \{x^1, x^2, \dots, x^n, x^{n+1}\}$ and consider the linearly independent probability densities $p_1, \dots, p_{n+1}$, defined by $p_j(x_i) = \delta_{ij}$. Consider their weighted average

$$p(x;\xi) = \xi^i p_i(x) + \Big(1 - \sum_{i=1}^{n} \xi^i\Big) p_{n+1}(x).$$

We can easily check that $p(x;\xi)$ is the distribution given by Example 1.3.5. From Example 1.5.1, it follows that the densities $p(x;\xi)$ form a mixture family. ∎

## 1.6 Fisher Information

This section introduces the Fisher information and studies its main properties solely from the differential geometric point of view.[4]

A fundamental object in differential geometry is the Riemannian metric tensor $g$ (from tangent vectors $X, Y$ it produces a real number $g(X, Y)$). The metric tensor is used to define the length of, and angle between tangent vectors. For this reason the metric $g$ must be symmetric and positive definite (i.e., a Riemannian metric). The metric tensor $g$ determines the Levi–Civita connection (Christoffel symbols) and implicitly the Riemannian curvature tensor field.

In differential geometry, the parallel transport is a way of transporting geometrical data along smooth curves in a manifold. If the manifold is equipped with a symmetric affine connection $\nabla$ (a covariant derivative or connection on the tangent bundle), then this connection allows one to transport tangent vectors of the manifold along curves so that they stay parallel with respect to the connection. A connection $\nabla$ determines the curvature tensor field $R$ of type $(1,3)$. A manifold is called: (i) $\nabla$-flat if there is a coordinate system in which $\Gamma_{ij}^k = 0$; (ii) flat if $R = 0$. It is worth noting that (i) depends on the coordinate system while (ii) does not.

[4]This follows the 1940s idea of C. R. Rao and H. Jeffreys to use the Fisher information to define a Riemannian metric.

### Definition and Properties

A metric structure, similar to the Riemannian structure on a hypersurface, can be introduced on a statistical model, based on the parameter $\xi = (\xi_1, \dots, \xi_n) \in \mathbb{E}$ and the log-likelihood function $\ell(\xi) = \ln p_\xi(x)$. The *Fisher information matrix* is defined by

$$g_{ij}(\xi) = E_\xi[\partial_i \ell(\xi) \partial_j \ell(\xi)], \quad \forall i, j \in \{1, \dots, n\}, \tag{1.6.16}$$

which can be written explicitly as

$$g_{ij}(\xi) = E_\xi[\partial_{\xi^i} \ln p_\xi \cdot \partial_{\xi^j} \ln p_\xi] = \int_{\mathcal{X}} \partial_{\xi^i} \ln p_\xi(x) \cdot \partial_{\xi^j} \ln p_\xi(x) \cdot p_\xi(x)\, dx.$$

We shall assume that the previous integral exists. There are several equivalent formulas for $g_{ij}(\xi)$, which will be presented in the following.

**Proposition 1.6.1** *The Fisher information matrix can be represented in terms of the square root of probability densities as*

$$g_{ij}(\xi) = 4 \int_{\mathcal{X}} \partial_{\xi^i} \sqrt{p_\xi(x)} \cdot \partial_{\xi^j} \sqrt{p_\xi(x)}\, dx.$$

*Proof:* A straightforward computation yields

$$\begin{aligned}
g_{ij}(\xi) &= \int_{\mathcal{X}} \partial_{\xi^i} \ln p_\xi(x) \cdot \partial_{\xi^j} \ln p_\xi(x) \cdot p_\xi(x)\, dx \\
&= \int_{\mathcal{X}} \frac{\partial_{\xi^i} p_\xi(x)}{p_\xi(x)} \cdot \frac{\partial_{\xi^j} p_\xi(x)}{p_\xi(x)} \cdot p_\xi(x)\, dx \\
&= 4 \int_{\mathcal{X}} \frac{\partial_{\xi^i} p_\xi(x)}{2\sqrt{p_\xi(x)}} \cdot \frac{\partial_{\xi^j} p_\xi(x)}{2\sqrt{p_\xi(x)}}\, dx \\
&= 4 \int_{\mathcal{X}} \partial_{\xi^i} \sqrt{p_\xi(x)} \cdot \partial_{\xi^j} \sqrt{p_\xi(x)}\, dx.
\end{aligned}$$

■

We note that the discrete version of Proposition 1.6.1 states

$$g_{ij}(\xi) = 4 \sum_k \partial_{\xi^i} \sqrt{p_\xi(x_k)}\, \partial_{\xi^j} \sqrt{p_\xi(x_k)}. \tag{1.6.17}$$

**Example 1.6.1** Consider the case of a two-dimensional discrete probability model given by

| $X$ | $x_1$ | $x_2$ | $x_3$ |
|---|---|---|---|
| $P(X = x_k)$ | $\xi^1$ | $\xi^2$ | $1 - \xi^1 - \xi^2$ |

with $\xi^i \in (0,1)$. The previous table denotes a random variable $X$ with three outcomes $x_1, x_2, x_3$, which occur with probabilities $\xi_1$, $\xi_2$, and $1 - \xi^1 - \xi^2$. This is a statistical model depending on two parameters, $\xi^1$ and $\xi^2$. Formula (1.6.17) provides the Fisher information matrix

$$\begin{aligned} g_{ij}(\xi) &= 4\Big(\partial_{\xi^i}\sqrt{\xi^1}\cdot\partial_{\xi^j}\sqrt{\xi^1} + \partial_{\xi^i}\sqrt{\xi^2}\cdot\partial_{\xi^j}\sqrt{\xi^2}\Big) \\ &\quad +4\Big(\partial_{\xi^i}\sqrt{1-\xi^1-\xi^2}\cdot\partial_{\xi^j}\sqrt{1-\xi^1-\xi^2}\Big) \\ &= \frac{\delta_{i_1 j_1}}{\xi^1} + \frac{\delta_{i_2 j_2}}{\xi^2} + \frac{1}{\sqrt{1-\xi^1-\xi^2}}. \\ (g_{ij}) &= \begin{pmatrix} \frac{1}{\xi^1} + \frac{1}{\sqrt{1-\xi^1-\xi^2}} & \frac{1}{\sqrt{1-\xi^1-\xi^2}} \\ \frac{1}{\sqrt{1-\xi^1-\xi^2}} & \frac{1}{\xi^2} + \frac{1}{\sqrt{1-\xi^1-\xi^2}} \end{pmatrix}. \end{aligned}$$

The matrix is non-degenerate since

$$\det(g_{ij}) = \frac{1}{\xi^1\xi^2} + \frac{1}{\sqrt{1-\xi^1-\xi^2}}\Big(\frac{1}{\xi^1} + \frac{1}{\xi^2}\Big) \neq 0, \quad \forall \xi^i \in (0,1).$$

In fact, this property holds for all Fisher information matrices, as the next result shows.

**Proposition 1.6.2** *The Fisher information matrix on any statistical model is symmetric, positive definite and non-degenerate (i.e., a Riemannian metric).*

*Proof:* The symmetry follows from formula (1.6.16).
Forall $\xi$, $\forall v \in T_\xi\mathcal{S}$, $v \neq 0$, we find using Proposition 1.6.1

$$\begin{aligned} g(v,v) &= \sum_{i,j} g_{ij}v^i v^j = 4\sum_{i,j}\Big(\int_{\mathcal{X}} v^i\partial_{\xi^i}\sqrt{p_{\xi^i}(x)}v^j\partial_{\xi^j}\sqrt{p_{\xi^j}(x)}\Big)\,dx \\ &= 4\int_{\mathcal{X}}\Big(\sum_i v^i\partial_{\xi^i}\sqrt{p}\Big)\Big(\sum_j v^j\partial_{\xi^j}\sqrt{p}\Big)\,dx \\ &= 4\int_{\mathcal{X}}\Big(\sum_i v^i\partial_{\xi^i}\sqrt{p}\Big)^2\,dx \geq 0, \end{aligned}$$

so $(g_{ij})$ is non-negative definite.

Next we shall show that $g$ is non-degenerate:

$$\begin{aligned}
g(v,v) &= 0 \Leftrightarrow \int_{\mathcal{X}} \Big(\sum_i v^i \partial_{\xi^i}\sqrt{p}\Big)^2 dx = 0 \Leftrightarrow \\
\Big(\sum_i v^i \partial_{\xi^i}\sqrt{p}\Big)^2 &= 0 \Leftrightarrow \sum_i v^i \partial_{\xi^i}\sqrt{p} = 0 \Leftrightarrow \\
\sum_i v^i \partial_{\xi^i} p &= 0 \Leftrightarrow v^i = 0, \qquad \forall i = 1,\dots,n,
\end{aligned}$$

since $\{\partial_{\xi^i} p\}$ are linear independent, which is an assumption made previously.

Since $(g_{ij})$ is non-degenerate, we have in fact that

$$4\int_{\mathcal{X}} \Big(\sum_i v^i \partial_{\xi^i}\sqrt{p}\Big)^2 dx > 0,$$

and hence $(g_{ij})$ is positive definite. ■

Hence the Fisher information matrix provides the coefficients of a Riemannian metric on the surface $\mathcal{S}$. This allows us to measure distances, angles and define connections on statistical models.

The next formula is useful in practical applications.

**Proposition 1.6.3** *The Fisher information matrix can be written as the negative expectation of the Hessian of the log-likelihood function*

$$g_{ij}(\xi) = -E_\xi[\partial_{\xi^i}\partial_{\xi^j}\ell(\xi)] = -E_\xi[\partial_{\xi^i}\partial_{\xi^j}\ln p_\xi]. \qquad (1.6.18)$$

*Proof:* Differentiating in $\int_{\mathcal{X}} p(x,\xi)\,dx = 1$ yields

$$\int_{\mathcal{X}} \partial_{\xi^i} p(x,\xi)\,dx = 0,$$

which can be also written as

$$E_\xi[\partial_{\xi^i}\ln p_\xi] = \int_{\mathcal{X}} \partial_{\xi^i}\ln p(x,\xi)\; \cdot p_\xi(x)\,dx = 0.$$

Differentiating again with respect to $\xi^j$, we obtain

$$\begin{aligned}
\int_{\mathcal{X}} \partial_{\xi^j}\partial_{\xi^i}\ln p(x,\xi)\;\cdot p(x,\xi)\,dx + \int_{\mathcal{X}} \partial_{\xi^i}\ln p(x,\xi)\cdot \partial_{\xi^j} p(x,\xi)\,dx &= 0 \Leftrightarrow \\
E_\xi[\partial_{\xi^j}\partial_{\xi^i}\ln p_\xi] + \int_{\mathcal{X}} \partial_{\xi^i}\ln p(x,\xi)\cdot \partial_{\xi^j}\ln p(x,\xi)\cdot p(x,\xi)\,dx &= 0 \Leftrightarrow \\
E_\xi[\partial_{\xi^j}\partial_{\xi^i}\ln p_\xi] + g_{ij}(\xi) &= 0.
\end{aligned}$$

This relation implies (1.6.18). ■

**Example 1.6.2 (Exponential Distribution)** In this case $p_\xi(x) = \xi e^{-\xi x}$ and then $\partial_\xi^2 \ln p_\xi = -\dfrac{1}{\xi^2}$. Therefore

$$g = g_{11} = -E[\partial_\xi^2 \ln p_\xi] = \int_0^\infty \frac{1}{\xi^2} p_\xi(x)\,dx = \frac{1}{\xi^2}.$$

The length of a curve $\xi = \xi(t)$ joining the points $\xi_0 = \xi(0)$ and $\xi_1 = \xi(1)$ is

$$\int_0^1 \sqrt{g(\dot\xi(t),\dot\xi(t))}\,dt = \int_0^1 \sqrt{g_{11}\dot\xi^2(t)}\,dt = \int_0^1 \Big|\frac{\dot\xi(t)}{\xi(t)}\Big|\,dt = \Big|\ln\frac{\xi_1}{\xi_0}\Big|.$$

This induces on the one-dimensional statistical model $\mathcal{X} = (0,\infty)$ the following hyperbolic distance function

$$d(x_1,x_2) = |\ln x_2 - \ln x_1|.$$

**Example 1.6.3 (Exponential Family)** Recall the exponential family (1.4.7)

$$p(x;\xi) = e^{C(x)+\xi^i F_i(x)-\psi(\xi)}. \tag{1.6.19}$$

Using (1.4.11) and (1.4.12) we can represent the Fisher information matrix either in terms of the functions $F_i(x)$, or in terms of $\psi$

$$g_{ij}(\xi) = E_\xi[\partial_i \ell_x(\xi)\partial_j \ell_x(\xi)] = Cov_\xi(F_i,F_j), \tag{1.6.20}$$

$$g_{ij}(\xi) = -E_\xi[\partial_i\partial_j \ell_x(\xi)] = E_\xi[\partial_i\partial_j\psi(\xi)] = \partial_i\partial_j\psi(\xi). \tag{1.6.21}$$

It is worthy to note the role of convexity of function $\psi$ in the positive definiteness of $g_{ij}$.

**Example 1.6.4 (Mixture Family)** Recall the probability density family (1.5.15)

$$p(x;\xi) = C(x) + \xi^i F_i(x). \tag{1.6.22}$$

From Proposition 1.5.1, part *(ii)*, we obtain

$$g_{ij}(\xi) = -E_\xi[\partial_i\partial_j \ell_x(\xi)] = E_\xi\Big[\frac{F_i(x)F_j(x)}{p(x;\xi)^2}\Big] = \int_{\mathcal{X}} \frac{F_i(x)F_j(x)}{p(x;\xi)}\,dx.$$

The next two results deal with important invariance properties of the Fisher information metric.

**Theorem 1.6.4** *The Fisher metric is invariant under reparametrizations of the sample space.*

*Proof:* Let $X$ be a random variable with the sample space $\mathcal{X} \subseteq \mathbb{R}^n$, whose associated density function $p_\xi$ form a statistical model. Consider the invertible transform of sample spaces $f : \mathcal{X} \to \mathcal{Y} \subseteq \mathbb{R}^n$, and denote by $\tilde{p}_\xi(y)$ the density function associated with the random variable $Y = f(X)$. The relation between the foregoing densities and the Jacobian of $f$ is given by

$$p_\xi(x) = \tilde{p}_\xi(y)\left|\frac{df(x)}{dx}\right|. \tag{1.6.23}$$

Since the log-likelihood functions are given by

$$\begin{aligned}\tilde{\ell}(\xi) &= \ln \tilde{p}_\xi(y) = \ln \tilde{p}_\xi\big(f(x)\big) \\ \ell(\xi) &= \ln p_\xi(x) = \tilde{\ell}(\xi) + \ln\left|\frac{df(x)}{dx}\right|,\end{aligned}$$

and $f$ does not depend on the parameter $\xi$, we have

$$\partial_{\xi^i}\ell(\xi) = \partial_{\xi^i}\tilde{\ell}(\xi). \tag{1.6.24}$$

Therefore, using (1.6.23) and (1.6.24), the Fisher information transforms as

$$\begin{aligned}g_{ij}(\xi) &= \int_{\mathcal{X}} \partial_{\xi^i}\ell(\xi)\,\partial_{\xi^j}\ell(\xi)\,p_\xi(x)\,dx \\ &= \int_{\mathcal{X}} \partial_{\xi^i}\tilde{\ell}(\xi)\,\partial_{\xi^j}\tilde{\ell}(\xi)\,\tilde{p}_\xi(f(x))\left|\frac{df(x)}{dx}\right|dx \\ &= \int_{\mathcal{Y}} \partial_{\xi^i}\tilde{\ell}(\xi)\,\partial_{\xi^j}\tilde{\ell}(\xi)\,\tilde{p}_\xi(y)\,dy \\ &= \tilde{g}_{ij}(\xi),\end{aligned}$$

which proves the desired invariance property. ■

**Theorem 1.6.5** *The Fisher metric is covariant under reparametrizations of the parameters space.*

*Proof:* Consider two sets of coordinates $\xi = (\xi^1, \ldots, \xi^n)$ and $\theta = (\theta^1, \ldots, \theta^n)$ related by the invertible relationship $\xi = \xi(\theta)$, i.e., $\xi^j = \xi^j(\theta^1, \ldots, \theta^n)$. Let $\tilde{p}_\theta(x) = p_{\xi(\theta)}(x)$. By chain rule we have

$$\partial_{\theta^i}\tilde{p}_\theta = \frac{\partial \xi^k}{\partial \theta^i}\,\partial_{\xi^k}p_\xi, \qquad \partial_{\theta^j}\tilde{p}_\theta = \frac{\partial \xi^r}{\partial \theta^j}\,\partial_{\xi^r}p_\xi,$$

and then

$$\begin{aligned}\tilde{g}_{ij}(\theta) &= \int_{\mathcal{X}} \frac{1}{\tilde{p}_\theta(x)} \partial_{\theta^i} p_\theta(x)\, \partial_{\theta^j} p_\theta(x)\, dx \\ &= \left[ \int_{\mathcal{X}} \frac{1}{p_{\xi(\theta)}(x)} \partial_{\xi^k} p_\xi(x)\, \partial_{\xi^r} p_\xi(x)\, dx \right] \frac{\partial \xi^k}{\partial \theta^i} \frac{\partial \xi^r}{\partial \theta^j} \\ &= g_{kr}(\xi)\Big|_{\xi=\xi(\theta)} \frac{\partial \xi^k}{\partial \theta^i} \frac{\partial \xi^r}{\partial \theta^j}.\end{aligned}$$

■

Theorems 1.6.4 and 1.6.5 state that the Fisher metric has two distinguished properties:

1. $g_{ij}$ is invariant under reparametrizations of the sample space $\mathcal{X}$.
2. $g_{ij}$ is covariant under reparametrizations of the parameters space $\mathbb{E}$ (i.e., transforms as a 2-covariant tensor).

It is important to note that a metric satisfying the abovementioned properties is unique, and hence equal to the Fisher metric; this result can be found in Corcuera and Giummolé [30].

## 1.7 Christoffel Symbols

The most simple connection on the statistical model $\mathcal{S}$ is defined by the Christoffel symbols. If $g_{ij}$ denotes a Riemannian metric, particularly the Fisher information matrix, then the *Christoffel symbols of first kind* are given by

$$\Gamma_{ij,k} = \frac{1}{2}\Big( \partial_i g_{jk} + \partial_j g_{ik} - \partial_k g_{ij} \Big), \tag{1.7.25}$$

where we used the abbreviation $\partial_i = \partial_{\xi^i}$. Before computing the Christoffel symbols, a few equivalent expressions for the derivative of the Fisher information matrix are needed.

**Proposition 1.7.1** *With the notation $\ell = \ell_x(\xi)$, we have*

(*i*) $\partial_k g_{ij}(\xi) = -E_\xi[\partial_i \partial_j \partial_k \ell] - E_\xi[(\partial_i \partial_j \ell)(\partial_k \ell)]$;

(*ii*) $\partial_k g_{ij}(\xi) = E_\xi[(\partial_k \partial_i \ell)(\partial_j \ell)] + E_\xi[(\partial_k \partial_j \ell)(\partial_i \ell)] + E_\xi[(\partial_i \ell)(\partial_j \ell)(\partial_k \ell)]$;

(*iii*) $\partial_k g_{ij}(\xi) = 4 \int_{\mathcal{X}} \partial_i \partial_j \sqrt{p} \cdot \partial_k \sqrt{p}\, dx + 4 \int_{\mathcal{X}} \partial_i \partial_k \sqrt{p} \cdot \partial_j \sqrt{p}\, dx.$

*Proof:*

(*i*) Differentiating in (1.6.18), we get

$$\begin{aligned}\partial_k g_{ij}(\xi) &= -\partial_k E_\xi[\partial_i\partial_j\ell] \\ &= -\int_{\mathcal{X}}(\partial_k\partial_i\partial_j\ell)p(x;\xi)\,dx - \int(\partial_i\partial_j\ell)\partial_k p(x;\xi)\,dx \\ &= -\int_{\mathcal{X}}(\partial_k\partial_i\partial_j\ell)p(x;\xi)\,dx - \int(\partial_i\partial_j\ell)\partial_k\ell\, p(x;\xi)\,dx \\ &= -E_\xi[\partial_i\partial_j\partial_k\ell] - E_\xi[(\partial_i\partial_j\ell)(\partial_k\ell)].\end{aligned}$$

(*ii*) It follows from a direct application of the product rule in the definition relation (1.6.16) and the use of $\partial_k p(x;\xi) = \partial_k\ell(\xi)p(x;\xi)$:

$$\begin{aligned}\partial_k g_{ij}(\xi) =& \int_{\mathcal{X}}\partial_k\partial_i\ell\,\partial_j\ell\,p(x,\xi)\,dx + \int_{\mathcal{X}}\partial_i\ell\,\partial_k\partial_j\ell\,p(x,\xi)\,dx \\ &+ \int_{\mathcal{X}}\partial_i\ell\,\partial_j\ell\,\partial_k p(x,\xi)\,dx \\ =& E_\xi[\partial_k\partial_i\ell\,\partial_j\ell] + E_\xi[\partial_i\ell\,\partial_k\partial_j\ell] + E_\xi[\partial_i\ell\,\partial_j\ell\,\partial_k\ell].\end{aligned}$$

(*iii*) A computation using Proposition 1.6.1 shows

$$\begin{aligned}\partial_k g_{ij} &= 4\partial_k\int_{\mathcal{X}}\partial_i\sqrt{p}\cdot\partial_j\sqrt{p}\,dx \\ &= 4\int_{\mathcal{X}}\partial_k\partial_i\sqrt{p}\cdot\partial_j\sqrt{p}\,dx + 4\int_{\mathcal{X}}\partial_k\partial_j\sqrt{p}\cdot\partial_i\sqrt{p}\,dx.\end{aligned}$$

■

**Proposition 1.7.2** *The following equivalent expressions of the Christoffel symbols of first type hold:*

(*i*) $2\Gamma_{ij,k}(\xi) = E_\xi[(\partial_i\partial_j\ell)(\partial_k\ell)] - E_\xi[(\partial_k\partial_j\ell)(\partial_i\ell)] - E_\xi[(\partial_k\partial_i\ell)(\partial_j\ell)] - E_\xi[\partial_i\partial_j\partial_k\ell]$;

(*ii*) $\Gamma_{ij,k}(\xi) = E_\xi[(\partial_i\partial_j\ell + \frac{1}{2}\partial_i\ell\,\partial_j\ell)\partial_k\ell]$;

(*iii*) $\Gamma_{ij,k}(\xi) = 4\int_{\mathcal{X}}\partial_i\partial_j\sqrt{p(x;\xi)}\cdot\partial_k\sqrt{p(x;\xi)}\,dx.$

*Proof:* It follows from relations (*i*), (*ii*), and (*iii*), supplied by Proposition 1.7.1, after substituting in (1.7.25) and performing pair cancelations. ∎

It is worth noting that the aforementioned relations work also in the case of a discrete sample space. For instance, formula (*iii*) has the following discrete analog

$$\Gamma_{ij,k}(\xi) = 4\sum_{r=1}^{n} \partial_{\xi^i}\partial_{\xi^j}\sqrt{p_\xi(x_r)} \cdot \partial_{\xi^k}\sqrt{p_\xi(x_r)}. \tag{1.7.26}$$

**Example 1.7.1** We shall work out the Christoffel coefficients in the case of the discrete probability model given by Example 1.6.1 using (1.7.26).

If $i \neq j$, then

$$\begin{aligned}
\Gamma_{ij,k}(\xi) &= 4\sum_{k=1}^{2}\Big(\underbrace{\partial_{\xi^i}\partial_{\xi^j}\sqrt{\xi^r}\cdot\partial_{\xi^k}\sqrt{\xi^r}}_{=0}\Big) \\
&\quad +4\partial_{\xi^i}\partial_{\xi^j}\sqrt{1-\xi_1-\xi_2}\cdot\partial_{\xi^k}\sqrt{1-\xi_1-\xi_2} \\
&= \partial_{\xi^i}(1-\xi^1-\xi^2)^{-1/2}\cdot\frac{1}{\sqrt{1-\xi^1-\xi^2}} = \frac{1}{2(1-\xi^1-\xi^2)^2}.
\end{aligned}$$

Let $i = j$. Since the second term of the sum is the same as above, we have

$$\begin{aligned}
\Gamma_{ii,k}(\xi) &= 4\sum_{r=1}^{2}\partial^2_{\xi^i}\sqrt{\xi^r}\cdot\partial_{\xi^k}\sqrt{\xi^r} + \frac{1}{2(1-\xi^1-\xi^2)^2} \\
&= -\frac{1}{2}\sum_{r=1}^{2}\frac{\delta_i^r\delta_r^k}{\xi^{r2}} + \frac{1}{2(1-\xi^1-\xi^2)^2}.
\end{aligned}$$

If $i = j \neq k$ the first term of the above formula vanishes, so

$$\Gamma_{ii,k}(\xi) = \frac{1}{2(1-\xi^1-\xi^2)^2}.$$

**Example 1.7.2 (Exponential Family)** This refers to the family of distributions (1.4.7). We shall work out the Christoffel coefficients in terms of the function $\psi(\xi)$. Using relations (1.4.9), (1.4.10), and (1.4.12) yields

$$\begin{aligned}
E_\xi[(\partial_i\partial_j\ell)(\partial_k\ell)] &= -E_\xi[(\partial_i\partial_j\psi)(F_k(x)-\partial_k\psi)] \\
&= -(\partial_i\partial_j\psi)E_\xi[F_k(x)] + (\partial_i\partial_j\psi)(\partial_k\psi) \\
&= (\partial_i\partial_j\psi)\Big(\partial_k\psi - E_\xi[F_k(x)]\Big) = 0.
\end{aligned}$$

Substituting in Proposition 1.7.2, part $(i)$, we find

$$\Gamma_{ij,k} = \frac{1}{2}E_\xi[\partial_i\partial_j\partial_k\ell] = \frac{1}{2}E_\xi[\partial_i\partial_j\partial_k\psi] = \frac{1}{2}\partial_i\partial_j\partial_k\psi(\xi).$$

**Example 1.7.3 (Mixture Family)** Consider the family of distributions (1.5.15). From Propositions 1.5.1 and 1.7.2, we have

$$\begin{aligned}\Gamma_{ij,k}(\xi) &= E_\xi\big[(\partial_i\partial_j\ell + \frac{1}{2}\partial_i\ell\,\partial_j\ell)\partial_k\ell\big] = E_\xi\big[(-\partial_i\ell\,\partial_j\ell + \frac{1}{2}\partial_i\ell\,\partial_j\ell)\partial_k\ell\big] \\ &= -\frac{1}{2}E_\xi[\partial_i\ell\,\partial_j\ell\,\partial_k\ell] = -\frac{1}{2}\int\frac{F_i(x)F_j(x)F_k(x)}{p(x;\xi)^2}\,dx.\end{aligned}$$

## 1.8 Levi–Civita Connection

The coefficients (1.6.16) induce a Riemannian metric on $\mathcal{S}$. This is a 2-covariant tensor $g$ defined locally by

$$g(X_\xi, Y_\xi) = \sum_{i,j=1}^n g_{ij}(\xi)a^i(\xi)b^j(\xi), \qquad p_\xi \in \mathcal{S}$$

where $X_\xi = \sum_{i=1}^n a^i(\xi)\partial_i$ and $Y_\xi = \sum_{i=1}^n b^i(\xi)\partial_i$ are vector fields in the 0-representation on $\mathcal{S}$. Here, for the sake of simplicity we used the notation abuse $\partial_i = \frac{\partial p_\xi}{\partial \xi^i}$. The tensor $g$ is called the *Fisher–Riemann metric*. Its associated *Levi–Civita connection* is denoted by $\nabla^{(0)}$ and is defined by

$$g(\nabla^{(0)}_{\partial_i}\partial_j, \partial_k) = \Gamma_{ij,k},$$

with $\Gamma_{ij,k}$ given by (1.7.25). The fact that $\nabla^{(0)}$ is metrical connection can be written locally as

$$\partial_k g_{ij} = \Gamma_{ki,j} + \Gamma_{kj,p}.$$

This can be checked also directly from Propositions 1.7.1 and 1.7.2.

It is worth noting that the superscript of $\nabla^{(0)}$ denotes a parameter of the connection. The same superscript convention will be also employed for Christoffel symbols. In the next section we shall introduce several connections parameterized by the parameter $\alpha$. The case $\alpha = 0$ corresponds to the Levi–Civita connection induced by the Fisher metric.

## 1.9 $\nabla^{(1)}$-Connection

We shall introduce a new linear connection $\nabla^{(1)}$ on the statistical model $S$. Generally, to define a linear connection, it suffices to know its components on a basis, and to have a metric $g$ that rises and lowers indices. Using the Fisher metric $g$, the $\nabla^{(1)}$-*connection* is defined by

$$g(\nabla^{(1)}_{\partial_i}\partial_j, \partial_k) = E_\xi[(\partial_i\partial_j\ell)\,(\partial_k\ell)]. \tag{1.9.27}$$

It can also be expressed equivalently by stating directly the Christoffel coefficients

$$\Gamma^{(1)}_{ij,k}(\xi) = E_\xi[(\partial_i\partial_j\ell)\,(\partial_k\ell)]. \tag{1.9.28}$$

In the following the $\nabla$-flatness is considered with respect to the system of coordinates $\xi$. The next result shows the importance of the $\nabla^{(1)}$-connection.

**Proposition 1.9.1** *The statistical model given by the exponential family (1.4.7) is $\nabla^{(1)}$-flat.*

*Proof:* We need to show that $\Gamma^{(1)}_{ij,k}(\xi) = 0$. Using (1.4.12) and Proposition 1.3.2 we have

$$\begin{aligned}
\Gamma^{(1)}_{ij,k}(\xi) &= E_\xi[(\partial_i\partial_j\ell)\,(\partial_k\ell)] = -E_\xi[\partial_i\partial_j\psi(\xi)\,(\partial_k\ell)] \\
&= -\partial_i\partial_j\psi(\xi)E_\xi[(\partial_k\ell)] = 0.
\end{aligned}$$

■

As a consequence, the torsion and curvature of an exponential family with respect to connection $\nabla^{(1)}$ vanish everywhere.

**Proposition 1.9.2** *The mixture family (1.5.15) is $\nabla^{(1)}$-flat if and only if is $\nabla^{(0)}$-flat.*

*Proof:* Since

$$\begin{aligned}
\Gamma^{(0)}_{ij,k}(\xi) &= -\frac{1}{2}E_\xi[((\partial_i\ell)\,(\partial_j\ell)\,(\partial_k\ell)] \\
\Gamma^{(1)}_{ij,k}(\xi) &= E_\xi[(\partial_i\partial_j\ell)\,(\partial_k\ell)] = -E_\xi[((\partial_i\ell)\,(\partial_j\ell)\,(\partial_k\ell)],
\end{aligned}$$

we have $\Gamma^{(0)}_{ij,k}(\xi) = 0$ if and only if $\Gamma^{(1)}_{ij,k}(\xi) = 0$. The first identity is provided by Example 1.7.3, and the second uses Proposition 1.5.1, part $(iii)$. ■

## 1.10 $\nabla^{(-1)}$-Connection

Using the Fisher metric $g$, the $\nabla^{(-1)}$-connection on a statistical model $S$ is defined by

$$g(\nabla^{(-1)}_{\partial_i}\partial_j, \partial_k) = \Gamma^{(-1)}_{ij,k} = E_\xi[(\partial_i\partial_j\ell + \partial_i\ell\,\partial_j\ell)\,(\partial_k\ell)], \qquad (1.10.29)$$

where $\ell$ is the log-likelihood function.

**Proposition 1.10.1** *The mixture family (1.5.15) is $\nabla^{(-1)}$-flat.*

*Proof:* It follows from the fact that in any mixture family $\partial_i\partial_j\ell = -\partial_i\ell\,\partial_j\ell$. ■

The $\nabla^{(-1)}$-connection is related to the $\nabla^{(0)}$ and $\nabla^{(1)}$ connections.

**Proposition 1.10.2** *The relation between the foregoing three connections is given by*

$$\nabla^{(0)} = \frac{1}{2}\Big(\nabla^{(-1)} + \nabla^{(1)}\Big). \qquad (1.10.30)$$

*Proof:* It suffices to show

$$\Gamma^{(0)}_{ij,k} = \frac{1}{2}\Big(\Gamma^{(-1)}_{ij,k} + \Gamma^{(1)}_{ij,k}\Big).$$

Using (1.9.28) and (1.10.29) and Proposition 1.7.2, part (*ii*), we find

$$\begin{aligned}
\Gamma^{(-1)}_{ij,k} + \Gamma^{(1)}_{ij,k} &= E_\xi[(\partial_i\partial_j\ell + \partial_i\ell\,\partial_j\ell)\,(\partial_k\ell)] + E_\xi[(\partial_i\partial_j\ell)(\partial_k\ell)] \\
&= E_\xi[(2\partial_i\partial_j\ell + \partial_i\ell\,\partial_j\ell)\,(\partial_k\ell)] \\
&= 2E_\xi[(\partial_i\partial_j\ell + \frac{1}{2}\partial_i\ell\,\partial_j\ell)\,(\partial_k\ell)] \\
&= 2\Gamma^{(0)}_{ij,k}.
\end{aligned}$$

■

**Corollary 1.10.3** *An exponential family is $\nabla^{(-1)}$-flat if and only if is $\nabla^{(0)}$-flat.*

None of the connections $\nabla^{(-1)}$ and $\nabla^{(1)}$ are metrical. However, they are related to the Fisher metric $g$ by the following duality relation.

**Proposition 1.10.4** *For any vector fields $X$, $Y$, $Z$ on the statistical model $\mathcal{S} = \{p_\xi\}$, we have*

$$Zg(X,Y) = g(\nabla_Z^{(1)} X, Y) + g(X, \nabla_Z^{(-1)} Y). \tag{1.10.31}$$

*Proof:* It suffices to prove the relation on a basis. Choosing $X = \partial_i$, $Y = \partial_j$ and $Z = \partial_k$, the relation we need to show becomes

$$\partial_k g_{ij} = \Gamma^{(1)}_{ki,j} + \Gamma^{(-1)}_{kj,i}. \tag{1.10.32}$$

Using (1.9.28) and (1.10.29) we get

$$\begin{aligned}
\Gamma^{(1)}_{ki,j} + \Gamma^{(-1)}_{kj,i} &= E_\xi[(\partial_k\partial_i\ell)\,(\partial_j\ell)] + E_\xi[(\partial_k\partial_j\ell + \partial_k\ell\,\partial_j\ell)\,(\partial_i\ell)] \\
&= E_\xi[(\partial_k\partial_i\ell)(\partial_j\ell)] + E_\xi[(\partial_k\partial_j\ell)(\partial_i\ell)] \\
&\quad + E_\xi[(\partial_i\ell)(\partial_j\ell)(\partial_k\ell)] \\
&= \partial_k g_{ij}(\xi),
\end{aligned}$$

where the last identity follows from Proposition 1.7.1, part (*ii*). ∎

## 1.11 $\nabla^{(\alpha)}$-Connection

The $\nabla^{(-1)}$ and $\nabla^{(1)}$ are two special connections on $S$ with respect to which the mixture family and the exponential family are, respectively, flat. Moreover, they are related by the duality condition (1.10.31). Midway between these connections there is the $\nabla^{(0)}$-connection, which is the Levi–Civita connection with respect to the Fisher metric. Interpolating, we define the following 1-parameter family of connections

$$\nabla^{(\alpha)} = \frac{1+\alpha}{2}\nabla^{(1)} + \frac{1-\alpha}{2}\nabla^{(-1)}, \tag{1.11.33}$$

with $\alpha$ real parameter, on the statistical model $S$. For $\alpha = -1, 0, 1$ we obtain, respectively, the connections $\nabla^{(-1)}$, $\nabla^{(0)}$ and $\nabla^{(1)}$. Using the Fisher metric $g$, the connection components

$$\Gamma^{(\alpha)}_{ij,k} = g(\nabla^{(\alpha)}_{\partial_i}\partial_j, \partial_k)$$

are given by the following result.

**Proposition 1.11.1** *The components $\Gamma^{(\alpha)}_{ij,k}$ can be written as*

$$\Gamma^{(\alpha)}_{ij,k} = E_\xi\Big[\Big(\partial_i\partial_j\ell + \frac{1-\alpha}{2}\partial_i\ell\partial_j\ell\Big)\partial_k\ell\Big]. \tag{1.11.34}$$

*Proof:* From (1.11.33) written on components and using (1.9.28) and (1.10.29), we obtain

$$\begin{aligned}
\Gamma_{ij,k}^{(\alpha)} &= \frac{1+\alpha}{2}\Gamma_{ij,k}^{(1)} + \frac{1-\alpha}{2}\Gamma_{ij,k}^{(-1)} \\
&= \frac{1+\alpha}{2}E_\xi[(\partial_i\partial_j\ell)(\partial_k\ell)] + \frac{1-\alpha}{2}E_\xi[(\partial_i\partial_j\ell + \partial_i\ell\partial_j\ell)(\partial_k\ell)] \\
&= E_\xi[(\partial_i\partial_j\ell)(\partial_k\ell) + \frac{1-\alpha}{2}\partial_i\ell\partial_j\ell\partial_k\ell] \\
&= E_\xi\Big[\Big(\partial_i\partial_j\ell + \frac{1-\alpha}{2}\partial_i\ell\partial_j\ell\Big)\partial_k\ell\Big].
\end{aligned}$$

■

The connection $\nabla^{(\alpha)}$ is symmetric. This follows directly from $\Gamma_{ij,k}^{(\alpha)} = \Gamma_{ji,k}^{(\alpha)}$. However, $\nabla^{(\alpha)}$ is not metrical for $\alpha \neq 0$. The pair connections $\nabla^{(\alpha)}$ and $\nabla^{(-\alpha)}$ are in the same duality relation as connections $\nabla^{(1)}$ and $\nabla^{(-1)}$.

**Proposition 1.11.2** *For any vector fields $X, Y, Z$ on the statistical model $\mathcal{S} = \{p_\xi\}$, we have*

$$Zg(X,Y) = g(\nabla_Z^{(\alpha)}X, Y) + g(X, \nabla_Z^{(-\alpha)}Y). \tag{1.11.35}$$

*Proof:* It suffices to prove the local version of (1.11.35)

$$\partial_k g_{ij} = \Gamma_{ki,j}^{(\alpha)} + \Gamma_{kj,i}^{(-\alpha)}. \tag{1.11.36}$$

Using (1.11.34), we write

$$\begin{aligned}
\Gamma_{ki,j}^{(\alpha)} &= E_\xi\Big[\Big(\partial_k\partial_i\ell + \frac{1-\alpha}{2}\partial_k\ell\partial_i\ell\Big)\partial_j\ell\Big] \\
\Gamma_{kj,i}^{(-\alpha)} &= E_\xi\Big[\Big(\partial_k\partial_j\ell + \frac{1+\alpha}{2}\partial_k\ell\partial_j\ell\Big)\partial_i\ell\Big].
\end{aligned}$$

Taking the sum and reducing terms yields

$$\begin{aligned}
\Gamma_{ki,j}^{(\alpha)} + \Gamma_{kj,i}^{(-\alpha)} &= E_\xi[(\partial_k\partial_i\ell)(\partial_j\ell)] + E_\xi[(\partial_k\partial_j\ell)(\partial_i\ell)] \\
&\quad + E_\xi[(\partial_i\ell)(\partial_j\ell)(\partial_k\ell)] \\
&= \partial_k g_{ij}(\xi),
\end{aligned}$$

see Proposition 1.7.1, part *(ii)*. ■

Other relations among connections $\nabla^{(0)}$, $\nabla^{(1)}$ and $\nabla^{(-1)}$ are given in the following result. The proof follows from a direct application of relation (1.11.33).

**Proposition 1.11.3** *The following relations hold*

$$\begin{aligned}\nabla^{(\alpha)} &= (1-\alpha)\nabla^{(0)} + \alpha\nabla^{(1)} \\ &= (1+\alpha)\nabla^{(0)} - \alpha\nabla^{(-1)} \\ &= \nabla^{(0)} + \frac{\alpha}{2}(\nabla^{(1)} - \nabla^{(-1)}); \\ \nabla^{(0)} &= \frac{1}{2}\Big(\nabla^{(-\alpha)} + \nabla^{(\alpha)}\Big).\end{aligned}$$

**Corollary 1.11.4** *Let $\mathcal{S}$ be a statistical model. Then the following statements are equivalent:*

*(i) $\mathcal{S}$ is $\nabla^{(\alpha)}$-flat for all $\alpha \in \mathbb{R}$;*

*(ii) $\mathcal{S}$ is both $\nabla^{(1)}$ and $\nabla^{(-1)}$-flat;*

*(iii) $\mathcal{S}$ is both $\nabla^{(0)}$ and $\nabla^{(1)}$-flat;*

*(iv) $\mathcal{S}$ is both $\nabla^{(0)}$ and $\nabla^{(-1)}$-flat.*

In general, if a model $\mathcal{S}$ is flat with respect to two distinct $\alpha$-connections, then is flat with respect to all $\alpha$-connections.

## 1.12 Skewness Tensor

Generally, it is well known that the difference of two linear connections is a tensor field. For two connections, $\nabla^{(\alpha)}$ and $\nabla^{(\beta)}$, on the statistical model $\mathcal{S}$, we define the *generalized difference tensor* as the $(2,1)$-type tensor

$$K^{(\alpha,\beta)}(X,Y) = \nabla_X^{(\beta)}Y - \nabla_X^{(\alpha)}Y. \tag{1.12.37}$$

**Proposition 1.12.1** *There is a 3-covariant, totally symmetric tensor $T$ such that for any distinct $\alpha, \beta$ we have*

$$g\big(K^{(\alpha,\beta)}(X,Y), Z\big) = \frac{\alpha-\beta}{2}T(X,Y,Z), \tag{1.12.38}$$

*where $g$ is the Fisher metric.*

*Proof:* Using (1.11.34)

$$\begin{aligned}\Gamma_{ij,k}^{(\beta)}(\xi) - \Gamma_{ij,k}^{(\alpha)}(\xi) &= E_\xi\Big[\Big(\partial_i\partial_j\ell + \frac{1-\beta}{2}\partial_i\ell\partial_j\ell\Big)\partial_k\ell\Big] \\ &\quad -E_\xi\Big[\Big(\partial_i\partial_j\ell + \frac{1-\alpha}{2}\partial_i\ell\partial_j\ell\Big)\partial_k\ell\Big] \\ &= \frac{\alpha-\beta}{2}E_\xi[\partial_i\ell\,\partial_j\ell\,\partial_k\ell],\end{aligned}$$

Choosing the components

$$T_{ijk}(\xi) = E_\xi[\partial_i \ell \, \partial_j \ell \, \partial_k \ell],$$

we find

$$\Gamma^{(\beta)}_{ij,k}(\xi) - \Gamma^{(\alpha)}_{ij,k}(\xi) = \frac{\alpha - \beta}{2} T_{ijk}(\xi), \tag{1.12.39}$$

which is the local coordinates version of equation (1.12.38). We note that $T$ is symmetric in any of the indices $i, j, k$, i.e., it is totally symmetric. ■

The fact that $T$ is covariant under parametrizations is the subject of Problem 1.16.

The 3-covariant, symmetric tensor $T$ with components

$$T(\partial_i, \partial_j, \partial_k) = T_{ijk} = E_\xi[\partial_i \ell \, \partial_j \ell \, \partial_k \ell]$$

is called the *skewness tensor*. This measures the expectation of the third-order cummulants of the variations of log-likelihood function. It is worth noting the similarity with the Fisher metric, which measures the expectation of the second-order cummulants.

**Example 1.12.1** In the case of an exponential family we find

$$\begin{aligned}
\Gamma^{(\alpha)}_{ij,k} &= E_\xi\Big[\Big(\partial_i\partial_j \ell + \frac{1-\alpha}{2}\partial_i \ell \partial_j \ell\Big)\partial_k \ell\Big] \\
&= E_\xi\Big[\Big(-\partial_i\partial_j \psi(\xi) + \frac{1-\alpha}{2}\partial_i \ell \partial_j \ell\Big)\partial_k \ell\Big] \\
&= -\partial_i\partial_j\psi(\xi) E_\xi[\partial_k \ell] + \frac{1-\alpha}{2} E_\xi[\partial_i \ell \partial_j \ell \partial_k \ell] \\
&= \frac{1-\alpha}{2} T_{ijk}.
\end{aligned}$$

We note that $\Gamma^{(1)}_{ij,k} = 0$, and hence an exponential family is $\nabla^{(1)}$-flat.

**Example 1.12.2** In the case of a mixture family we have

$$\begin{aligned}
\Gamma^{(\alpha)}_{ij,k} &= E_\xi\Big[\Big(\partial_i\partial_j \ell + \frac{1-\alpha}{2}\partial_i \ell \partial_j \ell\Big)\partial_k \ell\Big] \\
&= E_\xi\Big[\Big(-\partial_i \ell \, \partial_j \ell + \frac{1-\alpha}{2}\partial_i \ell \partial_j \ell\Big)\partial_k \ell\Big] \\
&= -E_\xi\Big[\frac{1+\alpha}{2}\partial_i \ell \partial_j \ell \partial_k \ell\Big] \\
&= -\frac{1+\alpha}{2} T_{ijk}.
\end{aligned}$$

We have that $\Gamma^{(-1)}_{ij,k} = 0$, i.e., a mixture family is $-1$-flat.

It is worth noting that there are statistical models, that are neither exponential nor mixture families, which are $\nabla^{(\alpha)}$-flat, for any $\alpha$, see Problem 1.15.

**Example 1.12.3** Next we shall compute the skewness tensor for the exponential distribution. Consider $p(x;\xi) = \xi e^{-\xi x}$, $\xi > 0$ and $x \geq 0$. Then $\partial_\xi \ell = \dfrac{1}{\xi} - x$ and the skewness tensor is given by the following component

$$\begin{aligned} T_{111} &= E_\xi[(\partial_\xi \ell)^2] = \int_0^\infty \Big(\frac{1}{\xi} - x\Big)^2 p(x;\xi)\, dx \\ &= \frac{1}{\xi^2}\int_0^\infty (1-\xi x)^3 e^{-\xi x}\, dx = -\frac{2}{\xi^3}. \end{aligned}$$

The next result deals with a few useful formulas for the skewness tensor.

**Proposition 1.12.2** *The skewness tensor on the statistical model $\mathcal{S} = \{p_\xi(x); x \in \mathcal{X}, \xi \in \mathbb{E}\}$ is given by:*

$$\begin{aligned} (a)\quad & T_{ijk}(\xi) = 27\int_{\mathcal{X}} \partial_i(p_\xi^{1/3}(x))\partial_j(p_\xi^{1/3}(x))\partial_k(p_\xi^{1/3}(x))\, dx; \\ (b)\quad & T_{ijk}(\xi) = -E_\xi[\partial_i\partial_j\partial_k \ell] - E_\xi[(\partial_j\partial_k\ell)(\partial_i\ell)] \\ & \qquad - E_\xi[(\partial_k\partial_i\ell)(\partial_j\ell)] - E_\xi[(\partial_i\partial_j\ell)(\partial_k\ell)]; \\ (c)\quad & T_{ijk}(\xi) = 2E_\xi[\partial_i\partial_j\partial_k\ell] + \partial_i g_{jk}(\xi) + \partial_j g_{ki}(\xi) + \partial_k g_{ij}(\xi). \end{aligned}$$

*Proof:* $(a)$ We have

$$\begin{aligned} T_{ijk}(\xi) &= E_\xi[(\partial_i\ell)(\partial_j\ell)(\partial_k\ell)] \\ &= \int_{\mathcal{X}} (\partial_i \ln p(x))(\partial_j \ln p(x))(\partial_k \ln p(x))p(x)\, dx \\ &= \int_{\mathcal{X}} \frac{\partial_i p(x)}{p(x)}\frac{\partial_j p(x)}{p(x)}\frac{\partial_k p(x)}{p(x)} p(x)\, dx \\ &= 3^3\int_{\mathcal{X}} \partial_i(p(x)^{1/3})\partial_j(p(x)^{1/3})\partial_k(p(x)^{1/3})\, dx. \end{aligned}$$

$(b)$ Equating the two formulas of the Fisher information $g_{ij}(\xi) = E_\xi[\partial_i\ell\partial_j\ell]$ and $g_{ij}(\xi) = -E_\xi[\partial_i\partial_j\ell]$, we obtain

$$\int_{\mathcal{X}} \partial_i\ell\, \partial_j\ell\, p(x)\, dx = -\int_{\mathcal{X}} \partial_i\partial_j\ell\, p(x)\, dx.$$

Differentiate with respect to $\xi^k$:

$$\int_{\mathcal{X}} \partial_i\partial_k \ell\, \partial_j \ell\, p(x)\, dx + \int_{\mathcal{X}} \partial_i \ell\, \partial_j\partial_k \ell\, p(x)\, dx + \int_{\mathcal{X}} \partial_i \ell\, \partial_j \ell\, \partial_k p(x)\, dx$$
$$= -\int_{\mathcal{X}} \partial_i\partial_j\partial_k \ell\, p(x)\, dx - \int_{\mathcal{X}} \partial_i\partial_j \ell\, \partial_k p(x)\, dx \iff$$
$$E_\xi[(\partial_i\partial_k\ell)(\partial_j\ell)] + E_\xi[(\partial_i\ell)(\partial_j\partial_k\ell)] + \int_{\mathcal{X}} \partial_i \ell\, \partial_j \ell\, \partial_k \ell\, p(x)\, dx$$
$$= E_\xi[\partial_i\partial_j\partial_k\ell] - \int_{\mathcal{X}} \partial_i\partial_j \ell\, \partial_k \ell\, p(x)\, dx \iff$$
$$E_\xi[(\partial_i\partial_k\ell)(\partial_j\ell)] + E_\xi[(\partial_j\partial_k\ell)(\partial_i\ell)] + T_{ijk}(\xi)$$
$$= -E_\xi[\partial_i\partial_j\partial_k\ell] - E_\xi[(\partial_i\partial_j\ell)(\partial_k\ell)]$$

Solving for $T_{ijk}(\xi)$ leads to the desired formula.

($c$) Applying Proposition 1.7.1 parts ($i$) and ($ii$), we have

$$\begin{aligned} T_{ijk}(\xi) &= E_\xi[(\partial_i\ell)(\partial_j\ell)(\partial_k\ell)] \\ &= \partial_k g_{ij}(\xi) - E_\xi[(\partial_k\partial_i\ell)(\partial_j\ell)] - E_\xi[(\partial_k\partial_j\ell)(\partial_i\ell)] \\ &= \partial_k g_{ij}(\xi) + \partial_j g_{ki}(\xi) + E_\xi[\partial_i\partial_j\partial_k\ell] + \partial_i g_{jk}(\xi) + E_\xi[\partial_i\partial_j\partial_k\ell] \\ &= 2E_\xi[\partial_i\partial_j\partial_k\ell] + \partial_i g_{jk}(\xi) + \partial_j g_{ki}(\xi) + \partial_k g_{ij}(\xi). \end{aligned}$$

These formulas can be used to find the skewness tensor for the cases of exponential and mixture families, see Problems 1.20. and 1.23.

It is worth noting that $T_{ijk}$ is covariant under reparametrizations (see Problem 1.16.) and it is invariant under transformations of the random variable (see Problem 1.18.). ■

An equivalent way of defining the skewness tensor $T$ is to choose $\beta = 0$ in (1.12.38)

$$\frac{\alpha}{2} T(X, Y, Z) = g\big(\nabla^{(0)}_X Y - \nabla^{(\alpha)}_X Y, Z\big). \tag{1.12.40}$$

Until now, all definitions of the skewness tensor involved two connections. The following result derives this tensor from only one connection and the Fisher metric. Recall the following derivative formula of a 2-covariant tensor $h$ with respect to a linear connection $\nabla$

$$(\nabla h)(X, Y, Z) = Xh(Y, Z) - h(\nabla_X Y, Z) - h(Y, \nabla_X Z).$$

We note that $\nabla h$ is a 3-covariant tensor (it acts on three vector fields).

**Theorem 1.12.3** *The metric $g$ and the tensor $T$ are related by*

$$\alpha T = \nabla^{(\alpha)} g.$$

*Proof:* We need to show that for any vector fields $X, Y, Z$ tangent to $\mathcal{S}$, we have

$$\alpha T(X, Y, Z) = Xg(Y, Z) - g(\nabla_X^{(\alpha)} Y, Z) - g(Y, \nabla_X^{(\alpha)} Z).$$

This can be written in local coordinates as

$$\alpha T_{ijk} = \partial_i g_{jk} - \Gamma_{ij,k}^{(\alpha)} - \Gamma_{ik,j}^{(\alpha)}. \tag{1.12.41}$$

We start working out the right side using (1.11.34) and Proposition 1.7.1, part *(ii)*

$$\begin{aligned}
\partial_i g_{jk} - \Gamma_{ij,k}^{(\alpha)} - \Gamma_{ik,j}^{(\alpha)} &= E_\xi[(\partial_i \partial_j \ell)(\partial_k \ell)] \\
&\quad + E_\xi[(\partial_i \partial_k \ell)(\partial_j \ell)] + E_\xi[(\partial_i \ell)(\partial_j \ell)(\partial_k \ell)] \\
&\quad - E_\xi[(\partial_i \partial_j \ell)\partial_k \ell] - \frac{1-\alpha}{2} E_\xi[\partial_i \ell\, \partial_j \ell\, \partial_k \ell] \\
&\quad - E_\xi[(\partial_i \partial_k \ell)\partial_j \ell] - \frac{1-\alpha}{2} E_\xi[\partial_i \ell\, \partial_j \ell\, \partial_k \ell] \\
&= \alpha E_\xi[\partial_i \ell\, \partial_j \ell\, \partial_k \ell] \\
&= \alpha T_{ijk},
\end{aligned}$$

which is (1.12.41). ■

In particular, for $\alpha = 1$, we obtain

$$T = \nabla^{(1)} g,$$

i.e., the skewness tensor is the $\nabla^{(1)}$-derivative of the Fisher metric $g$.

**Corollary 1.12.4** *For any $\epsilon > 0$, the tensor $T$ is given by*

$$T = \frac{\nabla^{(\alpha+\epsilon)} g - \nabla^{(\alpha)} g}{\epsilon}.$$

## 1.13 Autoparallel Curves

Consider a smooth curve $\xi(s)$ in the parameter space $\mathbb{E} \subset \mathbb{R}^n$, so each component of $\xi(s) = \big(\xi^1(s), \dots, \xi^n(s)\big)$ is smooth. A curve on the statistical model $\mathcal{S} = \{p_\xi\}$ is defined via

$$\gamma(s) = \iota\big(\xi(s)\big) = p_{\xi(s)}, \qquad s \in [0, T].$$

The velocity of the curve $\gamma(s)$ is given by

$$\dot{\gamma}(s) = \iota_*\big(\dot{\xi}(s)\big) = \frac{d}{ds}\iota\big(\xi(s)\big) = \frac{d}{ds}p_{\xi(s)}.$$

A curve $\gamma : [0,T] \to \mathcal{S}$ is called $\nabla^{(\alpha)}$-*autoparallel* if $\dot{\gamma}(s)$ is parallel transported along $\gamma(s)$, that is, the *acceleration* with respect to the $\nabla^{(\alpha)}$-connection vanishes

$$\nabla^{(\alpha)}_{\dot{\gamma}(s)}\dot{\gamma}(s) = 0, \qquad \forall s \in [0,T]. \tag{1.13.42}$$

In local coordinates $\gamma(s) = (\gamma^k(s))$, the autoparallelism means

$$\ddot{\gamma}^k(s) + \Gamma^{(\alpha)\,k}_{ij}\dot{\gamma}^i(s)\dot{\gamma}^j(s) = 0, \tag{1.13.43}$$

where $\Gamma^{(\alpha)\,k}_{ij} = \Gamma^{(\alpha)}_{ij,l}g^{lk}$ is evaluated along $\gamma(s)$, and $g$ denotes the Fisher metric. It is worth noting that this is just a Riccati system of ODEs. This system of equations states that the velocity vector $\dot{\gamma}(s)$ is moving parallel to itself with respect to the connection $\nabla^{(\alpha)}$.

If $\alpha = 0$, the autoparallel curves (1.13.42) become geodesics with respect to the Fisher metric $g$. In general, the system of equations (1.13.43) is nonlinear and hence, hard to solve explicitly. However, there are a few particular cases when explicit solutions are possible.

**Example 1.13.1** Since any exponential family model is $\nabla^{(1)}$-flat, the $\nabla^{(1)}$-autoparallel curves satisfy $\ddot{\gamma}(s) = 0$, so $\gamma_k(s) = c_k s + \gamma_k(0)$, and hence they are straight lines. Using $\Gamma^{(\alpha)}_{ij,k} = \frac{1-\alpha}{2}T_{ijk}$, the equation of $\nabla^{(\alpha)}$-autoparallel curves becomes, after lowering the indices using the Fisher metric $g$,

$$g_{rk}\ddot{\gamma}^k(s) + \Gamma^{(\alpha)}_{ij,r}\dot{\gamma}^i(s)\dot{\gamma}^j(s) = 0 \Leftrightarrow$$

$$g_{rk}\ddot{\gamma}^k(s) + \frac{1-\alpha}{2}T_{ijr}\dot{\gamma}^i(s)\dot{\gamma}^j(s) = 0.$$

If $X$ is a vector field along $\gamma(s)$, then multiplying by $X^r$ and summing over $r$ yields

$$g(\ddot{\gamma}, X) + \frac{1-\alpha}{2}T(\dot{\gamma}, \dot{\gamma}, X) = 0.$$

Two cases are of particular importance. If $X = \dot{\gamma}(s)$, then

$$g(\ddot{\gamma}(s), \dot{\gamma}(s)) + \frac{1-\alpha}{2}T\big(\dot{\gamma}(s), \dot{\gamma}(s), \dot{\gamma}(s)\big) = 0.$$

For $\alpha = 1$, we get $g(\ddot{\gamma}(s), \dot{\gamma}(s)) = 0$, i.e., the velocity is perpendicular to acceleration. This always holds true in the case of geodesic curves. If $X = \ddot{\gamma}(s)$, then

$$\|\ddot{\gamma}(s)\|_g^2 + \frac{1-\alpha}{2} T\big(\dot{\gamma}(s), \dot{\gamma}(s), \ddot{\gamma}(s)\big) = 0.$$

If $s$ denotes the arc length along $\gamma$, then the term $\kappa(s) = \|\ddot{\gamma}(s)\|$ has the interpretation of curvature. Hence the previous relation provides a relationship between the curvature and skewness tensor.

**Proposition 1.13.1** *Let $\alpha \neq \beta$. If a curve $\gamma(s)$ is both $\nabla^{(\alpha)}$ and $\nabla^{(\beta)}$-autoparallel, then*

$$T(\dot{\gamma}(s), \dot{\gamma}(s), X) = 0,$$

*for any vector field $X$ along the curve $\gamma(s)$.*

*Proof:* Since $\nabla^{(\alpha)}_{\dot{\gamma}(s)} \dot{\gamma}(s) = 0$ and $\nabla^{(\alpha)}_{\dot{\gamma}(s)} \dot{\gamma}(s) = 0$, then subtracting yields

$$K^{(\alpha,\beta)}(\dot{\gamma}(s), \dot{\gamma}(s)) = 0,$$

and hence $T(\dot{\gamma}(s), \dot{\gamma}(s), X) = 0$ by (1.12.38). ■

## 1.14 Jeffrey's Prior

The Fisher metric induces a distribution that has proved useful in the study of universal data compression, see Clarke and Barron [28].

Let $\mathcal{S} = \{p_\xi; \xi \in \mathbb{E}\}$ be a statistical model, and $G(\xi) = \det g_{ij}(\xi)$ denote the determinant of the Fisher information matrix. Assume the volume

$$Vol(\mathcal{S}) = \int_{\mathbb{E}} \sqrt{G(\xi)}\, d\xi < \infty.$$

Then

$$Q(\xi) = \frac{1}{Vol(\mathcal{S})} \sqrt{G(\xi)}$$

defines a probability distribution on $\mathbb{E}$, called the *Jeffrey prior.*

It is worth noting that this distribution is invariant under reparametrizations of the parameters space, see Problem 1.26.

## 1.15 Problems

**1.1.** Prove that the normalization function $\psi(\xi)$ used in the definition of an exponential family model, see formula (1.4.6), is convex.

**1.2.** A tangent vector field $A$ to a statistical model $\mathcal{S} = \{p_\xi\}$ in the 0-representation is a mapping $\xi \to A_\xi$, with $A_\xi = A^j_\xi \, \partial_j p_\xi$. The set of all these vector fields forms the tangent space in the 0-representation to $\mathcal{S}$ at $p_\xi$ and is denoted by $T^{(0)}_\xi \mathcal{S}$. The vector $A_\xi(\cdot)$ can be also considered as a function defined on the sample space $\mathcal{X}$. However, when taking the expectation $E_\xi[A_\xi]$, we consider $A_\xi(x)$ to be a random variable defined on $\mathcal{X}$.

A tangent vector field $B$ to a statistical model $\mathcal{S} = \{p_\xi\}$ in the 1-representation is a mapping $\xi \to B_\xi$, with $B_\xi = B^j_\xi \, \partial_j \ell(\xi)$. The set of all these vector fields forms the tangent space in the 1-representation to $\mathcal{S}$ at $p_\xi$ and is denoted by $T^{(1)}_\xi \mathcal{S}$. Observe that $\{\partial_j(\ln p_\xi(x))\}_{j=1,n}$ and $\{\partial_j p_\xi(x)\}_{j=1,n}$ are two distinct bases of the tangent space $T_\xi \mathcal{S}$, corresponding to the 1-representation and 0-representation, respectively.

(*a*) Show that the tangent space in the 0-representation is given by

$$T^{(0)}_\xi \mathcal{S} = \{A_\xi : \mathcal{X} \to \mathbb{R}, \int_{\mathcal{X}} A_\xi(x)\, dx = 0\}.$$

(*b*) Prove that the vector spaces $T^{(0)}_\xi \mathcal{S}$ and $T^{(1)}_\xi \mathcal{S}$ are isomorphic and find an isomorphism.

**1.3.** Let $\mathcal{S} = \{p_\xi\}$ be a statistical model with Fisher metric $g$. Consider $A_\xi, B_\xi \in T^{(1)}_\xi \mathcal{S}$ vectors in the 1-representation. Show the following:

(*a*) $g(A_\xi, B_\xi) = E_\xi[A_\xi, B_\xi]$;

(*b*) $E_\xi[A_\xi] = E_\xi[B_\xi] = 0$;

(*c*) $|A_\xi|_g = \sqrt{Var(A_\xi)}$, where $|\cdot|_g$ denotes the vector length with respect to $g$;

(*d*) $|g(A_\xi, B_\xi)| \le \sqrt{Var(A_\xi)}\sqrt{Var(B_\xi)}$.

**1.4.** Do the results of the previous problem hold if the vector fields are in the 0-representation, i.e., $A_\xi, B_\xi \in T_\xi^{(0)}\mathcal{S}$?

**1.5.** Consider the log-likelihood function $\ell(\xi) = \ln p_\xi$. Show the following:

(*a*) $\partial_i\partial_j\ell(\xi) = \dfrac{\partial_i\partial_j p_\xi}{p_\xi} - \partial_i\ell(\xi)\partial_j\ell(\xi)$;

(*b*) $E_\xi[\partial_i\partial_j\ell(\xi)] = -E_\xi[\partial_i\ell(\xi)\,\partial_i\ell(\xi)]$;

(*c*) The random variable $A_\xi = \partial_i\partial_j\ell(\xi) + \partial_i\ell(\xi)\,\partial_j\ell(\xi)$ is a tangent vector to $\mathcal{S} = \{p_\xi\}$ in the 1-representation, i.e., $A_\xi \in T_\xi^{(1)}\mathcal{S}$.

(*d*) Let $B_\xi = \partial_k\ell(\xi) \in T_\xi^{(1)}\mathcal{S}$. Using $g(A_\xi, B_\xi) = E_\xi[A_\xi B_\xi]$, show that

$$\Gamma_{ij,k}^{(-1)}(\xi) = g\Big(\partial_k\ell(\xi), \partial_i\partial_j\ell(\xi) + \partial_i\ell(\xi)\,\partial_j\ell(\xi)\Big).$$

(*e*) Write the Fisher metric $g(\xi)$ both as $-E_\xi[\partial_i\partial_j\ell(\xi)]$ and as $E_\xi[\partial_i\ell(\xi)\,\partial_j\ell(\xi)]$ and then prove (*b*).

**1.6.** Use the following linear approximation

$$\partial_j\ell(\xi + \delta\xi) = \partial_j\ell(\xi) + \partial_i\partial_j\ell(\xi)\,\delta\xi^i + O(\delta\xi^i\,\delta\xi^j)$$

to prove the following formulas:

(*a*) $E_\xi[\partial_j\ell(\xi + \delta\xi) - \partial_j\ell(\xi)] = -g_{ij}(\xi)\delta\xi^i + O(\delta\xi^i\,\delta\xi^j)$;

(*b*) $E_\xi[\partial_j\ell(\xi+\delta\xi)+(\delta\xi)_j] = O((\delta\xi^j)^2)$, where $(\delta\xi)_j = g_{ij}(\xi)\delta\xi^i$;

(*c*) $E_\xi[\partial_j\ell(\xi + \delta\xi) + \partial_i\ell(\xi)\,\partial_j\ell(\xi)\delta\xi^i] = O(\delta\xi^i\,\delta\xi^j)$;

(*d*) Let $A_\xi = \partial_j\ell(\xi) + \big(\partial_i\partial_j\ell(\xi) + g_{ij}(\xi)\big)\delta\xi^i$. Show that $A_\xi \in T_\xi^{(1)}\mathcal{S}$;

(*e*) If $A_\xi = A_\xi^r\partial_r\ell(\xi)$, show the following linear approximation $A_\xi^r = \delta_j^r + \Gamma_{ij}^{(1)}(\xi)\,\delta\xi^i$, where $\delta_j^r$ is the Kronecker's symbol.

**1.7.** Let $\mathcal{S} = \Big\{p_{\mu,\sigma}(x) = \frac{1}{\sqrt{2\pi}\sigma}e^{-\frac{(x-\mu)^2}{2\sigma^2}}, \mu \in \mathbb{R}, \sigma > 0\Big\}$ be the statistical model defined by a family of normal distributions. Consider two vector fields in the 1-representation

$$A_\xi = A^1(\xi)\partial_\mu\ell + A^2(\xi)\partial_\sigma\ell$$
$$B_\xi = B^1(\xi)\partial_\mu\ell + B^2(\xi)\partial_\sigma\ell,$$

where $\xi = (\mu, \sigma)$. Show that the vectors $A_\xi$ and $B_\xi$ are perpendicular if and only if

$$A^1(\xi)B^1(\xi) + 2A^2(\xi)B^2(\xi) = 0.$$

**1.8.** Consider the statistical model given by the exponential distribution $\mathcal{S} = \{p_\xi(x) = \xi e^{-\xi x}, \xi > 0, x \geq 0\}$.

(*a*) Show that $\partial_1 \ell(\xi) = \frac{1}{\xi} - x$, $\partial_1 p_\xi = e^{-\xi x}(1 - \xi x)$;

(*b*) If $A_\xi \in T_\xi^{(1)}\mathcal{S}$ is a vector in the 1-representation, show that

$$A_\xi = A^1(\xi)\Big(\frac{1}{\xi} - x\Big).$$

(*c*) If $B_\xi \in T_\xi^{(0)}\mathcal{S}$ is a vector in the 0-representation, show that

$$B_\xi = B^1(\xi)e^{-\xi x}\Big(1 - \xi x\Big).$$

(*d*) Let $f$ be a smooth function on $\mathcal{S}$. Find explicit formulas for $\partial_1 p_\xi(f)$ and $\partial_1 \ell(\xi)(f)$.

(*e*) Solve part (*d*) in the cases when $f(p) = \ln p$, $f(p) = e^p$ and $f(p) = p$.

(*f*) Let $A_\xi, B_\xi \in T_\xi^{(1)}(\mathcal{S})$. Find $g(A_\xi, B_\xi)$, where $g$ is the Fisher–Riemann metric.

**1.9.** Consider the normal family $p(x; \mu, \sigma) = \frac{1}{\sqrt{2\pi}\sigma} e^{-\frac{(x-\mu)^2}{2\sigma^2}}$. Show that this is an exponential family, $p(x; \theta) = e^{\theta^i F_i(x) - \psi(x)}$, with $\psi(\theta) = \dfrac{\mu^2}{2\sigma^2} + \ln(\sqrt{2\pi}\sigma)$, $\theta^1 = \dfrac{\mu}{\sigma^2}$, $\theta^2 = -\dfrac{1}{2\sigma^2}$, $F_1(x) = x$, and $F_2(x) = x^2$.

**1.10.** The exponential distribution determines the Fisher–Riemann manifold

$$\Big(\mathbb{R}_+, g(x) = \frac{1}{x^2}\Big)$$

(*a*) Compute the Christoffel symbols.

(*b*) Find the geodesics and the induced distance.

**1.11.** The two-dimensional discrete probability model determines the Fisher–Riemann manifold $(M, g(x,y))$, where

$$M = \{(x,y) \in \mathbb{R}^2_+;\ x + y < 1\}$$

and

$$g(x,y) = \begin{pmatrix} \frac{1}{x} + \frac{1}{\sqrt{1-x-y}} & \frac{1}{\sqrt{1-x-y}} \\ \frac{1}{\sqrt{1-x-y}} & \frac{1}{y} + \frac{1}{\sqrt{1-x-y}} \end{pmatrix}.$$

Using MAPLE facilities, compute the Christoffel symbols and find the geodesics.

**1.12.** The geometric probability distribution determines the Riemann manifold

$$\left(M, g(x) = \frac{1}{x^2(1-x)}\right),$$

where $M = \{x \in \mathbb{R}_+; x < 1\}$.

(*a*) Compute the Christoffel symbols.

(*b*) Find the geodesics and the induced distance.

(*c*) Solve the ODE $Hess_g(f) = 0$.

**1.13.** Find the subset of $\mathbb{R}^3$ on which the $(0,2)$-tensor

$$g_{ij}(x) = \left(\frac{\delta_{ij}}{x^i x^j} + \frac{1}{1 - x^1 - x^2}\right), \quad x = (x^1, x^2, x^3),\ i,j = 1,2,3$$

is a Riemannian metric.

**1.14.** Consider the product $q^{(n)}(x) = \prod_{i=1}^{n} q_i(x;\xi^i)$ where $q_i(x) = \xi^i e^{-\xi^i x}$, $\xi^i > 0$, $x \geq 0$.

(*a*) Show that $\mathcal{S} = \{q^{(n)}(x,\xi), x \geq 0; \xi \in \mathbb{R}^n_+\}$ is an exponential family. Specify the expressions of $C(x)$, $F_i(x)$ and $\psi(\xi)$.

(*b*) Find the Fisher–Riemann metric on $\mathcal{S}$.

(*c*) Find the distance induced by the Fisher information between $q^{(n)}(x,\xi)$ and $q^{(n)}(x,\theta)$.

**1.15.** Let $p(x)$ be a probability density function on $\mathbb{R}$ and consider $p^{(n)}(\mathbf{x}) = p(x_1)\dots p(x_n)$, $\mathbf{x} = (x_1,\dots,x_n) \in \mathbb{R}^n$. Consider a nonsingular matrix $A \in \mathbb{R}^n \times \mathbb{R}^n$ and define

$$q(\mathbf{x}; A, \mu) = \frac{1}{|\det A|} p^{(n)}\big(A^{-1}(\mathbf{x} - \mu)\big),$$

with $\mu = (\mu_1,\dots,\mu_n) \in \mathbb{R}^n$.

(*a*) Show that $q(\mathbf{x}; A, \mu)$ is a probability density on $\mathbb{R}^n$.

(*b*) The statistical model $\mathcal{S} = \{q(\mathbf{x}; A, \mu)\}$ is $\alpha$-flat for any $\alpha$ real.

(*c*) Show that the Fisher metric on $\mathcal{S} = \{q(\mathbf{x}; A, \mu)\}$ is the Euclidean metric.

(*d*) Choose a probability density $p(x)$ such that $\mathcal{S}=\{q(\mathbf{x}; A, \mu)\}$ is neither an exponential, nor a mixture family.

**1.16.** Prove that the skewness tensor $T_{ijk} = E[\partial_i \ell\, \partial_j \ell\, \partial_k \ell]$ is covariant under reparametrizations, i.e., if $\zeta^a = \zeta^a(\theta^1, \cdots, \theta^n)$ is a reparametrization, then

$$\tilde{T}_{ijk}(\theta) = T_{abc}(\zeta) \frac{\partial \zeta^a}{\partial \theta^i} \frac{\partial \zeta^b}{\partial \theta^j} \frac{\partial \zeta^c}{\partial \theta^k}.$$

**1.17.** Consider the reparametrization $\zeta^a = \zeta^a(\theta^1, \cdots, \theta^n)$. Find a formula relating $\Gamma^{(\alpha)}_{ijk}(\zeta)$ and $\tilde{\Gamma}^{(\alpha)}_{abc}(\theta)$. Is $\Gamma^{(\alpha)}_{ijk}(\zeta)$ covariant under reparametrizations?

**1.18.** Prove that the skewness tensor $T_{ijk}$ is invariant under transformations of the random variable, i.e., if $f : \mathcal{X} \to \mathcal{Y}$ is an invertible mapping (with $\mathcal{X}, \mathcal{Y} \subset \mathbb{R}^n$), and if $\tilde{p}_\xi(y) = p_\xi(x)$, with $y = f(x)$, and denote $\ell(\xi) = \ln p_\xi$, $\tilde{\ell}(\xi) = \ln p_\xi$, then

$$T_{ijk}(\xi) = \tilde{T}_{ijk}(\xi),$$

where $T_{ijk} = E_\xi[\partial_i \ell\, \partial_j \ell\, \partial_k \ell]$ and $\tilde{T}_{ijk} = E_\xi[\partial_i \tilde{\ell}\, \partial_j \tilde{\ell}\, \partial_k \tilde{\ell}]$.

**1.19.** (*a*) Prove that $\Gamma^{(\alpha)}_{ij,k}$ is invariant under transformations of the random variable.

(*b*) Use part (*a*) to show that the skewness tensor $T_{ijk}$ has the same property.

(*c*) Does $\Gamma^{(\alpha)\,k}_{ij}$ have the same invariance property?

**1.20.** Consider the exponential family

$$p(x;\xi) = e^{C(x)+\xi^i F_i(x)-\psi(\xi)}, \quad i = 1, \cdots, n.$$

(*a*) Show that $E_\xi[(\partial_i\partial_j\ell)(\partial_k\ell)] = 0$;

(*b*) Prove that the skewness tensor can be written in terms of $\psi(\xi)$ as $T^{(e)}_{ijk}(\xi) = \partial_i\partial_j\partial_k\psi(\xi)$.

**1.21.** Using Problem 1.20 (*b*), find the skewness tensor for the following statistical models:

(*a*) Exponential distribution;

(*b*) Normal distribution;

(*c*) Gamma distribution;

(*d*) Beta distribution.

**1.22.** Give an example of a nontrivial statistical model with a zero skewness tensor, $T_{ijk}(\xi) = 0$.

**1.23.** Consider the exponential family

$$p(x;\xi) = e^{C(x)+\xi^i F_i(x)-\psi(\xi)}, \quad i = 1, \cdots, n,$$

and denote the $\alpha$-connection by $\nabla^{(\alpha)}$.

(*a*) Show that $\Gamma^{(\alpha)}_{ij,k}(\xi) = \dfrac{1-\alpha}{2}\partial_i\partial_j\partial_k\psi(\xi)$;

(*b*) $\Gamma^{(1)}_{ij,k}(\xi) = 0$ and $\Gamma^{(-1)}_{ij,k} = \partial_i\partial_j\partial_k\psi(\xi)$;

(*c*) Show that the model is both $(\pm 1)$-flat, i.e., $R^{(1)} = 0$, $R^{(-1)} = 0$;

(*d*) Find the $\alpha$-curvature tensor $R^{(\alpha)}$;

(*e*) Assume the function $\psi(\xi)$ is quadratic, $\psi(\xi) = \sum_{i,j} a_{ij}\xi^i\xi^j$, with $(a_{ij})$ positive definite matrix. Show that $\Gamma^{(\alpha)}_{ij,k} = 0$. Find the geodesics.

**1.24.** Consider the mixture model

$$p(x;\xi) = C(x) + \xi^i F_i(x), \quad i = 1, \ldots, n.$$

(*a*) Show that the skewness tensor is

$$T^{(m)}_{ijk}(\xi) = \frac{1}{2} E_\xi[\partial_i \partial_j \partial_k \ell].$$

(*b*) Verify the relation

$$T^{(m)}_{ijk}(\xi) = \int_{\mathcal{X}} \frac{F_i(x)F_j(x)F_k(x)}{p^2(x;\xi)}\, dx.$$

**1.25.** Let $\mathcal{S} \times \mathcal{P}$ be the product of two statistical models. Denote by $g^{(\mathcal{S})}$, $g^{(\mathcal{P})}$, $g^{(\mathcal{S}\times\mathcal{P})}$ and $T^{(\mathcal{S})}$, $T^{(\mathcal{P})}$, $T^{(\mathcal{S}\times\mathcal{P})}$ the Fisher–Riemann metrics and the skewness tensors on the statistical models $\mathcal{S}$, $\mathcal{P}$ and $\mathcal{S} \times \mathcal{P}$, respectively.

(*a*) Prove that

$$g^{(\mathcal{S}\times\mathcal{P})} = \begin{pmatrix} g^{(\mathcal{S})} & 0 \\ 0 & g^{(\mathcal{P})} \end{pmatrix}$$

(*b*) Compute $T^{(\mathcal{S}\times\mathcal{P})}$ in terms of $T^{(\mathcal{S})}$ and $T^{(\mathcal{P})}$;

(*c*) Assume $T^{(\mathcal{S})} = 0$ and $T^{(\mathcal{P})} = 0$. Is $T^{(\mathcal{S}\times\mathcal{P})} = 0$?

**1.26.** Prove that Jeffrey's prior is invariant under reparametrizations of the parameters space $\mathbb{E}$.

**1.27.** Consider $\mathcal{S} = \{p_\xi(x)\}$, where $p_\xi(x) = \xi e^{-\xi x}$, $\xi > 0$, $x \geq 0$, is the exponential distribution model. Consider the submanifold $\mathcal{M}_c = \{p_\xi; \xi \geq c\}$. Find the Jeffrey's prior associated with $\mathcal{M}_c$.

**1.28.** Consider the statistical model $\mathcal{S} = \mathcal{P}(\mathcal{X})$, with the sample space $\mathcal{X} = \{0, 1, \ldots, n\}$, and denote the Fisher–Riemann metric by $g_{ij}$.

(*a*) Let $V = \int_{\mathbb{E}} \sqrt{\det g_{ij}(\xi)}\, d\xi$ be the volume of $\mathcal{S}$. Show that $V = \dfrac{\pi^{(n+1)/2}}{\Gamma\left(\frac{n+1}{2}\right)}$.

(*b*) Show that the associate Jeffrey prior is the uniform distribution on an $n$-dimensional sphere.

**1.29.** Consider the functions $F_1(x) = \sin x$, $F_2(x) = \cos x$ on $\mathbb{R}$.

(*a*) Show that $\{1, F_1(x), F_2(x)\}$ are linearly independent on $\mathbb{R}$;

(b) Let $\psi(\xi) = \ln\Big(\int_{\mathbb{R}} e^{x^2+\xi^i F_i(x)}\,dx\Big)$, and consider the set $\mathbb{E} = \{\xi; \psi(\xi) < \infty\}$. Show that $\mathbb{E} = \emptyset$.

**1.30.** Consider the statistical model given by the Multivariate Normal Distribution, see Example 1.3.8, with the notations therein.

(a) Verify that

$$\partial_{\mu_r} \ell_x(\mu, A) = -\sum_{j=1}^{k} \frac{A^{rj} + A^{jr}}{2}(x_j - \mu_j).$$

(b) Show that

$$\frac{\partial(\det A^{-1})}{\partial A^{is}} = \frac{A_{ij}}{\det A}.$$

(c) Use (b) to prove that

$$\partial_{A^{\alpha\beta}} \ell_x(\mu, A) = \frac{1}{2}A_{\alpha\beta} - \frac{1}{2}(x_\alpha - \mu_\alpha)(x_\beta - \mu_\beta).$$

# Chapter 2

# Explicit Examples

This chapter presents a few examples of usual statistical models (normal, lognormal, beta, gamma, Bernoulli, and geometric) for which we provide the Fisher metric explicitly and, if possible, the geodesics and $\alpha$-autoparallel curves. Some Fisher metrics will involve the use of non-elementary functions, such as the digamma and trigamma functions.

A distinguished role is dedicated to the normal distribution, which is associated with a manifold of negative constant curvature (hyperbolic space) and to the multinomial geometry, which corresponds to a space with positive constant curvature (spherical space).

## 2.1 The Normal Distribution

In this section we shall determine the geodesics with respect to the Fisher information metric of a family of normal distributions. Given two distributions of the same family, the geodesics are curves of minimum information joining the distributions. We shall see that such a curve always exists between any two distributions on a normal family. This is equivalent with the possibility of deforming one distribution into the other by keeping the change of information to a minimum.

### 2.1.1 The Fisher Metric

Recall the formula for the density of a normal family

$$p(x, \xi) = \frac{1}{\sigma\sqrt{2\pi}} e^{-\frac{(x-\mu)^2}{2\sigma^2}}, \quad x \in \mathcal{X} = \mathbb{R},$$

O. Calin and C. Udrişte, *Geometric Modeling in Probability and Statistics*,
DOI 10.1007/978-3-319-07779-6_2,

with parameters $(\xi^1, \xi^2) = (\mu, \sigma) \in \mathbb{R}\times(0,\infty)$. Using Proposition 1.6.3 we obtain the following components for the Fisher–Riemann metric.

**Proposition 2.1.1** *The Fisher information matrix for the normal distribution is given by*

$$g_{ij} = \begin{pmatrix} \frac{1}{\sigma^2} & 0 \\ 0 & \frac{2}{\sigma^2} \end{pmatrix}. \tag{2.1.1}$$

For the computation details see Problem 2.1. It is worth noting that the metric does not depend on $\mu$, i.e., it is translation invariant. This metric is also very similar to the upper-half plane metric.

### 2.1.2 The Geodesics

A straightforward computation shows that the nonzero Christoffel symbols of first and second kind are:

$$\Gamma_{11,2} = \frac{1}{\sigma^3}, \quad \Gamma_{12,1} = -\frac{1}{\sigma^3}, \quad \Gamma_{22,2} = -\frac{2}{\sigma^3}$$

$$\Gamma^1_{ij} = \begin{pmatrix} 0 & -\frac{1}{\sigma} \\ -\frac{1}{\sigma} & 0 \end{pmatrix}, \qquad \Gamma^2_{ij} = \begin{pmatrix} \frac{1}{2\sigma} & 0 \\ 0 & -\frac{1}{\sigma} \end{pmatrix}.$$

Consequently, the geodesics equations (1.13.43) are solutions of a Riccati ODE system

$$\ddot{\mu} - \frac{2}{\sigma}\dot{\mu}\dot{\sigma} = 0 \tag{2.1.2}$$

$$\ddot{\sigma} + \frac{1}{2\sigma}(\dot{\mu})^2 - \frac{1}{\sigma}(\dot{\sigma})^2 = 0. \tag{2.1.3}$$

Separating and integrating in the first equation yields

$$\frac{\ddot{\mu}}{\dot{\mu}} = \frac{2\dot{\sigma}}{\sigma} \Longleftrightarrow \frac{d}{ds}\ln\dot{\mu} = 2\frac{d}{ds}\ln\sigma \Longleftrightarrow \dot{\mu} = c\sigma^2,$$

with $c$ constant. We solve the equation in the following two cases:

*1. The case* $c = 0$. It follows that $\mu =$ constant, which corresponds to vertical half lines. Then $\sigma$ satisfies the equation $\ddot{\sigma} = \frac{1}{\sigma}\dot{\sigma}^2$. Writing the equation as

$$\frac{\ddot{\sigma}}{\dot{\sigma}} = \frac{\dot{\sigma}}{\sigma}$$

and integrating yields $\ln \dot{\sigma} = \ln(C\sigma)$, with $C$ constant. Integrating again, we find $\sigma(s) = Ke^{Cs}$. Hence, the geodesics in this case have the following explicit equations

$$\mu = c \tag{2.1.4}$$

$$\sigma(s) = Ke^{Cs}, \tag{2.1.5}$$

with $c, C \in \mathbb{R}$, $K > 0$ constants.

*2. The case $c \neq 0$.* Substituting $\dot{\mu} = x\sigma^2$ in Eq. (2.1.3), we obtain the following equation in $\sigma$

$$\sigma\ddot{\sigma} + \frac{c^2}{2}\sigma^4 - (\dot{\sigma})^2 = 0. \tag{2.1.6}$$

Let $\dot{\sigma} = u$. Then $\ddot{\sigma} = \dfrac{du}{d\sigma}u$ and (2.1.6) becomes

$$\sigma\frac{du}{d\sigma}u + \frac{c^2}{2}\sigma^4 - u^2 = 0.$$

Multiplying by the integrant factor $\dfrac{1}{\sigma^3}$ leads to the exact equation

$$\underbrace{\frac{u}{\sigma^2}}_{=M}\,du + \underbrace{\Big(\frac{c^2}{2}\sigma - \frac{u^2}{\sigma^3}\Big)}_{N}d\sigma = 0,$$

since

$$\frac{\partial M}{\partial \sigma} = \frac{\partial N}{\partial u} = -2u\sigma^{-3}.$$

Then there is a function $f(\sigma, u)$ such that $df = 0$, with

$$\frac{\partial f}{\partial u} = M, \qquad \frac{\partial f}{\partial \sigma} = N.$$

Integrating in the first equation yields

$$f(\sigma, u) = \frac{u^2}{2\sigma^2} + h(\sigma),$$

with function $h$ to be determined in the following. Differentiating with respect to $\sigma$ in the above equation,

$$\frac{\partial f}{\partial \sigma} = -\frac{u^2}{\sigma^3} + h'(\sigma),$$

and comparing with

$$\frac{\partial f}{\partial \sigma} = N = \frac{c^2}{2}\sigma - \frac{u^2}{\sigma^3},$$

we get

$$h'(\sigma) = \frac{c^2}{2}\sigma \Longrightarrow h(\sigma) = \frac{c^2\sigma^2}{4} + c_0,$$

with $c_0$ constant. Hence, a first integral for the system is

$$f(\sigma, u) = \frac{u^2}{2\sigma^2} + \frac{c^2\sigma^2}{4} = \frac{E}{2},$$

with $E$ positive constant. Solving for $u$, we obtain

$$\begin{aligned}
\frac{u^2}{\sigma^2} + \frac{c^2\sigma^2}{2} &= E \Longleftrightarrow \\
\frac{\dot{\sigma}}{\sigma} &= \frac{c}{\sqrt{2}}\sqrt{C^2 - \sigma^2},
\end{aligned}$$

where $C^2 = 2E/c^2$. Separating and integrating, we find

$$\int \frac{d\sigma}{\sigma\sqrt{C^2 - \sigma^2}} = (s + s_0)\frac{c}{\sqrt{2}}.$$

Using the value of the integral

$$\int \frac{dx}{x\sqrt{C^2 - x^2}} = -\frac{1}{\sqrt{C}}\tanh^{-1}\sqrt{1 - \left(\frac{x}{C}\right)^2},$$

we obtain

$$-\frac{1}{\sqrt{C}}\tanh^{-1}\sqrt{1 - \left(\frac{\sigma}{C}\right)^2} = (s + s_0)\frac{c}{\sqrt{2}}.$$

Solving for $\sigma$, we get

$$\sigma = c\sqrt{1 - \tanh^2\left(\sqrt{E}(s + s_0)\right)} = \frac{c}{\cosh\left(\sqrt{E}(s + s_0)\right)}.$$

In order to find $\mu$ we integrate in $\dot{\mu} = c\sigma^2$ and obtain

$$\mu(s) = \int \frac{c^3}{\cosh^2\left(\sqrt{E}(s + s_0)\right)}\, ds = \frac{c^3}{\sqrt{E}}\tanh\left(\sqrt{E}(s + s_0)\right) + K.$$

Since we have

$$\sigma(s)^2 + (\mu(s) - K)^2 \frac{E}{c^4} = c^2,$$

the geodesics will be half-ellipses, with $\sigma > 0$.

In the case $c = 0$, the unknown $\sigma$ satisfies

$$\frac{\dot{\sigma}}{\sigma} = \sqrt{E} \Longleftrightarrow \frac{d}{ds} \ln \sigma = \sqrt{E}$$

with solution

$$\sigma(s) = \sigma(0) e^{\sqrt{E}s},$$

while $\mu$ is constant, $\mu = K$. The geodesics in this case are vertical half-lines.

**Proposition 2.1.2** *Consider two normal distributions with equal means, $\mu_0 = \mu_1$, and distinct standard deviations $\sigma_0$ and $\sigma_1$. Then the smallest information transform, which deforms the first distribution into the second one, is a normal distribution with constant mean and standard deviation*

$$\sigma(s) = \sigma_1^{s/\tau} \sigma_0^{1-s/\tau}, \quad s \in [0, \tau].$$

*Proof:* The geodesic in this case is a vertical half-line with constant mean and $\sigma(s) = \sigma(0)e^{\sqrt{E}s}$. The amount $\sqrt{E}$ can be found from the boundary condition $\sigma(\tau) = \sigma_1$. ∎

Let $x_0 = \ln \sigma_0$, $x_1 = \ln \sigma_1$, and $x(s) = \ln \sigma(s)$. Then $x(s) = \frac{s}{\tau} x_0 + \left(1 - \frac{s}{\tau}\right) x_1$, which corresponds to a line segment. The minimal information loss during the deformation occurs when the log-likelihood function describes a line segment.

### 2.1.3 $\alpha$-Autoparallel Curves

A straightforward computation, using (1.11.34), yields the following Christoffel coefficients of first kind

$$\Gamma_{11,1}^{(\alpha)} = \Gamma_{21,2}^{(\alpha)} = \Gamma_{12,2}^{(\alpha)} = \Gamma_{22,1}^{(\alpha)} = 0$$

$$\Gamma_{11,2}^{(\alpha)} = \frac{1-\alpha}{\sigma^3}, \quad \Gamma_{12,1}^{(\alpha)} = \Gamma_{21,1}^{(\alpha)} = -\frac{1+\alpha}{\sigma^3}, \quad \Gamma_{22,2}^{(\alpha)} = -\frac{2(1+2\alpha)}{\sigma^3}.$$

The Christoffel symbols of second kind are obtained by rising indices

$$\begin{aligned}\Gamma_{ij}^{1\,(\alpha)} &= g^{11}\Gamma_{ij,1}{}^{(\alpha)} + g^{12}\Gamma_{ij,2}{}^{(\alpha)} = \sigma^2 \Gamma_{ij,1}{}^{(\alpha)} \\ &= \sigma^2 \begin{pmatrix} 0 & -\frac{1+\alpha}{\sigma^3} \\ -\frac{1+\alpha}{\sigma^3} & 0 \end{pmatrix} = \begin{pmatrix} 0 & -\frac{1+\alpha}{\sigma} \\ -\frac{1+\alpha}{\sigma} & 0 \end{pmatrix}.\end{aligned}$$

$$\begin{aligned}\Gamma_{ij}^{2\,(\alpha)} &= g^{21}\Gamma_{ij,1}{}^{(\alpha)} + g^{22}\Gamma_{ij,2}{}^{(\alpha)} \\ &= \frac{\sigma^2}{2} \begin{pmatrix} \frac{1-\alpha}{\sigma^3} & 0 \\ 0 & -2\frac{1+2\alpha}{\sigma^3} \end{pmatrix} = \begin{pmatrix} \frac{1-\alpha}{2\sigma} & 0 \\ 0 & -\frac{1+2\alpha}{\sigma} \end{pmatrix}.\end{aligned}$$

The Riccati equations (1.13.43) for the $\alpha$-autoparallel curves are

$$\begin{aligned}\ddot{\mu} - \frac{2(1+\alpha)}{\sigma}\dot{\sigma}\dot{\mu} &= 0 \\ \ddot{\sigma} + \frac{1-\alpha}{2\sigma}\dot{\mu}^2 - \frac{1+2\alpha}{\sigma}\dot{\sigma}^2 &= 0.\end{aligned}$$

The first equation can be transformed as in the following

$$\begin{aligned}\frac{\ddot{\mu}}{\dot{\mu}} &= 2(1+\alpha)\frac{\dot{\sigma}}{\sigma} \Longleftrightarrow \\ \frac{d}{ds}\ln\dot{\mu} &= 2(1+\alpha)\frac{d}{ds}\ln\sigma \Longleftrightarrow \\ \ln\dot{\mu} &= 2(1+\alpha)\ln\sigma + c_0 \Longleftrightarrow \\ \dot{\mu} &= c\,\sigma^{2(1+\alpha)},\end{aligned}$$

with $c$ constant. Substituting in the second equation yields

$$\ddot{\sigma} + \frac{1-\alpha}{2\sigma}c^2\sigma^{4(1+\alpha)} - \frac{1+2\alpha}{\sigma}\dot{\sigma}^2 = 0,$$

which after the new substitution $u = \dot{\sigma}$ writes as

$$\frac{du}{d\sigma}u + \frac{1-\alpha}{2\sigma}c^2\sigma^{4(1+\alpha)} - \frac{1+2\alpha}{\sigma}u^2 = 0.$$

Multiplying the equation by an integral factor of the form $\sigma^{k+1}$, we obtain

$$\underbrace{\sigma^{k+1}u}_{=M}\,du + \Big(\underbrace{\frac{1-\alpha}{2}c^2\sigma^{4(\alpha+1)+k} - (1+2\alpha)\sigma^k u^2}_{=N}\Big)d\sigma = 0.$$

From the closeness condition

$$\frac{\partial M}{\partial \sigma} = \frac{\partial N}{\partial u},$$

we determine $k+1 = -(4\alpha+2)$. The exact equation we need to solve is

$$u\sigma^{-(4\alpha+2)}\,du + \left(\frac{1-\alpha}{2}c^2\sigma - (1+2\alpha)u^2\sigma^{-(4\alpha+3)}\right)d\sigma = 0.$$

We need to determine a function $f$ that satisfies the system

$$\begin{aligned}\frac{\partial f}{\partial u} &= u\sigma^{-(4\alpha+2)}\\ \frac{\partial f}{\partial \sigma} &= \frac{1-\alpha}{2}c^2\sigma - (1+2\alpha)u^2\sigma^{-(2\alpha+3)}.\end{aligned}$$

From the first equation, we have

$$f = \frac{u^2}{2}\sigma^{-(4\alpha+2)} + h(\sigma) \Longrightarrow \frac{\partial f}{\partial \sigma} = -(1+2\alpha)u^2\sigma^{-(4\alpha+3)} + h'(\sigma)$$

and comparing with the second equation yields

$$h'(\sigma) = \frac{1-\alpha}{2}c^2\sigma \Longrightarrow h(\sigma) = \frac{(1-\alpha)c^2}{4}\sigma^2 + \mathcal{C}.$$

Hence, a first integral of motion is given by

$$f = \frac{u^2}{2}\sigma^{-(4\alpha+2)} + \frac{1-\alpha}{4}c^2\sigma^2 = \frac{E}{2},$$

with $E$ constant. Next we shall solve for $\sigma$. Using that $u = \dot{\sigma}$, we have

$$\begin{aligned}\frac{u^2}{\sigma^{2(2\alpha+1)}} + \frac{1-\alpha}{2}c^2\sigma^2 &= E \Longleftrightarrow\\ \left(\frac{\dot{\sigma}}{\sigma^{2\alpha+1}}\right)^2 + \frac{1-\alpha}{2}c^2\sigma^2 &= E \Longleftrightarrow\\ \left(\frac{\dot{\sigma}}{\sigma^{2\alpha+1}}\right)^2 &= E - \frac{1-\alpha}{2}c^2\sigma^2 \Longleftrightarrow\\ \int \frac{d\sigma}{\sigma^{2\alpha+1}\sqrt{E - \frac{1-\alpha}{2}c^2\sigma^2}} &= \pm s + s_0 \Longleftrightarrow\\ \int \frac{d\sigma}{\sigma^{2\alpha+1}\sqrt{C^2-\sigma^2}} &= (\pm s + s_0)\sqrt{\frac{1-\alpha}{2}}\,c, \qquad (2.1.7)\end{aligned}$$

where

$$C = C_\alpha = \frac{2E}{c}\frac{1}{1-\alpha}.$$

The left side integral can be transformed using the substitutions $t=\sigma^2$, $v=\sqrt{C^2-t}$ as follows

$$\begin{aligned}\int\frac{d\sigma}{\sigma^{2\alpha+1}\sqrt{C^2-\sigma^2}} &= \int\frac{dt}{2\sigma^{2(\alpha+1)}\sqrt{C^2-\sigma^2}}=\int\frac{dt}{2t^{\alpha+1}\sqrt{C^2-t}}\\ &= \int\frac{-2v\,dv}{2t^{\alpha+1}v}=-\int\frac{dv}{(C^2-v^2)^{\alpha+1}},\end{aligned}$$

and hence (2.1.7) becomes

$$-\int\frac{dv}{(C^2-v^2)^{\alpha+1}}=(\pm s+s_0)\sqrt{\frac{1-\alpha}{2}}\,c. \tag{2.1.8}$$

The $\mu$-component is given by

$$\mu=c\int\sigma^{2(1+\alpha)}(s)ds. \tag{2.1.9}$$

There are a few particular values of $\alpha$ for which this equation can be solved explicitly.

**Case** $\alpha=-1$

Equation (2.1.8) becomes

$$-v-K=(\pm s+s_0)\sqrt{\frac{1-\alpha}{2}}\,c,$$

with solution

$$\sigma^2(s)=C^2-\left((\pm s+s_0)\sqrt{\frac{1-\alpha}{2}}\,c+K\right)^2,$$

for $K$ constant. Equation (2.1.3) easily yields

$$\mu(s)=cs+\mu(0).$$

**Case** $\alpha=1/2$

Since

$$\int\frac{dv}{\left(C^2-v^2\right)^{3/2}}=\frac{v}{C^2\sqrt{C^2-v^2}},$$

we solve

$$-\frac{v}{C^2\sqrt{C^2-v^2}}=(\pm s+s_0)\frac{c}{2}+K$$

and obtain

$$\sigma(s) = \frac{C}{\sqrt{1 + C^4\Big((\pm s + s_0)^{\frac{c}{2}} + K\Big)^2}}.$$

The $\mu$-component is given by the integral

$$\mu(s) = c \int \sigma^3(s)\,ds.$$

## 2.2 Jeffrey's Prior

In the following we shall compute the prior on the statistical model

$$\mathcal{S}_\mu = \{p_\xi; E[p_\xi = \mu], Var[p_\xi] > 1\} = \{p_{(\mu,\sigma)}; \sigma > 1\}$$

which represents a vertical half line in the upper-half plane. The determinant is

$$G(\xi) = \det g_{ij}(\xi) = \det \begin{pmatrix} \frac{1}{\sigma^2} & 0 \\ 0 & \frac{2}{\sigma^2} \end{pmatrix} = \frac{2}{\sigma^4}.$$

Then the volume is computed as

$$Vol(\mathcal{S}_\mu) \;=\; \int_1^\infty \sqrt{G(\xi)}\,d\sigma = \int_1^\infty \frac{\sqrt{2}}{\sigma^2}\,d\sigma = \sqrt{2} < \infty.$$

Therefore the prior on $\mathcal{S}_\mu$ is given by

$$Q(\sigma) = \frac{\sqrt{G(\sigma)}}{Vol(\mathcal{S}_\mu)} = \frac{1}{\sigma^2}.$$

## 2.3 Lognormal Distribution

In the case of lognormal distribution

$$p_{\mu,\sigma}(x) = \frac{1}{\sqrt{2\pi}\,\sigma x} e^{-\frac{(\ln x - \mu)^2}{2\sigma^2}}, \qquad x > 0,$$

the Fisher information matrix (Fisher–Riemann metric) is given by

$$g = \begin{pmatrix} g_{\mu\mu} & g_{\mu\sigma} \\ g_{\sigma\mu} & g_{\sigma\sigma} \end{pmatrix} = \begin{pmatrix} \frac{1}{\sigma^2} & 0 \\ 0 & \frac{2}{\sigma^2} \end{pmatrix}.$$

The computation details are left for the reader and are the subject of Problem 2.2. It is worth noting that this coincides with the Fisher metric of a normal distribution model. Hence, the associated geodesics are vertical half lines or halfs of ellipses.

## 2.4 Gamma Distribution

In this case the statistical model is defined by the following family of densities

$$p_\xi(x) = p_{\alpha,\beta}(x) = \frac{1}{\beta^\alpha \Gamma(\alpha)}\, x^{\alpha-1} e^{-x/\beta},$$

with $(\alpha, \beta) \in (0, \infty) \times (0, \infty)$, $x \in (0, \infty)$. In the study of this model we need some special functions. Let

$$\psi(\alpha) = \frac{\Gamma'(\alpha)}{\Gamma(\alpha)}, \qquad \psi_1(\alpha) = \psi'(\alpha) \tag{2.4.10}$$

be the *digamma* and the *trigamma* functions, respectively. Differentiating in the Dirichlet's integral representation (see Erdélyi [42] vol. I, p. 17)

$$\psi(\alpha) = \int_0^\infty [e^{-t} - (1+t)^{-\alpha}] t^{-1}\, dt, \qquad \alpha > 0$$

yields the following integral expression for the trigamma function

$$\psi_1(\alpha) = \psi'(\alpha) = \int_0^\infty \frac{\ln(1+t)}{t(1+t)^\alpha}\, dt. \tag{2.4.11}$$

Another interesting formula is the expression of the trigamma function as a *Hurwitz zeta function*

$$\psi_1(\alpha) = \zeta(2, \alpha) = \sum_{n \geq 0} \frac{1}{(\alpha+n)^2}, \tag{2.4.12}$$

which holds for $\alpha \notin \{0, -1, -2, -3, \dots\}$, relation obviously satisfied in our case since $\alpha > 0$.

Then the components of the Fisher–Riemann metric are obtained from Proposition 1.6.3, using the relations

$$\int_0^\infty p_\xi(x)\, dx = 1, \qquad \int_0^\infty x p_\xi(x)\, dx = \alpha\beta$$

and the derivatives of the log-likelihood function that are asked to be computed in Problem 2.2:

$$\begin{aligned}
g_{\alpha\alpha} &= -E[\partial_\alpha^2 \ell_x(\xi)] = \int_0^\infty \psi'(\alpha) p_\xi(x)\,dx = \psi'(\alpha) = \psi_1(\alpha),\\
g_{\beta\beta} &= -E[\partial_\beta^2 \ell_x(\xi)] = -\int_0^\infty \Big(\frac{\alpha}{\beta^2} - \frac{2x}{\beta^3}\Big) p_\xi(x)\,dx\\
&= -\frac{\alpha}{\beta^2} + \frac{2}{\beta^3}\int_0^\infty x p_\xi(x)\,dx = \frac{\alpha}{\beta^2},\\
g_{\alpha\beta} &= -E[\partial_{\alpha\beta}\ell_x(\xi)] = \int_0^\infty \frac{1}{\beta} p_\xi(x)\,dx = \frac{1}{\beta}.
\end{aligned}$$

**Proposition 2.4.1** *The Fisher information matrix (Fisher–Riemann metric) for the gamma distribution is*

$$g = \begin{pmatrix} \psi_1(\alpha) & \frac{1}{\beta} \\ \frac{1}{\beta} & \frac{\alpha}{\beta^2} \end{pmatrix} = \begin{pmatrix} \displaystyle\sum_{n\geq 0} \frac{1}{(\alpha+n)^2} & \frac{1}{\beta} \\ \frac{1}{\beta} & \frac{\alpha}{\beta^2} \end{pmatrix}.$$

It is worth noting that here $\alpha$ is the parameter for the gamma distribution and it has nothing to do with $\alpha$-connections.

## 2.5 Beta Distribution

The Fisher information metric for the beta distribution

$$p_{a,b} = \frac{1}{B(a,b)}\, x^{a-1}(1-x)^{b-1}, \qquad a,b>0, x\in[0,1]$$

will be worked in terms of trigamma functions. Since the beta function

$$B(a,b) = \int_0^1 x^{a-1}(1-x)^{b-1}\,dx$$

can be expressed in terms of gamma functions as

$$B(a,b) = \frac{\Gamma(a)\Gamma(b)}{\Gamma(a+b)},$$

then its partial derivatives can be written in terms of digamma functions, using relation (2.11.17), see Problem 2.4, part $(a)$.

The log-likelihood function and its partial derivatives are left for the reader as an exercise in Problem 2.4, parts $(b)$ and $(c)$. Since the second partial derivatives of $\ell(a,b)$ do not depend on $x$, they will coincide with their own expected values. It follows the next components for the Fisher–Riemann metric:

$$\begin{aligned} g_{aa} &= -E[\partial_a^2 \ell(a,b)] = \psi_1(a) - \psi_1(a+b) \\ g_{bb} &= -E[\partial_b^2 \ell_x(a,b)] = \psi_1(b) - \psi_1(a+b) \\ g_{ab} &= g_{ba} = -E[\partial_a \partial_b \ell_x(a,b)] = -\psi_1(a+b). \end{aligned}$$

**Proposition 2.5.1** *The Fisher information matrix (Fisher–Riemann metric) for the beta distribution is given by*

$$g = \begin{pmatrix} \psi_1(a) - \psi_1(a+b) & -\psi_1(a+b) \\ -\psi_1(a+b) & \psi_1(b) - \psi_1(a+b) \end{pmatrix},$$

*where $\psi_1$ stands for the trigamma function.*

## 2.6 Bernoulli Distribution

Consider the sample space $\mathcal{X} = \{0, 1, \dots, n\}$ and parameter space $\mathbb{E} = [0,1]$. The Bernoulli, or binomial distribution, is given by

$$p(k;\xi) = \binom{n}{k} \xi^k (1-\xi)^{n-k},$$

where the parameter $\xi$ denotes the success probability. Then $\mathcal{S} = \{p_\xi; \xi \in [0,1]\}$ becomes a one-dimensional statistical model. The derivatives of the log-likelihood function $\ell_k(\xi) = \ln p(k;\xi)$ are proposed as an exercise in Problem 2.5. Then the Fisher information is given by the function

$$\begin{aligned} g_{11}(\xi) &= -E_\xi[\partial_\xi^2 \ell(\xi)] = \sum_{k=0}^{n} p(k;\xi) \partial_\xi^2 \ell_k(\xi) \\ &= \sum_{k=0}^{n} \frac{k}{\xi^2} p(k;\xi) + \sum \frac{(n-k)}{(1-\xi)^2} p(k;\xi) \\ &= \frac{n}{\xi} + \frac{n(1-\xi)}{(1-\xi)^2} = \frac{n}{\xi(1-\xi)}, \end{aligned}$$

where we used that the mean of a Bernoulli distribution is $n\xi$. Using that the variance is $n\xi(1-\xi)$, it follows that

$$g_{11}(\xi) = \frac{n^2}{Var(p_\xi)},$$

which is a Cramér–Rao type identity corresponding to an efficient estimator.

## 2.7 Geometric Probability Distribution

Let $\mathcal{X} = \{1, 2, 3, \dots\}$, $\mathbb{E} = [0, 1]$ and consider $p(k; \xi) = (1 - \xi)^{k-1}\xi$, $k \in \mathcal{X}$, $\xi \in \mathbb{E}$. The formulas for the partial derivatives of the log-likelihood function are left as an exercise for the reader in Problem 2.6. Then the Fisher information becomes

$$\begin{aligned} g_{11}(\xi) &= -E_{\xi}[\partial_{\xi}^{2}\ell(\xi)] \\ &= \sum_{k \geq 1} \frac{(k-1)p(k;\xi)}{(\xi - 1)^2} + \sum_{k \geq 1} \frac{1}{\xi^2} p(k;\xi) \\ &= \frac{1}{\xi^2(1-\xi)}, \end{aligned}$$

where we used the expression for the mean $\sum_{k \geq 1} k\, p(k; \xi) = \frac{1}{\xi}$.

## 2.8 Multinomial Geometry

In this section we investigate the geometry associated with the multinomial probability distribution. The computation performed here is inspired from Kass and Vos [49]. Consider $m$ independent, identical trials with $n$ possible outcomes. The probability that a single trial falls into class $i$ is $p_i$, $i = 1, 2, \dots, n$, and remains the same from trial to trial. Since $p_1 + \cdots + p_n = 1$, the parameter space is given by the $(n-1)$-dimensional simplex

$$\mathbb{E} = \{(p_1, \dots, p_{n-1}); 0 \leq p_i \leq 1, \sum_{i=1}^{n-1} p_i = 1\}.$$

It is advantageous to consider the new parameterization

$$z_i = 2\sqrt{p_i}, \qquad i = 1, \dots, n.$$

Then $\sum_{i=1}^{n} z_i^2 = 4$, and hence

$$z \in \mathbb{S}_{2,+}^{n-1} = \{z \in \mathbb{R}^n; \|z\|^2 = 4, z_i \geq 0\}.$$

Therefore, the statistical manifold of multinomial probability distributions can be identified with $\mathbb{S}_{2,+}^{n-1}$, the positive portion of the $(n-1)$-dimensional sphere of radius 2. The Fisher information matrix with respect to a local coordinate system $(\xi^i)$ is

$$\begin{aligned} g_{rs}(\xi) &= 4\sum_{i=1}^{n} \partial_r \sqrt{p_i(\xi)}\partial_s \sqrt{p_i(\xi)} \\ &= \sum_{i=1}^{n} \partial_r z_i(\xi)\, \partial_s z_i(\xi) \\ &= \langle \partial_r z, \partial_s z\rangle, \end{aligned}$$

where $\partial_s = \partial_{\xi^s}$. Therefore, the Fisher metric is the natural metric induced from the Euclidean metric of $\mathbb{R}^n$ on the sphere $\mathbb{S}^{n-1}_{2,+}$. We note that $\partial_r z$ is a tangent vector to the sphere in the direction of $\xi^r$.

To find the information distance between two multinomial distributions $p$ and $q$, we need to find the length of the shortest curve on the sphere $\mathbb{S}^{n-1}_{2,+}$, joining $p$ and $q$. The curve that achieves the minimum is an arc of great circle passing through $p$ and $q$, and this curve is unique.

Let $z_p$ and $z_q$ denote the points on the sphere corresponding to the aforementioned distributions. The angle $\alpha$ made by the unit vectors $z_p/2$ and $z_q/2$ satisfies $\cos\alpha = \langle z_p/2, z_q/2\rangle$. Since the distance on the sphere is the product between the radius and the central angle, we have

$$\begin{aligned} d(p,q) &= 2\alpha = 2\arccos\Big(\sum_{i=1}^{n} \frac{z_p^i}{2}\frac{z_q^i}{2}\Big) \\ &= 2\arccos\Big(\sum_{i=1}^{n} (p_i q_i)^{1/2}\Big). \end{aligned}$$

It is worthy to note that the Euclidean distance between $p$ and $q$ can be written as

$$\begin{aligned} \|z_p - z_q\|^2 &= \Big(\sum_{i=1}^{n}(z_p^i - z_q^i)^2\Big)^{1/2} = 2\Big(\sum_{i=1}^{n}\Big(\frac{z_p^i}{2} - \frac{z_q^i}{2}\Big)^2\Big)^{1/2} \\ &= 2\Big(\sum_{i=1}^{n}(\sqrt{p_i} - \sqrt{q_i})^2\Big)^{1/2} = d_H(p,q), \end{aligned}$$

which is called the *Hellinger distance* between $p$ and $q$. We shall discuss about this distance in more detail later.

The foregoing computation of the Fisher metric was exploiting geometric properties. In the following we shall provide a direct computation. We write the statistical model of multinomial distributions by $\mathcal{S} = \{p(k;\xi)\}$, with

$$p(k;\xi) = \frac{n!}{k_1! \dots k_m!} p_1^{k_1} \dots p_{m-1}^{k_{m-1}} p_m^{k_m},$$

where

$$\mathcal{X} = \{k = (k_1, \dots k_m) \in \mathbb{N}^m; k_1 + \dots k_m = n\},$$

and $\xi = (\xi^1, \dots, \xi^{m-1}) \in \mathbb{E} = [0,1]^{m-1}$, with $\xi^i = p_i$, $i = 1, \dots m-1$, and $p_m = 1 - p_1 - \dots - p_{m-1}$. Then a straightforward computation shows

$$\begin{aligned} \partial_i \ell(k;\xi) &= \frac{k_i}{p_i} - \frac{k_m}{p_m} \\ \partial_j \partial_i \ell(k;\xi) &= -\Big[\frac{k_i \delta_{ij}}{p_i^2} + \frac{k_m}{p_m^2}\Big]. \end{aligned}$$

Using the formula for the marginal probability

$$\sum_k k_i p(k;\xi) = n p_i,$$

we have

$$\begin{aligned} g_{ij}(\xi) &= -E_\xi[\partial_i \partial_j \ell(k;\xi)] = E\Big[\frac{k_i \delta_{ij}}{p_i^2} + \frac{k_m}{p_m^2}\Big] \\ &= \frac{\delta_{ij}}{p_i^2} \sum_k k_i p(k;\xi) + \frac{1}{p_m^2} \sum_k k_m p(k;\xi) \\ &= n\Big[\frac{\delta_{ij}}{p_i} + \frac{1}{p_m}\Big] = n\Big[\frac{\delta_{ij}}{\xi^i} + \frac{1}{1 - \xi^1 - \dots - \xi^{m-1}}\Big]. \end{aligned}$$

## 2.9 Poisson Geometry

Consider $m$ independent Poisson distributions with parameters $\lambda_i$, $i = 1, \dots, m$. The joint probability function is given by the product

$$p(x;\lambda) = \prod_{i=1}^m p_{\lambda_i}(x_i) = \prod_{i=1}^m e^{-\lambda_i} \frac{\lambda_i^{x_i}}{x_i!},$$

with $\lambda = (\lambda_1, \dots, \lambda_{m+1}) \in \mathbb{E} = (0,\infty)^m$, and $x = (x_1, \dots, x_m) \in \mathcal{X} = (\mathbb{N} \cup \{0\})^m$. The log-likelihood function and its derivatives with respect to $\partial_j = \partial_{\lambda_j}$ are

$$\begin{aligned} \ell(x;\lambda) &= -\lambda_i + x_i \ln \lambda_i - \ln(x_i!) \\ \partial_j \ell(x;\lambda) &= -1 + \frac{x_j}{\lambda_j} \\ \partial_k \partial_j \ell(x;\lambda) &= -\frac{x_j}{\lambda_j^2} \delta_{kj}. \end{aligned}$$

Then the Fisher information is obtained as

$$\begin{aligned} g_{jk}(\lambda) &= E\Big[\frac{x_j}{\lambda_j^2}\delta_{kj}\Big] = \frac{1}{\lambda_j^2}\delta_{kj}E[x_j] \\ &= \frac{1}{\lambda_j^2}\delta_{kj}\sum_x x_j p(x;\lambda) = \frac{1}{\lambda_j}\delta_{kj}. \end{aligned}$$

Therefore the Fisher matrix has a diagonal form with positive entries.

## 2.10 The Space $\mathcal{P}(\mathcal{X})$

Let $\mathcal{X} = \{x_1, \dots, x_n\}$ and consider the statistical model $\mathcal{P}(\mathcal{X})$ of all discrete probability densities on $\mathcal{X}$. The space $\mathcal{P}(\mathcal{X})$ can be imbedded into the function space $\mathbb{R}^{\mathcal{X}} = \{f; f : \mathcal{X} \to \mathbb{R}\}$ in several ways, as we shall describe shortly. This study can be found in Nagaoka and Amari [61].

For any $\alpha \in \mathbb{R}$ consider the function $\varphi_\alpha : (0, \infty) \to \mathbb{R}$

$$\varphi_\alpha(u) = \begin{cases} \dfrac{2}{1-\alpha}u^{\frac{1-\alpha}{2}}, & \text{if } \alpha \neq 1 \\ \ln u, & \text{if } \alpha = 1. \end{cases}$$

The imbedding

$$\mathcal{P}(\mathcal{X}) \ni p(x;\xi) \to \varphi_\alpha\big(p(x;\xi)\big) \in \mathbb{R}^{\mathcal{X}}$$

is called the *$\alpha$-representation* of $\mathcal{P}(\mathcal{X})$. A distinguished role will be played by the *$\alpha$-likelihood functions*

$$\ell^{(\alpha)}(x;\xi) = \varphi_\alpha\big(p(x;\xi)\big).$$

The coordinate tangent vectors in this representation are given by

$$\partial_i \ell^{(\alpha)}(x;\xi) = \partial_{\xi^i}\varphi_\alpha\big(p(x;\xi)\big).$$

The $\alpha$-representation can be used to define the Fisher metric and the $\nabla^{(\alpha)}$-connection on $\mathcal{P}(\mathcal{X})$.

**Proposition 2.10.1** *The Fisher metric can be written in terms of the $\alpha$-likelihood functions as in the following* (*i*) $g_{ij}(\xi) = \sum_{k=1}^{n} \partial_i \ell^{(\alpha)}(x_k;\xi) \partial_j \ell^{(-\alpha)}(x_k;\xi)$;

(*ii*) $g_{ij}(\xi) = -\frac{2}{1+\alpha} \sum_{k=1}^{n} p(x_k;\xi)^{\frac{1+\alpha}{2}} \partial_i \partial_j \ell^{(\alpha)}(x_k;\xi)$.

*Proof:* Differentiating yields

$$\partial_i \ell^{(\alpha)} = p^{\frac{1-\alpha}{2}} \partial_i \ell; \tag{2.10.13}$$

$$\partial_i \ell^{(-\alpha)} = p^{\frac{1+\alpha}{2}} \partial_i \ell; \tag{2.10.14}$$

$$\partial_i \partial_j \ell^{(\alpha)} = p^{\frac{1-\alpha}{2}} \Big( \partial_i \partial_j \ell + \frac{1-\alpha}{2} \partial_i \ell \partial_j \ell \Big), \tag{2.10.15}$$

where $\ell(x;\xi) = \ln p(x;\xi)$.

(*i*) The previous computations and formula (1.6.16) provide

$$\begin{aligned} \sum_{k=1}^{n} \partial_i \ell^{(\alpha)}(x_k;\xi) \partial_j \ell^{(-\alpha)}(x_k;\xi) &= \sum_{k=1}^{n} p^{\frac{1-\alpha}{2}} \partial_i \ell(x_k) p^{\frac{1+\alpha}{2}} \partial_j \ell(x_k) \\ &= \sum_{k=1}^{n} p(x_k;\xi) \partial_i \ell(x_k) \partial_j \ell(x_k) \\ &= E_\xi[\partial_i \ell \, \partial_j \ell] = g_{ij}(\xi). \end{aligned}$$

(*ii*) Relation (2.10.15) implies

$$p^{\frac{1+\alpha}{2}} \partial_i \partial_j \ell^{(\alpha)}(x;\xi) = p(x;\xi) \partial_i \partial_j \ell(x;\xi) + \frac{1-\alpha}{2} p(x;\xi) \partial_i \ell \partial_j \ell(x;\xi).$$

Summing and using (1.6.16) and (1.6.18), we have

$$\begin{aligned} \sum_{k=1}^{n} p(x_k;\xi)^{\frac{1+\alpha}{2}} \partial_i \partial_j \ell^{(\alpha)}(x_k;\xi) &= E_\xi[\partial_i \partial_j \ell] + \frac{1-\alpha}{2} E_\xi[\partial_i \ell \, \partial_j \ell] \\ &= -g_{ij}(\xi) + \frac{1-\alpha}{2} g_{ij}(\xi) \\ &= -\frac{1+\alpha}{2} g_{ij}(\xi). \end{aligned}$$

■

The symmetry of relation (*i*) implies that the Fisher metric induced by both $\alpha$ and $-\alpha$-representations are the same.

**Proposition 2.10.2** *The components of the $\alpha$-connection are given in terms of the $\alpha$-representation as*

$$\Gamma^{(\alpha)}_{ij,k} = \sum_{r=1}^{n} \partial_i \partial_j \ell^{(\alpha)}(x_r;\xi)\, \partial_k \ell^{(-\alpha)}(x_r;\xi). \tag{2.10.16}$$

*Proof:* Combining relations (2.10.14) and (2.10.15)

$$\begin{aligned}\sum_{r=1}^{n} \partial_i \partial_j \ell^{(\alpha)}\, \partial_k \ell^{(-\alpha)} &= \sum_{r=1}^{n} p(x_r;\xi)\Big(\partial_i\partial_j\ell + \frac{1-\alpha}{2}\partial_i\ell\partial_j\ell\Big)\partial_k\ell(x_r;\xi) \\ &= E_\xi\Big[\Big(\partial_i\partial_j\ell + \frac{1-\alpha}{2}\partial_i\ell\partial_j\ell\Big)\partial_k\ell\Big] \\ &= \Gamma^{(\alpha)}_{ij,k},\end{aligned}$$

by (1.11.34). ∎

The particular values $\alpha = -1, 0, 1$ provide distinguished important cases of representations of $\mathcal{P}(\mathcal{X})$.

### 2.10.1 −1-Representation

If $\alpha = -1$, then $\varphi_{-1}(u) = u$, and $\ell^{(-1)}(p(x;\xi)) = p(x;\xi)$ is the identical imbedding of $\mathcal{P}(\mathcal{X})$ into $\mathbb{R}^{\mathcal{X}}$. Thus $\mathcal{P}(\mathcal{X})$ is an open set of the affine space $\mathcal{A}_1 = \{f : \mathcal{X} \to \mathbb{R}; \sum_{k=1}^{n} f(x_k) = 1\}$. Therefore, the tangent space at any point $p_\xi$ can be identified with the following affine variety

$$T^{(-1)}_\xi(\mathcal{P}(\mathcal{X})) = \mathcal{A}_0 = \{f : \mathcal{X} \to \mathbb{R}; \sum_{k=1}^{n} f(x_k) = 0\}.$$

The coordinate vector fields in this representation are given by

$$(\partial_i^{-1})_\xi = \partial_i p_\xi.$$

We can easily check that

$$\sum_{k=1}^{n} (\partial_i^{-1})_\xi(x_k) = \sum_{k=1}^{n} \partial_i p_\xi(x_k) = \partial_i(1) = 0,$$

so $(\partial_i^{-1})_\xi \in T_\xi(\mathcal{P})$, for any $\xi$.

### 2.10.2 0-Representation

This is also called the *square root representation.* In this case $\varphi_0(u) = 2\sqrt{u}$, and the imbedding $\varphi_0 : \mathcal{P}(\mathcal{X}) \to \mathbb{R}^{\mathcal{X}}$ is

$$p(x;\xi) \to \varphi_0(p(x;\xi)) = \ell^{(0)}(x;\xi) = 2\sqrt{p(x;\xi)} = \theta(x) \in \mathbb{R}^{\mathcal{X}}.$$

Since $\sum_{k=1}^{n} \theta(x_k)^2 = 4$, the image of the imbedding $\varphi_0$ is an open subset of the sphere of radius 2,

$$\varphi_0\big(\mathcal{P}(\mathcal{X})\big) \subset \{\theta; \theta : \mathcal{X} \to \mathbb{R}; \sum_k \theta(x_k)^2 = 4\}.$$

The induced metric from the natural Euclidean metric of $\mathbb{R}^{\mathcal{X}}$ on this sphere is

$$\begin{aligned} \langle \partial_i \theta, \partial_j \theta \rangle &= \sum_{k=1}^{n} \partial_i \theta(x_k) \partial_j \theta(x_k) \\ &= 4 \sum_{k=1}^{n} \partial_i \sqrt{p(x_k;\xi)} \partial_j \sqrt{p(x_k;\xi)} \\ &= g_{ij}(\xi), \end{aligned}$$

i.e., the Fisher metric on the statistical model $\mathcal{P}(\mathcal{X})$.

The coordinate vector fields are given by

$$(\partial_i^0)_\xi = \partial_i \ell^{(0)}(x;\xi) = \frac{1}{\sqrt{p(x;\xi)}} \partial_i p(x;\xi).$$

The next computation deals with the tangent space generated by $(\partial_i^0)_\xi$. We have

$$\begin{aligned} \langle \theta, (\partial_i^0)_\xi \rangle &= \sum_{k=1}^{n} \theta(x_k) \frac{1}{p(x_k;\xi)} \partial_i p(x_k;\xi) \\ &= \sum_{k=1}^{n} 2\sqrt{p(x_k;\xi)} \frac{1}{p(x_k;\xi)} \partial_i p(x_k;\xi) \\ &= 2\partial_i \sum_{k=1}^{n} p(x_k;\xi) = 0, \end{aligned}$$

so that the vector $(\partial_i^0)_\xi$ is perpendicular on the vector $\theta$, and hence belongs to the tangent plane to the sphere at $\theta$. This can be identified with the tangent space $T_\xi^{(0)} \mathcal{P}(\mathcal{X})$ in the 0-representation.

### 2.10.3 1-Representation

This is also called the *exponential (or the logarithmic) representation*, because each distribution $p(x;\xi) \in \mathcal{P}(\mathcal{X})$ is identified with $\ln p(x;\xi) \in \mathbb{R}^{\mathcal{X}}$. In this case the 1-likelihood function becomes $\ell^{(1)}(x;\xi) = \ell(x;\xi) = \ln p(x;\xi)$, i.e., the usual likelihood function.

The coordinate vector fields are given by

$$(\partial_i^1)_\xi = \partial_i \ell^{(1)}(x;\xi) = \frac{1}{p(x;\xi)} \partial_i p(x;\xi).$$

In the virtue of the computation

$$E_p[(\partial_i^1)_\xi] = E_p[\partial_i \ell^{(1)}(x;\xi)] = \sum_{k=1}^{n} \partial_i p(x_k;\xi) = \partial_i(1) = 0,$$

it follows that the tangent space in this representation is given by

$$T_p^{(1)}(\mathcal{P}(\mathcal{X})) = \{f; f \in \mathbb{R}^{\mathcal{X}}\}, E_p[f] = 0\}.$$

It is worth noting that tangent spaces are invariant objects, that do not depend on any representation. However, when considering different system of parameters, tangent vectors can be described by some particular relations, like in the cases of $\pm 1$ and 0 representations.

### 2.10.4 Fisher Metric

Let $\xi^i = p(x_i;\xi)$, $i = 1, \ldots, n-1$, be the coordinates on $\mathcal{P}(\mathcal{X})$. Since $p(x_n;\xi) = 1 - \sum_{j=1}^{n-1} \xi^j$, then the partial derivatives with respect to $\xi^j$ are

$$\partial_i p(x_k;\xi) = \begin{cases} \delta_{ik}, & \text{if } k = 1, \ldots, n-1 \\ -1, & \text{if } k = n. \end{cases}$$

Then the Fisher metric is given by

$$\begin{aligned} g_{ij}(\xi) &= E_p[\partial_i \ell \, \partial_j \ell] = \sum_{k=1}^{n} p(x_k;\xi) \partial_i \ln p(x_k;\xi) \partial_j \ln p(x_k;\xi) \\ &= \sum_{k=1}^{n} \frac{\partial_i p(x_k;\xi) \partial_j p(x_k;\xi)}{p(x_k;\xi)} \\ &= \sum_{k=1}^{n-1} \frac{\delta_{ik}\delta_{jk}}{\xi^k} + \frac{1}{1 - \sum_{j=1}^{n-1} \xi^j} \\ &= \frac{\delta_{ij}}{\xi^j} + \frac{1}{1 - \sum_{j=1}^{n-1} \xi^j}. \end{aligned}$$

## 2.11 Problems

**2.1.** Consider the statistical model given by the densities of a normal family

$$p(x, \xi) = \frac{1}{\sigma\sqrt{2\pi}} e^{-\frac{(x-\mu)^2}{2\sigma^2}}, \quad x \in \mathcal{X} = \mathbb{R},$$

with parameters $(\xi^1, \xi^2) = (\mu, \sigma) \in \mathbb{R} \times (0, \infty)$.

(*a*) Show that the log-likelihood function and its derivatives are given by

$$\begin{aligned}
\ell_x(\xi) = \ln p(x, \xi) &= -\frac{1}{2}\ln(2\pi) - \ln\sigma - \frac{(x-\mu)^2}{2\sigma^2} \\
\partial_\sigma \ell_x(\xi) = \partial_\sigma \ln p(x, \xi) &= -\frac{1}{\sigma} + \frac{1}{\sigma^3}(x-\mu)^2 \\
\partial_\sigma\partial_\sigma \ell_x(\xi) = \partial_\sigma\partial_\sigma \ln p(x, \xi) &= \frac{1}{\sigma^2} - \frac{3}{\sigma^4}(x-\mu)^2 \\
\partial_\mu \ell_x(\xi) = \partial_\mu \ln p(x, \xi) &= \frac{1}{\sigma^2}(x-\mu) \\
\partial_\mu\partial_\mu \ell_x(\xi) = \partial_\mu\partial_\mu \ln p(x, \xi) &= -\frac{1}{\sigma^2} \\
\partial_\sigma\partial_\mu \ell_x(\xi) = \partial_\sigma\partial_\mu \ln p(x, \xi) &= -\frac{2}{\sigma^3}(x-\mu).
\end{aligned}$$

(*b*) Show that the Fisher–Riemann metric components are given by

$$g_{11} = \frac{1}{\sigma^2}, \qquad g_{12} = g_{21} = 0, \qquad g_{22} = \frac{2}{\sigma^2}.$$

**2.2.** Consider the statistical model defined by the lognormal distribution

$$p_{\mu,\sigma}(x) = \frac{1}{\sqrt{2\pi}\,\sigma x} e^{-\frac{(\ln x-\mu)^2}{2\sigma^2}}, \qquad x > 0.$$

(*a*) Show that the log-likelihood function and its derivatives are given by

$$\begin{aligned}
\ell(\mu, \sigma) &= -\ln\sqrt{2\pi} - \ln\sigma - \ln x - \frac{1}{2\sigma^2}(\ln x - \mu)^2 \\
\partial_\mu^2 \ell(\mu, \sigma) &= -\frac{1}{\sigma^2} \\
\partial_\sigma^2 \ell(\mu, \sigma) &= \frac{1}{\sigma^2} - \frac{3}{\sigma^4}(\ln x - \mu)^2 \\
\partial_\mu\partial_\sigma \ell(\mu, \sigma) &= -\frac{2}{\sigma^3}(\ln x - \mu).
\end{aligned}$$

(b) Using the substitution $y = \ln x - \mu$, show that the components of the Fisher–Riemann metric are given by

$$g_{\sigma\sigma} = \frac{2}{\sigma^2}, \qquad g_{\mu\mu} = \frac{1}{\sigma^2}, \qquad g_{\mu\sigma} = g_{\sigma\mu} = 0.$$

**2.3.** Let

$$p_\xi(x) = p_{\alpha,\beta}(x) = \frac{1}{\beta^\alpha \Gamma(\alpha)}\, x^{\alpha-1} e^{-x/\beta},$$

with $(\alpha, \beta) \in (0, \infty) \times (0, \infty)$, $x \in (0, \infty)$ be the statistical model defined by the gamma distribution.

(a) Show that the log-likelihood function is

$$\ell_x(\xi) = \ln p_\xi = -\alpha \ln \beta - \ln \Gamma(\alpha) + (\alpha - 1) \ln x - \frac{x}{\beta}.$$

(b) Verify the relations

$$\begin{aligned}
\partial_\beta \ell_x(\xi) &= -\frac{\alpha}{\beta} + \frac{x}{\beta^2} \\
\partial_{\alpha\beta} \ell_x(\xi) &= -\frac{1}{\beta} \\
\partial_\beta^2 \ell_x(\xi) &= \frac{\alpha}{\beta^2} - \frac{2x}{\beta^3} \\
\partial_\alpha \ell_x(\xi) &= -\ln \beta - \psi(\alpha) + \ln x \\
\partial_\alpha^2 \ell_x(\xi) &= -\psi_1(\alpha),
\end{aligned}$$

where

$$\psi(\alpha) = \frac{\Gamma'(\alpha)}{\Gamma(\alpha)}, \qquad \psi_1(\alpha) = \psi'(\alpha) \tag{2.11.17}$$

are the *digamma* and the *trigamma* functions, respectively.

(c) Prove that for $\alpha > 0$, we have

$$\sum_{n \geq 0} \frac{\alpha}{(\alpha + n)^2} > 1.$$

**2.4.** Consider the beta distribution

$$p_{a,b} = \frac{1}{B(a, b)}\, x^{a-1} (1 - x)^{b-1}, \qquad a, b > 0, x \in [0, 1].$$

($a$) Using that the beta function

$$B(a,b) = \int_0^1 x^{a-1}(1-x)^{b-1}\,dx$$

can be expressed in terms of gamma function as

$$B(a,b) = \frac{\Gamma(a)\Gamma(b)}{\Gamma(a+b)},$$

show that its partial derivatives can be written in terms of digamma functions, as

$$\partial_a \ln B(a,b) = \psi(a) - \psi(a+b) \tag{2.11.18}$$

$$\partial_b \ln B(a,b) = \psi(b) - \psi(a+b). \tag{2.11.19}$$

($b$) Show that the log-likelihood function is given by

$$\ell(a,b) = \ln p_{a,b} = -\ln B(a,b) + (a-1)\ln x + (b-1)\ln(1-x).$$

($c$) Take partial derivatives and use formulas (2.11.18) and (2.11.19) to verify relations

$$\begin{aligned}
\partial_a \ell(a,b) &= -\partial_a \ln B(a,b) + \ln x = \psi(a+b) - \psi(a) + \ln x \\
\partial_b \ell(a,b) &= \psi(a+b) - \psi(b) + \ln(1-x) \\
\partial_a^2 \ell(a,b) &= \psi'(a+b) - \psi'(a) = \psi_1(a+b) - \psi_1(a) \\
\partial_b^2 \ell(a,b) &= \psi'(a+b) - \psi'(b) = \psi_1(a+b) - \psi_1(b) \\
\partial_a \partial_b \ell(a,b) &= \psi'(a+b) = \psi_1(a+b).
\end{aligned}$$

($c$) Using the expression of trigamma functions as a Hurwitz zeta function, show that the Fisher information matrix can be written as a series $g = \sum_{n \geq 0} g_n$, where

$$g_n = \begin{pmatrix} \frac{1}{(a+n)^2} - \frac{1}{(a+b+n)^2} & -\frac{1}{(a+b+n)^2} \\ -\frac{1}{(a+b+n)^2} & \frac{1}{(b+n)^2} - \frac{1}{(a+b+n)^2} \end{pmatrix}.$$

**2.5.** Let $\mathcal{S} = \{p_\xi; \xi \in [0,1]\}$ be a one-dimensional statistical model, where

$$p(k;\xi) = \binom{n}{k} \xi^k (1-\xi)^{n-k}$$

is the Bernoulli distribution, with $k \in \{0, 1, \dots, n\}$ and $\xi \in [0, 1]$. Show that the derivatives of the log-likelihood function $\ell_k(\xi) = \ln p(k; \xi)$ are

$$\begin{aligned}\partial_\xi \ell_k(\xi) &= \frac{k}{\xi} - (n-k)\frac{1}{1-\xi}\\ \partial_\xi^2 \ell_k(\xi) &= -\frac{k}{\xi^2} - (n-k)\frac{1}{(1-\xi)^2}.\end{aligned}$$

**2.6.** Consider the geometric probability distribution $p(k; \xi) = (1-\xi)^{k-1}\xi$, $k \in \{1, 2, 3, \dots\}$, $\xi \in [0, 1]$. Show that

$$\begin{aligned}\partial_\xi \ell_k(\xi) &= \frac{k-1}{\xi-1} + \frac{1}{\xi}\\ \partial_\xi^2 \ell_k(\xi) &= -\frac{(k-1)}{(\xi-1)^2} - \frac{1}{\xi^2}.\end{aligned}$$

**2.7.** Let $f$ be a density function on $\mathbb{R}$ and define the statistical model

$$\mathcal{S}_f = \Big\{p(x; \mu, \sigma) = \frac{1}{\sigma} f\Big(\frac{x-\mu}{\sigma}\Big); \mu \in \mathbb{R}, \sigma > 0\Big\}.$$

(*a*) Show that $\int_{\mathbb{R}} p(x; \mu, \sigma)\, dx = 1$.

(*b*) Verify the following formulas involving the log-likelihood function $\ell = \ln p(\,\cdot\,; \mu, \sigma)$:

$$\partial_\mu \ell = -\frac{1}{\sigma}\frac{f'}{f}, \qquad \partial_\sigma \ell = -\frac{1}{\sigma} - \frac{(x-\mu)}{\sigma^2}\frac{f'}{f}$$

$$\partial_\mu \partial_\sigma \ell = \frac{1}{f^2}\Big[\Big(\frac{f'}{\sigma^2} + \frac{1}{\sigma}\frac{x-\mu}{\sigma^2} f''\Big) f - \frac{1}{\sigma}\frac{x-\mu}{\sigma^2}(f')^2\Big].$$

(*b*) Show that for any continuous function $h$ we have

$$E_{(\mu,\sigma)}\Big[h\Big(\frac{x-\mu}{\sigma}\Big)\Big] = E_{(0,1)}[h(x)].$$

(*c*) Assume that $f$ is an even function (i.e., $f(-x) = f(x)$). Show that the Fisher–Riemann metric, $g$, has a diagonal form (i.e., $g_{12} = 0$).

(*d*) Prove that the Riemannian space $(\mathcal{S}_f, g)$ has a negative, constant curvature.

(e) Consider $f(x) = \frac{1}{\sqrt{2\pi}} e^{-x^2/2}$. Use the aforementioned points to deduct the formula for $g_{ij}$ and to show that the curvature $K = -\frac{1}{2}$.

**2.8.** Search the movement of the curve

$$(\mu, \sigma) \to p_{\mu,\sigma}(x) = \frac{1}{\sigma\sqrt{2\pi}} e^{-\frac{(x-\mu)^2}{2\sigma^2}}, \ \mu^2 + \sigma^2 = 1$$

with $(\mu, \sigma, p) \in \mathbb{R} \times (0, \infty) \times (0, \infty)$, $x \in \mathbb{R}$, fixed, in the direction of the binormal vector field.

**2.9.** The graph of the normal density of probability

$$x \to p_{\mu,\sigma}(x) = \frac{1}{\sigma\sqrt{2\pi}} e^{-\frac{(x-\mu)^2}{2\sigma^2}}$$

is called Gauss bell. Find the equation of the surface obtained by revolving the Gauss bell about:

(a) $Ox$ axis;

(b) $Op$ axis.

**2.10.** Inspect the movement of the trajectories of the vector field $(y, z, x)$ after the direction of the vector field

$$\left(1, 1, \frac{1}{\sigma\sqrt{2\pi}} e^{-\frac{(x-\mu)^2}{2\sigma^2}}\right),$$

where $\mu$ and $\sigma$ are fixed.

**2.11.** The normal surface

$$(\mu, \sigma) \to p_{\mu,\sigma}(x) = \frac{1}{\sigma\sqrt{2\pi}} e^{-\frac{(x-\mu)^2}{2\sigma^2}},$$

$$(\mu, \sigma) \in \mathbb{R} \times (0, \infty); \ x \in \mathbb{R}$$

is deformed into $p_{\mu,\sigma}(tx)$, $t \in \mathbb{R}$. What happens with the Gauss curvature?

**2.12.** The gamma surface

$$(\alpha, \beta) \to p_{\alpha,\beta}(x) = \frac{1}{\beta^\alpha \Gamma(\alpha)} x^{\alpha-1} e^{-\frac{x}{\beta}}$$

$$(\alpha, \beta) \in (0, \infty) \times (0, \infty); \ x \in (0, \infty)$$

is deformed into $p_{t\alpha,\beta}(x)$, $t \in (0, \infty)$. What happens with the mean curvature?

# Chapter 3

# Entropy on Statistical Models

Entropy is a notion taken form Thermodynamics, where it describes the uncertainty in the movement of gas particles. In this chapter the entropy will be considered as a measure of uncertainty of a random variable.

Maximum entropy distributions, with certain moment constraints, will play a central role in this chapter. They are distributions with a maximal ignorance degree towards unknown elements of the distribution. For instance, if nothing is known about a distribution defined on the interval $[a, b]$, it makes sense to express our ignorance by choosing the distribution to be the uniform one. Sometimes the mean is known. In this case the maximum entropy decreases and the distribution is not uniform any more. More precisely, among all distributions $p(x)$ defined on $(0, \infty)$ with a given mean $\mu$, the one with the maximum entropy is the exponential distribution. Furthermore, if both the mean and the standard variation are given for a distribution $p(x)$ defined on $\mathbb{R}$, then the distribution with the largest entropy is the normal distribution.

Since the concept of entropy can be applied to any point of a statistical model, the entropy becomes a function defined on the statistical model. Then, likewise in Thermodynamics, we shall investigate the entropy maxima, as they have a distinguished role in the theory.

O. Calin and C. Udrişte, *Geometric Modeling in Probability and Statistics*,
DOI 10.1007/978-3-319-07779-6_3,

## 3.1 Introduction to Information Entropy

The notion of *entropy* comes originally from Thermodynamics. It is a quantity that describes the amount of disorder or randomness in a system bearing energy or information. In Thermodynamics the entropy is defined in terms of heat and temperature.

According to the second law of Thermodynamics, during any process the change in the entropy of a system and its surroundings is either zero or positive. The entropy of a free system tends to increase in time, towards a finite or infinite maximum. Some physicists define the arrow of time in the direction in which its entropy increases, see Hawking [43]. Most processes tend to increase their entropy in the long run. For instance, a house starts falling apart, an apple gets rotten, a person gets old, a car catches rust over time, etc.

Another application of entropy is in information theory, formulated by C. E. Shannon [73] in 1948 to explain aspects and problems of information and communication. In this theory a distinguished role is played by the *information source*, which produces a sequence of messages to be communicated to the receiver. The information is a measure of the freedom of choice with which a message can be selected from the set of all possible messages. The information can be measured numerically using the logarithm in base 2. In this case the resulting units are called binary digits, or *bits*. One bit measures a choice between two equally likely choices. For instance, if a coin is tossed but we are unable to see it as it lands, the landing information contains 1 bit of information. If there are $N$ equally likely choices, the number of bits is equal to the digital logarithm of the number of choices, $\log_2 N$. In the case when the choices are not equally probable, the situation will be described in the following.

Shannon defined a quantity that measures how much information, and at which rate this information is produced by an information source. Suppose there are $n$ possible elementary outcomes of the source, $A_1, \ldots, A_n$, which occur with probabilities $p_1 = p(A_1)$, ..., $p_n = p(A_n)$, so the source outcomes are described by the discrete probability distribution

| event | $A_1$ | $A_2$ | ... | $A_n$ |
|---|---|---|---|---|
| probability | $p_1$ | $p_2$ | ... | $p_n$ |

with $p_i$ given. Assume there is an uncertainty function, $H(p_1, \ldots, p_n)$, which "measures" how much "choice" is involved in selecting an event. It is fair to ask that $H$ satisfies the following properties (Shannon's axioms):

(*i*) $H$ is continuous in each $p_i$;

(*ii*) If $p_1 = \cdots = p_n = \dfrac{1}{n}$, then $H$ is monotonic increasing function of $n$ (i.e., for equally likely events there is more uncertainty when there are more possible events).

(*iii*) If a choice is broken down into two successive choices, then the initial $H$ is the weighted sum of the individual values of $H$:

$$H(p_1, p_2, \ldots, p_{n-1}, p_n', p_n'') = H(p_1, p_2, \ldots, p_{n-1}, p_n) + p_n H\Big(\frac{p_n'}{p_n}, \frac{p_n''}{p_n}\Big),$$

with $p_n = p_n' + p_n''$.

Shannon proved that the only function $H$ satisfying the previous three assumptions is of the form

$$H = -k \sum_{i=1}^{n} p_i \log_2 p_i,$$

where $k$ is a positive constant, which amounts to the choice of a unit of measure. The negative sign in front of the summation formula implies its non-negativity. This is the definition of the *information entropy* for discrete systems given by Shannon [73]. It is remarkable that this is the same expression seen in certain formulations of statistical mechanics.

Since the next sections involve integration and differentiation, it is more convenient to use the natural logarithm instead of the digital logarithm. The entropy defined by $H = -\sum_{i=1}^{n} p_i \ln p_i$ is measured in *natural units* instead of bits.[1] Sometimes this is also denoted by $H(p_1, \ldots, p_n)$.

We make some more remarks regarding notation. We write $H(X)$ to denote the entropy of a random variable $X$, $H(p)$ to denote the entropy of a probability density $p$, and $H(\xi)$ to denote the entropy $H(p_\xi)$ on a statistical model with parameter $\xi$. The joint entropy of two random variables $X$ and $Y$ will be denoted by $H(X, Y)$, while

[1] Since $\log_2 x = \ln x / \ln 2 = 1.44 \ln x$, a natural unit is about 1.44 bits.

$H(X|Y)$ will be used for the conditional entropy of $X$ given $Y$. These notations will be used interchangeably, depending on the context.

The entropy can be used to measure information in the following way. The information can be measured as a reduction in the uncertainty, i.e. entropy. If $X$ and $Y$ are random variables that describe an event, the initial uncertainty about the event is $H(X)$. After the random variable $Y$ is revealed, the new uncertainty is $H(X|Y)$. The reduction in uncertainty, $H(X) - H(X|Y)$, is called the information conveyed about $X$ by $Y$. Its symmetry property is left as an exercise in Problem 3.3, part $(d)$.

In the case of a discrete random variable $X$, the entropy can be interpreted as the weighted average of the numbers $-\ln p_i$, where the weights are the probabilities of the values of the associated random variable $X$. Equivalently, this can be also interpreted as the expectation of the random variable that assumes the value $-\ln p_i$ with probability $p_i$

$$H(X) = -\sum_{i=1}^{n} P(X = x_i) \ln P(X = x_i) = E[-\ln P(X)].$$

Extending the situation from the discrete case, the uncertainty of a continuous random variable $X$ defined on the interval $(a, b)$ will be defined by an integral. If $p$ denotes the probability density function of $X$, then the integral

$$H(X) = -\int_a^b p(x) \ln p(x)\, dx$$

defines the entropy of $X$, provided the integral is finite.

This chapter considers the entropy on statistical models as a function of its parameters. It provides examples of statistical manifolds and their associated entropies and deals with the main properties of the entropy regarding bounds, maximization and relation with the Fisher information metric.

## 3.2 Definition and Examples

Let $\mathcal{S} = \{p_\xi = p(x;\xi); \xi = (\xi^1, \dots, \xi^n) \in \mathbb{E}\}$ be a statistical model, where $p(\cdot, \xi) : \mathcal{X} \to [0, 1]$ is the probability density function which depends on parameter vector $\xi$. The entropy on the manifold $\mathcal{S}$ is a function $H : \mathbb{E} \to \mathbb{R}$, which is equal to the negative of the expectation of the log-likelihood function, $H(\xi) = -E_{p_\xi}[\ell_x(\xi)]$. More precisely,

$$H(\xi) = \begin{cases} -\displaystyle\int_{\mathcal{X}} p(x,\xi) \ln p(x,\xi)\,dx, & \text{if } \mathcal{X} \text{ is continuous;} \\ \\ -\displaystyle\sum_{x\in\mathcal{X}} p(x,\xi) \ln p(x,\xi), & \text{if } \mathcal{X} \text{ is discrete.} \end{cases}$$

Since the entropy is associated with each distribution $p(x,\xi)$, we shall also use the alternate notation $H\big(p(x,\xi)\big)$. Sometimes, the entropy in the continuous case is called *differential entropy*, while in the discrete case is called *discrete entropy*.

It is worth noting that in the discrete case the entropy is always positive, while in the continuous case might be zero or negative. Since a simple scaling of parameters will modify a continuous distribution with positive entropy into a distribution with a negative entropy (see Problem 3.4.), in the continuous case there is no canonical entropy, but just a relative entropy. In order to address this drawback, the entropy is modified into the *relative information entropy*, as we shall see in Chap. 4.

The entropy can be defined in terms of a base measure on the space $\mathcal{X}$, but for keeping the exposition elementary we shall assume that $\mathcal{X} \subseteq \mathbb{R}^n$ with the Lebesgue-measure $dx$.

The entropy for a few standard distributions is computed in the next examples.

**Example 3.2.1 (Normal Distribution)** In this case $\mathcal{X} = \mathbb{R}$, $\xi = (\mu, \sigma) \in \mathbb{R} \times (0, \infty)$ and

$$p(x;\xi) = \frac{1}{\sigma\sqrt{2\pi}} e^{-\frac{(x-\mu)^2}{2\sigma^2}}.$$

The entropy is

$$\begin{aligned} H(\mu,\sigma) &= -\int_{\mathcal{X}} p(x) \ln p(x)\,dx \\ &= -\int_{\mathcal{X}} p(x)\Big(-\frac{1}{2}\ln(2\pi) - \ln\sigma - \frac{(x-\mu)^2}{2\sigma^2}\Big)\,dx \\ &= \frac{1}{2}\ln(2\pi) + \ln\sigma + \frac{1}{2\sigma^2}\int_{\mathcal{X}} (x-\mu)^2 p\,dx \end{aligned}$$

$$
\begin{aligned}
&= \frac{1}{2}\ln(2\pi) + \ln\sigma + \frac{1}{2\sigma^2}\cdot\sigma^2 \\
&= \frac{1}{2}\ln(2\pi) + \ln\sigma + \frac{1}{2} \\
&= \ln(\sigma\sqrt{2\pi e}).
\end{aligned}
$$

It follows that the entropy does not depend on $\mu$, and is increasing logarithmically as a function of $\sigma$, with $\lim_{\sigma\searrow 0} H = -\infty$, $\lim_{\sigma\nearrow\infty} H = \infty$. Furthermore, the change of coordinates $\varphi : \mathbb{E} \to \mathbb{E}$ under which the entropy is invariant, i.e. $H(\xi) = H\big(\varphi(\xi)\big)$, are only the translations $\varphi(\mu,\sigma) = (\mu + k, \sigma)$, $k \in \mathbb{R}$.

**Example 3.2.2 (Poisson Distribution)** In this case the sample space is $\mathcal{X} = \mathbb{N}$, and the probability density

$$
p(n;\xi) = e^{-\xi}\frac{\xi^n}{n!}, \qquad n\in\mathbb{N},\ \xi\in\mathbb{R}
$$

depends only on one parameter, $\xi$. Using $\ln p(n,\xi) = -\xi + n\ln\xi - \ln(n!)$, we have

$$
\begin{aligned}
H(\xi) &= -\sum_{n\geq 0} p(n,\xi)\ln p(n,\xi) \\
&= -\sum_{n\geq 0}\Big(-\xi e^{-\xi}\frac{\xi^n}{n!} + n\ln\xi e^{-\xi}\frac{\xi^n}{n!} - \ln(n!)e^{-\xi}\frac{\xi^n}{n!}\Big) \\
&= \xi e^{-\xi}\underbrace{\sum_{n\geq 0}\frac{\xi^n}{n!}}_{=e^{\xi}} - \ln\xi\, e^{-\xi}\sum_{n\geq 0}\frac{n\xi^n}{n!} + e^{-\xi}\sum_{n\geq 0}\frac{\xi^n\ln(n!)}{n!} \\
&= \xi - \ln\xi\, e^{-\xi}\xi e^{\xi} + e^{-\xi}\sum_{n\geq 0}\frac{\ln(n!)}{n!}\xi^n \\
&= \xi(1-\ln\xi) + e^{-\xi}\sum_{n\geq 0}\frac{\ln(n!)}{n!}\xi^n.
\end{aligned}
$$

We note that $\lim_{\xi\searrow 0} H(\xi)=0$ and $H(x)<\infty$, since the series $\sum_{n\geq 0}\frac{\xi^n\ln(n!)}{n!}$ has an infinite radius of convergence, see Problem 3.21.

**Example 3.2.3 (Exponential Distribution)** Consider the exponential distribution

$$
p(x;\xi) = \xi e^{-\xi x}, \qquad x > 0,\ \xi > 0
$$

with parameter $\xi$. The entropy is

$$\begin{aligned}
H(\xi) &= -\int_0^\infty p(x)\ln p(x)\,dx = -\int_0^\infty \xi e^{-\xi x}(\ln\xi - \xi x)\,dx \\
&= -\xi\ln\xi\int_0^\infty e^{-\xi x}\,dx + \xi\int_0^\infty \xi e^{-\xi x}\,x\,dx \\
&= -\ln\xi\underbrace{\int_0^\infty p(x,\xi)\,dx}_{=1} + \xi\underbrace{\int_0^\infty xp(x,\xi)\,dx}_{=1/\xi} \\
&= 1-\ln\xi,
\end{aligned}$$

which is a decreasing function of $\xi$, with $H(\xi) > 0$ for $\xi \in (0, e)$. Making the parameter change $\lambda = \dfrac{1}{\xi}$, the model becomes $p(x;\lambda) = \frac{1}{\lambda}e^{-x/\lambda}$, $\lambda > 0$. The entropy $H(\lambda) = 1 + \ln\lambda$ increases logarithmically in $\lambda$. We note the fact that the entropy is parametrization dependent.

**Example 3.2.4 (Gamma Distribution)** Consider the family of distributions

$$p_\xi(x) = p_{\alpha,\beta}(x) = \frac{1}{\beta^\alpha\Gamma(\alpha)}\,x^{\alpha-1}e^{-x/\beta},$$

with positive parameters $(\xi^1, \xi^2) = (\alpha, \beta)$ and $x > 0$. We shall start by showing that

$$\int_0^\infty \ln x\; p_{\alpha,\beta}(x)\,dx = \ln\beta + \psi(\alpha), \tag{3.2.1}$$

where

$$\psi(\alpha) = \frac{\Gamma'(\alpha)}{\Gamma(\alpha)} \tag{3.2.2}$$

is the *digamma function*. Using that the integral of $p_{\alpha,\beta}(x)$ is unity, we have

$$\int_0^\infty x^{\alpha-1}\,e^{-\frac{x}{\beta}}\,dx = \beta^\alpha\,\Gamma(\alpha),$$

and differentiating with respect to $\alpha$, it follows

$$\int_0^\infty \ln x\, x^{\alpha-1}\,e^{-\frac{x}{\beta}}\,dx = \ln\beta\,\beta^\alpha\,\Gamma(\alpha) + \beta^\alpha\,\Gamma'(\alpha). \tag{3.2.3}$$

Dividing by $\beta^\alpha\Gamma(\alpha)$ yields relation (3.2.1).

Since

$$\ln p_{\alpha,\beta}(x) = -\alpha \ln \beta - \ln \Gamma(\alpha) + (\alpha - 1) \ln x - \frac{x}{\beta},$$

using $\int_0^\infty p_{\alpha,\beta}(x)\,dx = 1$, $\int_0^\infty x\,p_{\alpha,\beta}(x)\,dx = \alpha\beta$ and (3.2.1), the entropy becomes

$$\begin{aligned} H(\alpha,\beta) &= -\int_0^\infty p_{\alpha,\beta}(x) \ln p_{\alpha,\beta}(x)\,dx \\ &= \alpha \ln \beta + \ln \Gamma(\alpha) - (\alpha - 1) \int_0^\infty \ln x\, p_{\alpha,\beta}(x)\,dx \\ &\quad + \frac{1}{\beta} \int_0^\infty x\, p_{\alpha,\beta}(x)\,dx \\ &= \ln \beta + (1-\alpha)\psi(\alpha) + \ln \Gamma(\alpha) + \alpha. \end{aligned}$$

**Example 3.2.5 (Beta Distribution)** The beta distribution on $\mathcal{X} = [0,1]$ is defined by the density

$$p_{a,b}(x) = \frac{1}{B(a,b)}\, x^{a-1}(1-x)^{b-1},$$

with $a, b > 0$ and beta function given by

$$B(a,b) = \int_0^1 x^{a-1}(1-x)^{b-1}\,dx. \tag{3.2.4}$$

Differentiating with respect to $a$ and $b$ in (3.2.4) yields

$$\begin{aligned} \partial_a B(a,b) &= \int_0^1 \ln x\, x^{a-1}(1-x)^{b-1}\,dx \\ \partial_b B(a,b) &= \int_0^1 \ln(1-x)\, x^{a-1}(1-x)^{b-1}\,dx. \end{aligned}$$

Using

$$\ln p_{a,b} = -\ln B(a,b) + (a-1)\ln x + (b-1)\ln(1-x),$$

we find

$$\begin{aligned} H(a,b) &= -\int_0^1 p_{a,b}(x)\,\ln p_{a,b}(x)\,dx \\ &= \ln B(a,b) - \frac{a-1}{B(a,b)}\int_0^1 \ln x\, x^{a-1}(1-x)^{b-1}\,dx \\ &\quad -\frac{b-1}{B(a,b)}\int_0^1 \ln(1-x)\,x^{a-1}(1-x)^{b-1}\,dx \\ &= \ln B(a,b) - (a-1)\frac{\partial_a B(a,b)}{B(a,b)} - (b-1)\frac{\partial_b B(a,b)}{B(a,b)} \\ &= \ln B(a,b) - (a-1)\partial_a \ln B(a,b) - (b-1)\partial_b \ln B(a,b). \end{aligned} \tag{3.2.5}$$

We shall express the entropy in terms of digamma function (3.2.2). Using the expression of the beta function in terms of gamma functions

$$B(a,b) = \frac{\Gamma(a)\Gamma(b)}{\Gamma(a+b)},$$

we have

$$\ln B(a,b) = \ln\Gamma(a) + \ln\Gamma(b) - \ln\Gamma(a+b).$$

The partial derivatives of the function $B(a,b)$ are

$$\partial_a \ln B(a,b) = \psi(a) - \psi(a+b) \tag{3.2.6}$$

$$\partial_b \ln B(a,b) = \psi(b) - \psi(a+b). \tag{3.2.7}$$

Substituting in (3.2.5) yields

$$H(a,b) = \ln B(a,b) + (a+b-2)\psi(a+b) - (a-1)\psi(a) - (b-1)\psi(b). \tag{3.2.8}$$

For example

$$\begin{aligned} H(1/2,1/2) &= \ln\sqrt{2} + \ln\sqrt{2} - \psi(1) + \psi(1/2) \\ &= \ln 2 + \gamma - 2\ln 2 - \gamma = -\ln 2 < 0, \end{aligned}$$

where we used

$$\psi(1) = -\gamma = -0.5772\ldots, \qquad \psi(1/2) = -2\ln 2 - \gamma.$$

It can be shown that the entropy is always non-positive, see Problem 3.22. For $a = b = 1$ the entropy vanishes

$$H(1,1) = \ln\Gamma(1) + \ln\Gamma(1) - \ln\Gamma(2) = 0.$$

**Example 3.2.6 (Lognormal Distribution)** The lognormal distribution

$$p_{\mu,\sigma}(x) = \frac{1}{\sqrt{2\pi}\sigma x} e^{-\frac{(\ln x-\mu)^2}{2\sigma^2}}, \quad (\mu,\sigma) \in (0,\infty) \times (0,\infty)$$

defines a statistical model on the sample space $\mathcal{X} = (0,\infty)$. First, using the substitution $y = \ln x - \mu$, we have

$$\begin{aligned} \int_0^\infty \ln x \, p_{\mu,\sigma}(x)\, dx &= \int_0^\infty (\ln x - \mu)\, p_{\mu,\sigma}(x)\, dx + \mu \\ &= \int_{-\infty}^{+\infty} \frac{1}{\sqrt{2\pi} y\sigma} e^{-\frac{y^2}{2\sigma^2}}\, dy + \mu = \mu. \\ \int_0^\infty (\ln x - \mu)^2\, p_{\mu,\sigma}(x)\, dx &= \int_{-\infty}^{+\infty} \frac{1}{\sqrt{2\pi}\sigma} e^{-\frac{y^2}{2\sigma^2}} y^2\, dy = \sigma^2. \end{aligned}$$

Using

$$\ln p_{\mu,\sigma} = -\ln(\sqrt{2\pi}\sigma) - \ln x - (\ln x - \mu)^2 \frac{1}{2\sigma^2},$$

and the previous integrals, the entropy becomes

$$\begin{aligned} H(\mu,\sigma) &= -\int_0^\infty p_{\mu,\sigma}(x) \ln p_{\mu,\sigma}(x)\, dx \\ &= \ln(\sqrt{2\pi}\sigma) + \int_0^\infty \ln x \, p_{\mu,\sigma}(x)\, dx \\ &\quad + \frac{1}{2\sigma^2} \int_0^\infty (\ln x - \mu)^2 p_{\mu,\sigma}(x)\, dx \\ &= \ln(\sqrt{2\pi}) + \ln\sigma + \mu + \frac{1}{2}. \end{aligned}$$

**Example 3.2.7 (Dirac Distribution)** A Dirac distribution on $(a,b)$ centered at $x_0 \in (a,b)$ represents the density of an idealized point mass $x_0$. This can be thought of as an infinitely high, infinitely thin spike at $x_0$, with total area under the spike equal to 1. The Dirac distribution centered at $x_0$ is customarily denoted by $p(x) = \delta(x-x_0)$, and its relation with the integral can be written informally as

$(i)$ $\displaystyle\int_a^b p(x)\, dx = \int_a^b \delta(x - x_0)\, dx = 1;$

$(ii)$ $\displaystyle\int_a^b g(x)p(x)\, dx = \int_a^b g(x)\delta(x - x_0)\, dx = g(x_0),$

for any continuous function $g(x)$ on $(a, b)$.

The $k$-th moment is given by

$$m_k = \int_a^b x^k \delta(x - x_0)\, dx = x_0^k.$$

Then the mean of the Dirac distribution is $\mu = x_0$ and the variance is $Var = m_2 - (m_1)^2 = 0$. The underlying random variable, which is Dirac distributed, is a constant equal to $x_0$.

In order to compute the entropy of $\delta(x - x_0)$, we shall approximate the distribution by a sequence of distributions $\varphi_\epsilon(x)$ for which we can easily compute the entropy. For any $\epsilon > 0$, consider the distribution

$$\varphi_\epsilon(x) = \begin{cases} \dfrac{1}{\epsilon}, & \text{if } |x| < \epsilon/2 \\ 0, & \text{otherwise,} \end{cases}$$

with the entropy given by

$$\begin{aligned} H_\epsilon &= -\int_a^b \varphi_\epsilon(x) \ln \varphi_\epsilon(x)\, dx \\ &= -\int_{x_0 - \epsilon/2}^{x_2 + \epsilon/2} \frac{1}{\epsilon} \ln \frac{1}{\epsilon}\, dx \\ &= \ln \epsilon. \end{aligned}$$

Since $\lim\limits_{\epsilon \searrow 0} \varphi_\epsilon = \delta(x - x_0)$, by the Dominated Convergence Theorem the entropy of $\delta(x - x_0)$ is given by the limit

$$H = \lim_{\epsilon \searrow 0} H_\epsilon = \lim_{\epsilon \searrow 0} \ln \epsilon = -\infty.$$

In conclusion, the Dirac distribution has the lowest possible entropy. Heuristically, this is because of the lack of disorganization of the associated random variable, which is a constant.

## 3.3 Entropy on Products of Statistical Models

Consider the statistical manifolds $\mathcal{S}$ and $\mathcal{U}$ and let $\mathcal{S} \times \mathcal{U}$ be their product model, see Example 1.3.9. Any density function $f \in \mathcal{S} \times \mathcal{U}$, with $f(x,y) = p(x)q(y)$, $p \in \mathcal{S}$, $q \in \mathcal{U}$, has the entropy

$$\begin{aligned} H_{\mathcal{S}\times\mathcal{U}}(f) &= -\iint_{\mathcal{X}\times\mathcal{Y}} f(x,y) \ln f(x,y)\, dxdy \\ &\quad - \iint_{\mathcal{X}\times\mathcal{Y}} p(x)q(y)[\ln p(x) + \ln q(y)]\, dxdy \\ &= -\int_{\mathcal{Y}} q(y)\, dy \int_{\mathcal{X}} p(x) \ln p(x)\, dx \\ &\quad - \int_{\mathcal{X}} p(x)\, dx \int_{\mathcal{Y}} q(y) \ln q(y)\, dy \\ &= H_{\mathcal{S}}(p) + H_{\mathcal{U}}(q), \end{aligned}$$

i.e., the entropy of an element of the product model $\mathcal{S} \times \mathcal{U}$ is the sum of the entropies of the projections on $\mathcal{S}$ and $\mathcal{U}$. This can be also stated by saying that the joint entropy of two independent random variables $X$ and $Y$ is the sum of individual entropies, i.e.

$$H(X,Y) + H(X) + H(Y),$$

see Problem 3.5 for details.

## 3.4 Concavity of Entropy

**Theorem 3.4.1** *For any two densities $p, q : \mathcal{X} \to \mathbb{R}$ we have*

$$H(\alpha p + \beta q) \geq \alpha H(p) + \beta H(q), \tag{3.4.9}$$

*$\forall \alpha, \beta \in [0,1]$, with $\alpha + \beta = 1$.*

*Proof:* Using that $f(u) = -u \ln u$ is concave on $(0, \infty)$, we obtain

$$f(\alpha p + \beta q) \geq \alpha f(p) + \beta f(q).$$

Integrating (summing) over $\mathcal{X}$ leads to expression (3.4.9). ∎

With a similar proof we can obtain the following result.

**Corollary 3.4.2** *For any densities $p_1, \dots, p_n$ on $\mathcal{X}$ and $\lambda_i \in [0,1]$ with $\lambda_1 + \cdots + \lambda_n = 1$, we have*

$$H\Big(\sum_{i=1}^{n} \lambda_i p_i\Big) \geq \sum_{i=1}^{n} \lambda_i H(p_i).$$

The previous result suggests to look for the maxima of the entropy function on a statistical model.

## 3.5 Maxima for Entropy

Let $\mathcal{S} = \{p_\xi(x); x \in \mathcal{X}, \xi \in \mathbb{E}\}$ be a statistical model. We can regard the entropy $H$ as a function defined on the parameters space $\mathbb{E}$. We are interested in the value of the parameter $\xi$ for which the entropy $H(\xi)$ has a local maximum. This parameter value corresponds to a distinguished density $p_\xi$. Sometimes, the density $p_\xi$ satisfies some given constraints, which are provided by the given observations, and has a maximum degree of ignorance with respect to the unknown observations. This type of optimization problem is solved by considering the maximization of the entropy with constraints. In order to study this problem we shall start with the definition and characterization of critical points of entropy.

Let $f$ be a function defined on the statistical manifold $S = \{p_\xi\}$. If $\partial_i = \partial_{\xi^i}$ denotes the tangent vector field on $S$ in the direction of $\xi^i$, then

$$\partial_i f =: \partial_{\xi^i} f := \partial_{\xi^i}(f \circ p_{_\xi}).$$

In the following the role of the function $f$ is played by the entropy $H(\xi) = H(p_\xi)$.

**Definition 3.5.1** *A point $q \in S$ is a critical point for the entropy $H$ if*

$$X(H) = 0, \quad \forall X \in T_q S.$$

Since $\{\partial_i\}_i$ form a basis, choosing $X = \partial_i$, we obtain that the point $q = p_\xi \in S$ is a critical point for $H$ if and only if

$$\partial_i H(\xi) = 0, \qquad i = 1, 2, \dots, n.$$

A computation provides

$$\begin{aligned}
\partial_i H &= -\partial_i \int_{\mathcal{X}} p(x,\xi) \ln p(x,\xi)\, dx \\
&= -\int_{\mathcal{X}} \Big( \partial_i p(x,\xi)\, \ln p(x,\xi) + p(x,\xi) \frac{\partial_i p(x,\xi)}{p(x,\xi)} \Big)\, dx \\
&= -\int_{\mathcal{X}} \big( \ln p(x,\xi) + 1 \big) \partial_i p(x,\xi)\, dx \\
&= -\int_{\mathcal{X}} \ln p(x,\xi)\, \partial_i p(x,\xi)\, dx,
\end{aligned}$$

where we used that

$$\int_{\mathcal{X}} p(x,\xi)\, dx = 1$$

and

$$0 = \partial_i \int_{\mathcal{X}} p(x, \xi)\, dx = \int_{\mathcal{X}} \partial_i p(x, \xi)\, dx.$$

The previous computation can be summarized as in the following.

**Proposition 3.5.2** *The probability distribution $p_\xi$ is a critical point of the entropy $H$ if and only if*

$$\int_{\mathcal{X}} \ln p(x, \xi)\, \partial_{\xi^i} p(x, \xi)\, dx = 0, \quad \forall i = 1, \ldots, m. \tag{3.5.10}$$

*In the discrete case, when $\mathcal{X} = \{x^1, \ldots, x^n\}$, the Eq. (3.5.10) is replaced by the relation*

$$\sum_{k=1}^{n} \ln p(x^k, \xi)\, \partial_i p(x^k, \xi) = 0, \quad \forall i = 1, \ldots, m. \tag{3.5.11}$$

Observe that the critical points characterized by the previous result do not belong to the boundary. The entropy, which is a concave function, on a convex set (such as a mixture family) sometimes attains the local minima along the boundary. Even if these points are called critical by some authors, here we do not consider them as part of our analysis.

The first derivative of the entropy can be also expressed in terms of the log-likelihood function as in the following

$$\begin{aligned}
\partial_i H &= -\int_{\mathcal{X}} \ln p(x, \xi)\, \partial_{\xi^i} p(x, \xi)\, dx \\
&= -\int_{\mathcal{X}} p(x, \xi) \ln p(x, \xi)\, \partial_i \ln p(x, \xi)\, dx \\
&= -\int_{\mathcal{X}} p(x, \xi) \ell(\xi)\, \partial_i \ell(\xi)\, dx \\
&= -E_\xi[\ell(\xi)\, \partial_{\xi^i} \ell(\xi)].
\end{aligned} \tag{3.5.12}$$

The goal of this section is to characterize the distributions $p_\xi$ for which the entropy is maximum. Minima and maxima are among the set of critical points, see Definition 3.5.1. In order to deal with this issue we need to compute the Hessian of the entropy $H$.

The second order partial derivatives of the entropy $H$ are

$$\begin{aligned}\partial_{ji} H &= \partial_j \int_{\mathcal{X}} \ln p(x,\xi)\, \partial_i p(x,\xi)\, dx \\ &= -\int_{\mathcal{X}} \Big( \frac{\partial_j p(x)}{p(x)} \partial_i p(x) + \ln p(x)\, \partial_i \partial_j p(x) \Big)\, dx \\ &= -\int_{\mathcal{X}} \Big( \frac{1}{p(x)} \partial_i p(x)\, \partial_j p(x) + \ln p(x)\, \partial_{ji} p(x) \Big)\, dx.\end{aligned}$$

In the discrete case this becomes

$$\partial_{ji} H = -\sum_{k=1}^{n} \Big( \frac{\partial_i p(x^k,\xi)\, \partial_j p(x^k,\xi)}{p(x^k,\xi)} + \ln p(x_k,\xi)\, \partial_{ij} p(x^k,\xi) \Big). \tag{3.5.13}$$

We can also express the Hessian of the entropy in terms of the log-likelihood function only. Differentiating in (3.5.12) we have

$$\begin{aligned}\partial_{ji} H &= -\partial_j \int_{\mathcal{X}} p(x,\xi) \ell(\xi)\, \partial_i \ell(\xi)\, dx \\ &= -\int_{\mathcal{X}} \Big( \partial_j p(x,\xi) \ell(\xi)\, \partial_i \ell(\xi) + p(x,\xi) \partial_j \ell(\xi)\, \partial_i \ell(\xi) \\ &\qquad + p(x,\xi) \ell(\xi)\, \partial_i \partial_j \ell(\xi) \Big)\, dx \\ &= -E_\xi[\partial_i \ell\, \partial_j \ell] - E_\xi[(\partial_j \ell(\xi) \partial_i \ell(\xi) + \partial_i \partial_j \ell(\xi)) \ell(\xi)] \\ &= -g_{ij}(\xi) - h_{ij}(\xi).\end{aligned}$$

We arrived at the following result that relates the entropy and the Fisher information.

**Proposition 3.5.3** *The Hessian of the entropy is given by*

$$\partial_i \partial_j H(\xi) = -g_{ij}(\xi) - h_{ij}(\xi), \tag{3.5.14}$$

*where $g_{ij}(\xi)$ is the Fisher–Riemann metric and*

$$h_{ij}(\xi) = E_\xi[(\partial_j \ell(\xi) \partial_i \ell(\xi) + \partial_i \partial_j \ell(\xi)) \ell(\xi)].$$

**Corollary 3.5.4** *In the case of the mixture family (1.5.15)*

$$p(x;\xi) = C(x) + \xi^i F_i(x) \tag{3.5.15}$$

*the Fisher–Riemann metric is given by*

$$g_{ij}(\xi) = -\partial_i \partial_j H(\xi). \tag{3.5.16}$$

*Furthermore, any critical point of the entropy (see Definition 3.5.1) is a maximum point.*

*Proof:* From Proposition 1.5.1, part (*iii*) we have $\partial_i \partial_j \ell_x(\xi) = -\partial_i \ell_x(\xi)\, \partial_j \ell_x(\xi)$ which implies $h_{ij}(\xi) = 0$. Substituting in (3.5.14) yields (3.5.16). Using that the Fisher–Riemann matrix $g_{ij}(\xi)$ is positive definite at any $\xi$, it follows that $\partial_i \partial_j H(\xi)$ is globally negative definite, and hence all critical points must be maxima. We also note that we can express the Hessian in terms of $F_j$ as in the following

$$\partial_i \partial_j H(\xi) = -\int_{\mathcal{X}} \frac{F_i(x)F_j(x)}{p(x;\xi)}\, dx.$$

■

A Hessian $Hess(F) = (\partial_{ij} F)$ is called positive definite if and only if $\sum_{i,j} \partial_{ij} F\, v^i v^j > 0$, or, equivalently,

$$\langle Hess(F)v, v\rangle > 0, \qquad \forall v \in \mathbb{R}^m.$$

In the following we shall deal with the relationship between the Hessian and the second variation of the entropy $H$.

Consider a curve $\xi(s)$ in the parameter space and let $\big(\xi_u(s)\big)_{|u|<\epsilon}$ be a smooth variation of the curve with $\xi_u(s)_{|u=0} = \xi(s)$. Then $s \to p_{\xi_u(s)}$ is a variation of the curve $s \to p_{\xi(s)}$ on the statistical manifold $S$. Consider the variation

$$\xi_u(s) = \xi(s) + u\eta(s),$$

so $\partial_u \xi_u(s) = \eta(s)$ and $\partial_u^2 \xi_u(s) = 0$. The second variation of the entropy along the curve $s \to p_{\xi_u(s)}$ is

$$\begin{aligned}
\frac{d^2}{du^2} H\big(\xi_u(s)\big) &= \frac{d}{du}\langle \partial_\xi H, \partial_u \xi_u(s)\rangle \\
&= \langle \frac{d}{du}\partial_\xi H, \partial_u \xi(s)\rangle + \langle \partial_\xi H, \underbrace{\partial_u^2 \xi_u(s)}_{=0}\rangle \\
&= \frac{d}{du}(\partial_i H)\, \partial_u \xi^i(s) \\
&= \partial_i \partial_j H(\xi_u(s)) \cdot \partial_u \xi_u^i(s) \partial_u \xi_u^j(s).
\end{aligned}$$

Taking $u = 0$, we find

$$\begin{aligned}
\frac{d^2}{du^2} H\big(\xi_u(s)\big)_{|u=0} &= \partial_{ij} H\big(\xi(s)\big)\eta^i(s)\eta^j(s) \\
&= \langle Hess\, H\big(\xi(s)\big)\eta, \eta\rangle.
\end{aligned}$$

Hence $\frac{d^2}{du^2} H\big(\xi_u(s)\big)_{|u=0} < 0 (> 0)$ if and only if $Hess(H)$ is negative (positive) definite. Summarizing, we have:

**Theorem 3.5.5** *If $\xi$ is such that $p_\xi$ satisfies the critical point condition (3.5.10) (or condition (3.5.11) in the discrete case), and the Hessian $Hess(H(\xi))$ is negative definite at $\xi$, then $p_\xi$ is a local maximum point for the entropy.*

We shall use this result in the next section.

**Corollary 3.5.6** *Let $\xi_0$ be such that*

$$E_{\xi_0}[\ell(\xi_0)\partial_i \ell(\xi_0)] = 0 \tag{3.5.17}$$

*and $h_{ij}(\xi_0)$ is positive definite. Then $p(x, \xi_0)$ is a distribution for which the entropy reaches a local maximum.*

*Proof:* In the virtue of (3.5.12) the Eq. (3.5.17) is equivalent with the critical point condition $\partial_i H(\xi)_{|\xi=\xi_0} = 0$. Since $g_{ij}(\xi_0)$ is positive definite, then (3.5.14) implies that $\partial_i \partial_j H(\xi_0)$ is negative definite. Then applying Theorem 3.5.5 ends the proof. ■

## 3.6 Weighted Coin

Generally, for discrete distributions we may identify the statistical space $\mathcal{S}$ with the parameter space $\mathbb{E}$. We shall present next the case of a simple example where the entropy can be maximized. Flipping a weighted coin provides either heads with probability $\xi^1$, or tails with probability $\xi^2 = 1 - \xi^1$. The statistical manifold obtained this way depends on only one essential parameter $\xi := \xi^1$. Since $\mathcal{X} = \{x_1 = heads, x_2 = tails\}$, the manifold is just a curve in $\mathbb{R}^2$ parameterized by $\xi \in [0, 1]$. The probability distribution of the weighted coin is given by the table

| outcomes | $x_1$ | $x_2$ |
|---|---|---|
| probability | $\xi$ | $1-\xi$ |

We shall find the points of maximum entropy. First we write the Eq. (3.5.11) to determine the critical points

$$\begin{aligned} \ln p(x_1, \xi)\, \partial_\xi p(x_1, \xi) + \ln p(x_2, \xi)\, \partial_\xi p(x_2, \xi) &= 0 \Longleftrightarrow \\ \ln \xi - \ln(1-\xi) &= 0 \Longleftrightarrow \\ \xi &= 1 - \xi \end{aligned}$$

and hence there is only one critical point, $\xi = \frac{1}{2}$.

The Hessian has only one component, so formula (3.5.13) yields

$$\begin{aligned}
\partial_\xi^2 H &= -\Big(\frac{1}{p(x_1)}\big(\partial_\xi p(x_1)\big)^2 + \ln p(x_1)\partial_\xi^2 p(x_1)\Big) \\
&\quad -\Big(\frac{1}{p(x_2)}\big(\partial_\xi p(x_2)\big)^2 + \ln p(x_2)\partial_\xi^2 p(x_2)\Big) \\
&= -\Big(\frac{1}{\xi}\cdot 1 + \ln\xi \cdot 0\Big) \\
&\quad -\Big(\frac{1}{1-\xi}\big(\partial_\xi(1-\xi)\big)^2 + \ln(1-\xi)\ \partial_\xi^2(1-\xi)\Big) \\
&= -\Big(\frac{1}{\xi} + \frac{1}{1-\xi}\Big).
\end{aligned}$$

Evaluating at the critical point, we get

$$\partial_\xi^2 H_{|\xi=\frac{1}{2}} = -4 < 0,$$

and hence $\xi = \frac{1}{2}$ is a maximum point for the entropy. In this case $\xi^1 = \xi^2 = \frac{1}{2}$. This can be restated by saying that *the fair coin has the highest entropy among all weighted coins.*

## 3.7 Entropy for Finite Sample Space

Again, we underline that for discrete distributions we identify the statistical space $\mathcal{S}$ with the parameter space $\mathbb{E}$.

Consider a statistical model with a finite discrete sample space $\mathcal{X} = \{x^1, \dots, x^{n+1}\}$ and associated probabilities $p(x^i) = \xi^i$, $\xi^i \in [0,1]$, $i = 1, \dots n{+}1$. Since $\xi^{n+1} = 1 - \sum_{i=1}^n \xi^i$, the statistical manifold is described by $n$ essential parameters, and hence it has $n$ dimensions. The manifold can be also seen as a hypersurface in $\mathbb{R}^{n+1}$. The entropy function is

$$H = -\sum_{i=1}^{n+1} \xi^i \ln \xi^i. \qquad (3.7.18)$$

The following result deals with the maximum entropy condition. Even if it can be derived from the concavity property of $H$, see Theorem 3.4.1, we prefer to deduct it here in a direct way. We note that concavity is used as a tool to derive the case of continuous distributions, see Corollary 5.9.3.

**Theorem 3.7.1** *The entropy (3.7.18) is maximum if and only if*

$$\xi^1 = \cdots = \xi^{n+1} = \frac{1}{n+1}. \qquad (3.7.19)$$

*Proof:* The critical point condition (3.5.11) becomes

$$\begin{aligned}
\sum_{k=1}^{n} \ln p(x^k,\xi)\partial_{\xi^i} p(x^k,\xi) + \ln p(x^{n+1},\xi)\, \partial_{\xi^i} p(x^{n+1},\xi) &= 0 \iff \\
\sum_{k=1}^{n} \ln \xi^k\, \delta_{ik} + \ln \xi^{n+1}\, \partial_{\xi^{n+1}}(1-\xi^1-\cdots-\xi^n) &= 0 \iff \\
\ln \xi^i - \ln \xi^{n+1} &= 0 \iff \\
\xi^i &= \xi^{n+1},
\end{aligned}$$

$\forall i = 1,\ldots,n$. Hence condition (3.7.19) follows.

We shall investigate the Hessian at this critical point. Following formula (3.5.13) yields

$$\begin{aligned}
Hess(H)_{ij} &= -\sum_{k=1}^{n} \frac{\partial_i(\xi^k)\cdot\partial_j(\xi^k)}{\xi^k} - \frac{\partial_i(\xi^{n+1})\cdot\partial_j(\xi^{n+1})}{\xi^{n+1}} \\
&\quad -\sum_{k=1}^{n} \ln \xi^k\, \partial_i\partial_j(\xi^k) - \ln \xi^{n+1}\, \partial_i\partial_j(\xi^{n+1}) \\
&= -\Big(\sum_{k=1}^{n} \frac{\delta_{ik}\delta_{jk}}{\xi^k} - \frac{1}{\xi^{n+1}}\Big),
\end{aligned}$$

where we have used $\partial_i(\xi^{n+1}) = \partial_i(1-\xi^1-\cdots-\xi^n) = -1$, for $i = 1,\ldots,n$.

At the critical point the Hessian is equal to

$$Hess(H)_{ij}\big|_{\xi_k=\frac{1}{n+1}} = -(n+1)\Big(1+\sum_{k=1}^{n}\delta_{ik}\delta_{jk}\Big) = -2(n+1)I_n,$$

which shows that it is negative definite. Theorem 3.5.5 leads to the desired conclusion. ∎

**Example 3.7.2** Let $\xi^i$ be the probability that a die lands with the face $i$ up. This model depends on five essential parameters. According to the previous result, the fair die is the one which maximizes the entropy.

## 3.8 A Continuous Distribution Example

Let $p(x;\xi) = 2\xi x + 3(1-\xi)x^2$ be a continuous probability distribution function, with $x \in [0,1]$. The statistical manifold defined by the above probability distribution is one dimensional, since $\xi \in \mathbb{R}$. There is only one basic vector field equal to

$$\partial_\xi = 2x - 3x^2,$$

and which does not depend on $\xi$. In order to find the critical points, we follow Eq. (3.5.10)

$$\begin{aligned}
\int_0^1 p(x,\xi)\,\partial_\xi p(x,\xi)\,dx &= 0 \Longleftrightarrow \\
\int_0^1 (2x-3x^2)(2\xi x + 3(1-\xi)x^2)\,dx &= 0 \Longleftrightarrow \\
\frac{2}{15}\xi - \frac{3}{10} &= 0 \Longleftrightarrow \xi = \frac{9}{4}.
\end{aligned}$$

Before investigating the Hessian, we note that

$$\partial_\xi p(x;\xi) = 2x - 3x^2, \quad \partial_\xi^2 p(x;\xi) = 0, \quad p\Big(x;\frac{9}{4}\Big) = \frac{9}{4}x - \frac{15}{4}x^2,$$

so

$$\begin{aligned}
\partial_\xi^2 H_{|\xi=\frac{9}{4}} &= -\int_0^1 \Big(\frac{1}{p}(\partial_\xi p)^2 + \ln p\,\partial_\xi^2 p\Big)\,dx\Big|_{\xi=\frac{9}{4}} \\
&= -\int_0^1 \frac{(2x-3x^2)^2}{\frac{9}{2}x - \frac{15}{4}x^2}\,dx < 0,
\end{aligned}$$

because $\frac{9}{2}x - \frac{15}{4}x^2 < 0$ for $x \in (0,1]$.

Hence $\xi = \frac{9}{4}$ is a maximum point for the entropy. The maximum value of the entropy is

$$\begin{aligned}
H\Big(\frac{9}{4}\Big) &= -\int_0^1 \big(\frac{9}{2}x - \frac{15}{4}x^2\big)\ln\big(\frac{9}{2}x - \frac{15}{4}x^2\big)\,dx \\
&= -\frac{52}{25}\ln 3 + \frac{47}{30} + \frac{23}{25}\ln 2 \\
&= -0.807514878.
\end{aligned}$$

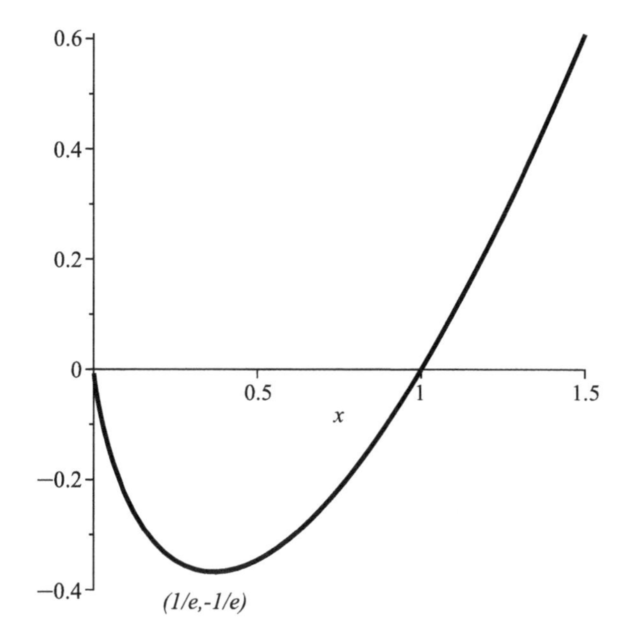


Figure 3.1: The function $x \to x \ln x$ has a global minimum value equal to $-1/e$ that is reached at $x = 1/e$

## 3.9 Upper Bounds for Entropy

We shall start with computing a rough upper bound for the entropy in the case when the sample space is a finite interval, $\mathcal{X} = [a, b]$. Consider the convex function

$$f : [0, 1] \to \mathbb{R}, \qquad f(u) = \begin{cases} u \ln u & if \quad u \in (0, 1] \\ 0 & if \quad u = 0. \end{cases}$$

Since $f'(u) = 1 + \ln u$, $u \in (0, 1)$, the function has a global minimum at $u = 1/e$, and hence $u \ln u \geq -1/e$, see Fig. 3.1.

Let $p : \mathcal{X} \to \mathbb{R}$ be a probability density. Substituting $u = p(x)$ yields $p(x) \ln p(x) \geq -1/e$. Integrating, we find

$$\int_a^b p(x) \ln p(x) \, dx \geq -\frac{b - a}{e}.$$

Using the definition of the entropy we obtain the following upper bound.

**Proposition 3.9.1** *The entropy $H(p)$ of a probability distribution $p : [a,b] \to [0,\infty)$ satisfies the inequality*

$$H(p) \leq \frac{b-a}{e}. \tag{3.9.20}$$

**Corollary 3.9.2** *The entropy $H(p)$ is smaller than half the length of the domain interval of the distribution $p$, i.e.,*

$$H(p) \leq \frac{b-a}{2}.$$

*This implies that the entropy $H(p)$ is smaller than the mean of the uniform distribution.*

We note that the inequality (3.9.20) becomes identity for the uniform distribution $p : [0,e] \to [0,\infty)$, $p(x) = 1/e$, see Problem 3.20. We shall present next another upper bound which is reached for all uniform distributions.

**Theorem 3.9.3** *The entropy of a smooth probability distribution $p : [a,b] \to [0,\infty)$ satisfies the inequality*

$$H(p) \leq \ln(b-a). \tag{3.9.21}$$

*Proof:* Since the function

$$f : [0,1] \to \mathbb{R}, \qquad f(u) = \begin{cases} u \ln u & if \quad u \in (0,1] \\ 0 & if \quad u = 0 \end{cases}$$

is convex on $[0,\infty)$, an application of Jensen integral inequality yields

$$\begin{aligned} f\Big(\frac{1}{b-a}\int_a^b p(x)\,dx\Big) &\leq \frac{1}{b-a}\int_a^b f\big(p(x)\big)\,dx \Longleftrightarrow \\ f\Big(\frac{1}{b-a}\Big) &\leq \frac{1}{b-a}\int_a^b p(x)\ln p(x)\,dx \Longleftrightarrow \\ \ln\Big(\frac{1}{b-a}\Big) &\leq \int_a^b p(x)\ln p(x)\,dx \Longleftrightarrow \\ -\ln(b-a) &\leq -H(p), \end{aligned}$$

which is equivalent to (3.9.21). The identity is reached for the uniform distribution $p(x) = 1/(b-a)$. ∎

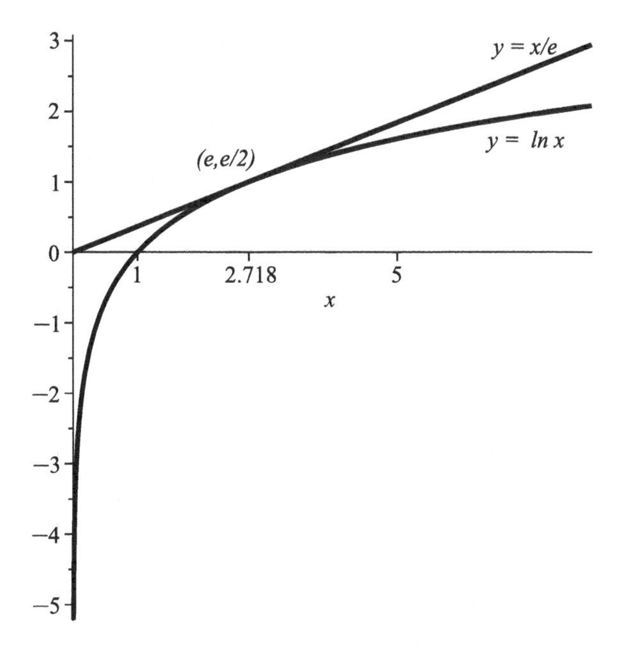


Figure 3.2: The inequality $\ln x \leq x/e$ is reached for $x = e$

The above result states that the maximum entropy is realized only for the case of the uniform distribution. In other words, the entropy measures the closeness of a distribution to the uniform distribution.

Since we have the inequality

$$\ln x \leq \frac{x}{e}, \qquad \forall x > 0$$

with equality only for $x = e$, see Fig. 3.2, it follows that the inequality (3.9.21) provides a better bound than (3.9.20).

In the following we shall present the bounds of the entropy in terms of the maxima and minima of the probability distribution. We shall use the following inequality involving the weighted average of $n$ numbers.

**Lemma 3.9.4** *If $\lambda_1, \dots, \lambda_n > 0$ and $\alpha_1, \dots, \alpha_n \in \mathbb{R}$, then*

$$\min_j\{\alpha_j\} \leq \frac{\sum_i \lambda_i \alpha_i}{\sum_i \lambda_i} \leq \max_j\{\alpha_j\}.$$

This says that if $\alpha_j$ are the coordinates of $n$ points of masses $\lambda_j$, then the coordinate of the center of mass of the system is larger than the smallest coordinate and smaller than the largest coordinate.

**Proposition 3.9.5** *Consider the discrete probability distribution* $p = \{p_j\}$*, with* $p_1 \le \cdots \le p_n$*. Then the entropy satisfies the double inequality*

$$-\ln p_n \le H(p) \le -\ln p_1.$$

*Proof:* Letting $\lambda_j = p_j$ and $\alpha_j = -\ln p_j$ in Lemma 3.9.4 and using

$$H(p) = -\sum_j p_j \ln p_j = \frac{\sum_i \lambda_i \alpha_i}{\sum_i \lambda_i},$$

we find the desired inequality. ∎

**Remark 3.9.6** The distribution $p = \{p_j\}$ is uniform with $p_j = \frac{1}{n}$ if and only if $p_1 = p_n$. In this case the entropy is given by

$$H(p) = -\ln p_1 = \ln p_n = -\ln \frac{1}{n} = \ln n.$$

The continuous analog of Proposition 3.9.5 is given below.

**Proposition 3.9.7** *Consider the continuous probability distribution* $p : \mathcal{X} \to [a,b] \subset [0,\infty)$*, with* $p_m = \min_{x \in \mathcal{X}} p(x)$ *and* $p_M = \max_{x \in \mathcal{X}} p(x)$*. Then the entropy satisfies the inequality*

$$-\ln p_M \le H(p) \le -\ln p_m.$$

*Proof:* The proof is using the following continuous analog of Lemma 3.9.4,

$$\min_{x \in \mathcal{X}} \alpha(x) \le \frac{\int_{\mathcal{X}} \lambda(x)\alpha(x)\,dx}{\int_{\mathcal{X}} \lambda(x)\,dx} \le \max_{x \in \mathcal{X}} \alpha(x),$$

where we choose $\alpha(x) = -\ln p(x)$ and $\lambda(x) = p(x)$. ∎

## 3.10 Boltzman–Gibbs Submanifolds

Let

$$\mathcal{S} = \{p_\xi : [0,1] \longrightarrow \mathbb{R}_+; \int_{\mathcal{X}} p_\xi(x)\,dx = 1\}, \quad \xi \in \mathbb{E},$$

be a statistical model with the state space $\mathcal{X} = [0,1]$. Let $\mu \in \mathbb{R}$ be a fixed constant and consider the set of elements of $\mathcal{S}$ with the mean $\mu$

$$\mathcal{M}_\mu = \{p_\xi \in \mathcal{S}; \int_{\mathcal{X}} x p_\xi(x)\,dx = \mu\}.$$

and assume that $\mathcal{M}_\mu$ is a submanifold of $\mathcal{S}$.

**Definition 3.10.1** *The statistical submanifold* $\mathcal{M}_\mu = \{p_\xi\}$ *defined above is called a Boltzman–Gibbs submanifold of* $\mathcal{S}$.

**Example 3.10.1** In the case of beta distribution, the Boltzman–Gibbs submanifold $\mathcal{M}_\mu = \{p_{a,ka}; a > 0, k = (1-\mu)/\mu\}$ is just a curve. In particular, $\mathcal{M}_1 = \{p_{a,0}; a > 0\}$, with $p_{a,0}(x) = \frac{1}{B(a,0)} x^{a-1}(1-x)^{-1}$.

One of the problems arised here is to find the distribution of maximum entropy on a Boltzman–Gibbs submanifold. Since the maxima are among critical points, which are introduced by Definition 3.5.1, we shall start the study with finding the critical points of the entropy

$$H(\xi) = H(p_\xi) = -\int_{\mathcal{X}} p_\xi(x) \ln p_\xi(x)\, dx$$

on a Boltzman–Gibbs submanifold $\mathcal{M}_\mu$. Differentiating with respect to $\xi^j$ in relations

$$\int_{\mathcal{X}} x p_\xi(x)\, dx = \mu, \qquad \int_{\mathcal{X}} p_\xi(x)\, dx = 1 \tag{3.10.22}$$

yields

$$\int_{\mathcal{X}} x\, \partial_j p(x,\xi)\, dx = 0, \qquad \int_{\mathcal{X}} \partial_j p(x,\xi)\, dx = 0. \tag{3.10.23}$$

A computation provides

$$\begin{aligned} -\partial_j H(\xi) &= \partial_j \int_{\mathcal{X}} p_\xi(x) \ln p_\xi(x)\, dx \\ &= \int_{\mathcal{X}} \Big( \partial_j p_\xi(x)\, \ln p_\xi(x) + p_\xi(x) \frac{\partial_j p_\xi(x)}{p_\xi(x)} \Big)\, dx \\ &= \int_{\mathcal{X}} \partial_j p(x)\, \ln p_\xi(x)\, dx + \underbrace{\int_{\mathcal{X}} \partial_j p_\xi(x)\, dx}_{=0 \text{ by } (3.10.23)}. \end{aligned}$$

Hence the critical points $p_\xi$ satisfying $\partial_j H(\xi) = 0$ are solutions of the integral equation

$$\int \partial_j p(x,\xi)\, \ln p(x,\xi)\, dx = 0, \tag{3.10.24}$$

subject to the constraint

$$\int_{\mathcal{X}} x \partial_j p(x,\xi)\, dx = 0. \tag{3.10.25}$$

Multiplying (3.10.25) by the Lagrange multiplier $\lambda = \lambda(\xi)$ and adding it to (3.10.24) yields

$$\int_{\mathcal{X}} \partial_j p(x,\xi)\Big( \ln p(x,\xi) + \lambda(\xi)x \Big)\, dx = 0.$$

Since $\displaystyle\int \partial_j p(x,\xi)\, dx = 0$, it makes sense to consider those critical points for which the term $\ln p(x,\xi) + \lambda(\xi)x$ is a constant function in $x$, i.e., depends only on $\xi$

$$\ln p(x,\xi) + \lambda(\xi)x = \theta(\xi).$$

Then the above equation has the solution

$$p(x,\xi) = e^{\theta(\xi)-\lambda(\xi)x}, \tag{3.10.26}$$

which is an exponential family. We still need to determine the functions $\theta$ and $\lambda$ such that the constraints (3.10.22) hold. This will be done explicitly for the case when the sample space is $\mathcal{X} = [0,1]$. From the second constraint we obtain a relation between $\theta$ and $\lambda$:

$$\int_0^1 p(x,\xi)\, dx{=}1 \Longrightarrow e^{\theta(\xi)} \int_0^1 e^{-\lambda(\xi)x}\, dx = 1 \Longleftrightarrow \frac{1-e^{-\lambda(\xi)}}{\lambda(\xi)} = e^{-\theta(\xi)},$$

which leads to

$$\theta(\xi) = \ln \frac{\lambda(\xi)}{1-e^{-\lambda(\xi)}}.$$

Substituting in (3.10.26) yields

$$p(x,\xi) = \frac{\lambda(\xi)}{1-e^{-\lambda(\xi)}} e^{-\lambda(\xi)x}. \tag{3.10.27}$$

Substituting in the constraint

$$\int_0^1 x p(x,\xi)\, dx = \mu,$$

we find

$$\begin{aligned}\frac{\lambda(\xi)}{1-e^{-\lambda(\xi)}}\int_0^1 xe^{-\lambda(\xi)x}\,dx &= \mu \Longleftrightarrow\\ \frac{1-\big(1+\lambda(\xi)\big)e^{-\lambda(\xi)}}{\lambda(\xi)(1-e^{-\lambda(\xi)})} &= \mu \Longleftrightarrow\\ \frac{e^{\lambda(\xi)}-\lambda(\xi)-1}{\lambda(\xi)(e^{\lambda(\xi)}-1)} &= \mu \Longleftrightarrow\\ \frac{1}{\lambda(\xi)}-\frac{1}{e^{\lambda(\xi)}-1} &= \mu.\end{aligned}$$

Given $\mu$, we need to solve the above equation for $\lambda(\xi)$. In order to complete the computation, we need the following result.

**Lemma 3.10.2** *The function*

$$f(x)=\frac{1}{x}-\frac{1}{e^x-1},\ x\in(-\infty,0)\cup(0,\infty),$$

*has the following properties*

*i)* $\displaystyle\lim_{x\searrow 0} f(x)=\lim_{x\nearrow 0} f(x)=\frac{1}{2}$,

*ii)* $\displaystyle\lim_{x\longrightarrow\infty} f(x)=0,\ \lim_{x\longrightarrow -\infty} f(x)=1$,

*iii)* *$f(x)$ is a strictly decreasing function of $x$.*

*Proof:* *i)* Applying l'Hôspital's rule twice, we get

$$\begin{aligned}\lim_{x\searrow 0} f(x) &= \lim_{x\searrow 0}\frac{e^x-1-x}{x(e^x-1)}=\lim_{x\searrow 0}\frac{e^x-1}{e^x-1+xe^x}\\ &= \lim_{x\searrow 0}\frac{e^x}{e^x+xe^x+e^x}=\lim_{x\searrow 0}\frac{1}{2+x}=\frac{1}{2}.\end{aligned}$$

*ii)* It follows easily from the properties of the exponential function. ■

Since the function $f$ is one-to-one, the equation $f(\lambda)=\mu$ has at most one solution, see Fig. 3.3. More precisely,

- if $\mu\geq 1$, the equation has no solution;

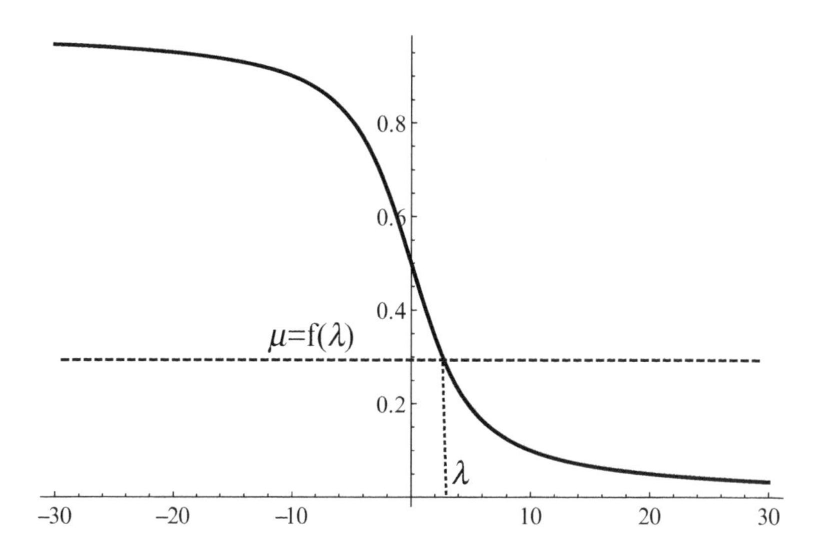


Figure 3.3: The graph of the decreasing function $f(x) = \frac{1}{x} - \frac{1}{e^x - 1}$ and the solution of the equation $f(\lambda) = \mu$ with $\mu \in (0, 1)$

- if $\mu \in (0, 1)$, the equation has a unique solution, for any $\xi$, i.e., $\lambda$ is constant, $\lambda = f^{-1}(\mu)$. For instance, if $\mu = 1/2$, then $\lambda = 0$.

It follows that $\theta$ is also constant,

$$\theta = \ln \frac{\lambda}{1 - e^{-\lambda}}.$$

Hence the distribution becomes

$$p(x) = e^{\theta - \lambda x}, \quad x \in (0, 1).$$

## 3.11 Adiabatic Flows

The entropy $H(\xi)$ is a real function defined on the parameter space $\mathbb{E}$ of the statistical model $\mathcal{S} = \{p_\xi\}$. The critical points of $H(\xi)$ are solutions of the system $\partial_i H(\xi) = 0$. Suppose that the set $C$ of critical points is void. Then the constant level sets $\sum_c := \{H(\xi) = c\}$ are hypersurfaces in $\mathbb{E}$. As usual, we accept the denomination of hypersurface for $\sum_c$ even if $\sum_c \cap C$ consists in a finite number of points.

Let $s \longrightarrow \xi(s)$, $\xi(s) \in \mathbb{E}$, be a curve situated in one of the hypersurfaces $\sum_c$. Since $H(\xi(s)) = c$, it follows

$$\frac{d}{ds}H\big(\xi(s)\big) = \partial_j H\big(\xi(s)\big)\,\dot{\xi}^j(s) = 0. \qquad (3.11.28)$$

Since $\dot{\xi}^j(s)$ is an arbitrary vector tangent to $\sum_c$, the vector field $\partial_i H$ is normal to $\sum_c$. Consequently, any vector field $X = (X^i)$ on $\mathbb{E}$ that satisfies

$$\partial_i H(\xi) X^i(\xi) = 0$$

is tangent to $\sum_c$.

Let $X = (X^i)$ be a vector field tangent to $\sum_c$. The flow $\xi(s)$ defined by

$$\dot{\xi}(s) = X^i(\xi(s)), \;\; i = 1, \ldots, n = \dim\, \mathrm{S}$$

is called *adiabatic flow* on $\sum_c$. This means $H(\xi) = c$, since the entropy is unchanged along the flow, i.e., $H(\xi)$ is a first integral, or $\sum_c$ is an invariant set with respect to this flow.

Suppose now that $S = \{p_\xi\}$ refers to a continuous distribution statistical model. Then

$$\begin{aligned} \partial_j H\big(\xi(s)\big) &= \int_{\mathcal{X}} \ln p\big(x, \xi(s)\big)\; \partial_j p\big(x, \xi(s)\big)\, dx \\ &= \int_{\mathcal{X}} \ell_x(\xi(s)) \partial_j \ell_x(\xi(s))\, dx, \end{aligned}$$

and combining with (3.11.28) we arrive at the following result:

**Proposition 3.11.1** *The flow* $\dot{\xi}^i(s) = X^i(\xi(s))$ *is adiabatic if and only if*

$$\int_{\mathcal{X}} \ell_x(\xi(s)) \frac{d}{ds} \ell_x(\xi(s))\, dx = 0.$$

**Example 3.11.1** If in the case of the normal distribution the entropy along the curve $s \longrightarrow p_{\sigma(s),\mu(s)}$ is constant, i.e.,

$$H\big(\sigma(s), \mu(s)\big) = \ln\big(\sigma(s)\sqrt{2\pi e}\big) = c$$

then $\sigma(s) = \dfrac{e^c}{\sqrt{2\pi e}}$, constant. Hence the adiabatic flow in this case corresponds to the straight lines

$$\{\sigma = constant,\; \mu(s)\},$$

with $\mu(s)$ arbitrary curve.

For more information regarding flows the reader is referred to Udriste [80, 82, 83].

## 3.12 Problems

**3.1.** Use the uncertainty function axioms to show the following relations:

(a) $H\left(\frac{1}{2}, \frac{1}{3}, \frac{1}{6}\right) = H\left(\frac{1}{2}, \frac{1}{2}\right) + \frac{1}{2}H\left(\frac{2}{3}, \frac{1}{3}\right)$.

(b) $H\left(\frac{1}{2}, \frac{1}{4}, \frac{1}{8}, \frac{1}{8}\right) = H\left(\frac{3}{4}, \frac{1}{4}\right) + \frac{3}{4}H\left(\frac{2}{3}, \frac{1}{3}\right) + \frac{1}{4}H\left(\frac{1}{2}, \frac{1}{2}\right)$.

(c) $H(p_1, \dots, p_n, 0) = H(p_1, \dots, p_n)$.

**3.2.** Consider two events $A = \{a_1, \dots, a_m\}$ and $B = \{b_1, \dots, b_n\}$, and let $p(a_i, b_j)$ be the probability of the joint occurrence of outcomes $a_i$ and $b_j$. The entropy of the joint event is defined by

$$H(A, B) = -\sum_{i,j} p(a_i, b_j) \log_2 p(a_i, b_j).$$

Prove the inequality

$$H(A, B) \leq H(A) + H(B),$$

with identity if and only if the events $A$ and $B$ are independent (i.e., $p(a_i, b_i) = p(a_i)p(b_j)$).

**3.3.** If $A = \{a_1, \dots, a_m\}$ and $B = \{b_1, \dots, b_n\}$ are two events, define the *conditional entropy* of $B$ given $A$ by

$$H(B|A) = -\sum_{i,j} p(a_i, b_j) \log_2 p_{a_i}(b_j),$$

and the information conveyed about $B$ by $A$ as

$$I(B|A) = H(B) - H(B|A),$$

where $p_{a_i}(b_j) = \dfrac{p(a_i, b_j)}{\sum_j p(a_i, b_j)}$ is the conditional probability of $b_j$ given $a_i$. Prove the following:

(a) $H(A, B) = H(A) + H(B|A)$;

(b) $H(B) \geq H(B|A)$. When does the equality hold?

(c) $H(B|A) - H(A|B) = H(B) - H(A)$;

(d) $I(B|A) = I(A|B)$.

**3.4.** Let $X$ be a real-valued continuous random variable on $\mathbb{R}^n$, with density function $p(x)$. Define the entropy of $X$ by

$$H(X) = -\int_{\mathbb{R}^n} p(x) \ln p(x) \, dx.$$

(*a*) Show that the entropy is translation invariant, i.e., $H(X) = H(X + c)$, for any constant $c \in \mathbb{R}$.

(*b*) Prove the formula $H(aX) = H(X) + \ln |a|$, for any constant $a \in \mathbb{R}$. Show that by rescaling the random variable the entropy can change from negative to positive and vice versa.

(*c*) Show that in the case of a vector valued random variable $Y : \mathbb{R}^n \to \mathbb{R}^n$ and an $n \times n$ matrix $A$ we have

$$H(AY) = H(Y) + \ln |\det A|.$$

(*d*) Use (*c*) to prove that the entropy is invariant under orthogonal transformations of the random variable.

**3.5.** The joint and conditional entropies of two continuous random variables $X$ and $Y$ are given by

$$H(X, Y) = -\iint p(x, y) \log_2 p(x, y) \, dxdy,$$

$$H(Y|X) = -\iint p(x, y) \log_2 \frac{p(x, y)}{p(x)} \, dxdy,$$

where $p(x) = \displaystyle\int p(x, y) \, dy$ is the marginal probability of $X$. Prove the following:

(*a*) $H(X, Y) = H(X) + H(Y|X) = H(Y) + H(X|Y)$;

(*b*) $H(Y|X) \le H(Y)$.

**3.6.** Let $\alpha(x, y)$ be a function with $\alpha(x, y) \ge 0$, $\displaystyle\int_{\mathbb{R}} \alpha(x, y) \, dx = \int_{\mathbb{R}} \alpha(x, y) \, dy = 1$. Consider the averaging operation

$$q(y) = \int_{\mathbb{R}} \alpha(x, y) p(x) \, dx.$$

Prove that the entropy of the averaged distribution $q(y)$ is equal to or greater than the entropy of $p(x)$, i.e., $H(q) \ge H(p)$.

**3.7.** Consider the two-dimensional statistical model defined by

$$p(x, \xi^1, \xi^2) = 2\xi^1 x + 3\xi^2 x^2 + 4(1 - \xi^1 - \xi^2)x^3, \qquad x \in (0, 1).$$

(*a*) Compute the Fisher metric $g_{ij}(\xi)$.

(*b*) Compute the entropy $H(p)$.

(*c*) Find $\xi$ for which $H$ is critical. Does it correspond to a maximum or to a minimum?

**3.8.** Find a generic formula for the informational entropy of the exponential family $p(\xi, x) = e^{C(x)+\xi^i F_i(x)-\phi(\xi)}$, $x \in \mathcal{X}$.

**3.9.** (The change of the entropy under a change of coordinates.) Consider the vector random variables $X$ and $Y$, related by $Y = \phi(X)$, with $\phi : \mathbb{R}^n \to \mathbb{R}^n$ invertible transformation.

(*a*) Show that

$$H(Y) = H(X) - E[\ln J_{\phi^{-1}}],$$

where $J_{\phi^{-1}}$ is the Jacobian of $\phi^{-1}$ and $E[\,\cdot\,]$ is the expectation with respect to the probability density of $X$.

(*b*) Consider the linear transformation $Y = AX$, with $A \in \mathbb{R}^{n\times n}$ nonsingular matrix. What is the relation expressed by part (*a*) in this case?

**3.10.** Consider the Gaussian distribution

$$p(x_1, \dots, x_n) = \frac{\sqrt{\det A}}{(2\pi)^{n/2}} e^{-\frac{1}{2}\langle Ax, x\rangle},$$

where $A$ is a symmetric $n \times n$ matrix. Show that the entropy of $p$ is

$$H = \frac{1}{2}\ln[(2\pi e)^n \det A].$$

**3.11.** Let $X = (X_1, \dots, X_n)$ be a random vector in $\mathbb{R}^n$, with $E[X_j] = 0$ and denote by $A = a_{ij} = E[X_i X_j]$ the associated covariance matrix. Prove that

$$H(X) \le \frac{1}{2}\ln[(2\pi e)^n \det A].$$

When is the equality reached?

**3.12.** Consider the density of an exponentially distributed random variable with parameter $\lambda > 0$

$$p(x, \lambda) = \lambda e^{-\lambda x}, \qquad x \geq 0.$$

Find its entropy.

**3.13.** Consider the Cauchy's distribution on $\mathbb{R}$

$$p(x, \xi) = \frac{\xi}{4\pi} \frac{1}{x^2 + \xi^2}, \qquad \xi > 0.$$

Show that its entropy is

$$H(\xi) = \ln(4\pi\xi).$$

**3.14.** Find a generic formula for the informational energy of the mixture family $p(\xi, x) = C(x) + \xi^i F_i(x)$, $x \in \mathcal{X}$.

**3.15.** Let $f(x) = \frac{x}{\sigma^2} e^{-\frac{x^2}{2\sigma^2}}$, $x \geq 0$, $\sigma > 0$, be the Rayleigh distribution. Prove that its entropy is given by

$$H(\sigma) = 1 + \ln \frac{\sigma}{\sqrt{2}} + \frac{\gamma}{2},$$

where $\gamma$ is Euler's constant.

**3.16.** Show that the entropy of the Maxwell–Boltzmann distribution

$$p(x, a) = \frac{1}{a^3} \sqrt{\frac{2}{\pi}} x^2 e^{-\frac{x^2}{2a^2}}, \qquad a > 0, \ x \in \mathbb{R}$$

is $H(a) = \frac{1}{2} - \gamma - \ln(a\sqrt{2\pi})$, where $\gamma$ is Euler's constant.

**3.17.** Consider the Laplace distribution

$$f(x, b, \mu) = \frac{1}{2b} e^{-|x-\mu|/b}, \qquad b > 0, \mu \in \mathbb{R}.$$

Show that its entropy is

$$H(b, \mu) = 1 + \ln(2b).$$

**3.18.** Let $\mu \in \mathbb{R}$. Construct a statistical model

$$\mathcal{S} = \{p_\xi(x);\ \xi \in \mathbb{E}, x \in \mathcal{X}\}$$

such that the functional $F : \mathcal{S} \longrightarrow \mathbb{R}$,

$$F(p(\cdot)) = \int_{\mathcal{X}} xp(x)\, dx - \mu$$

has at least one critical point. Is $\mathcal{M}_\mu = F^{-1}(0)$ a submanifold of $\mathcal{S}$?

**3.19.** Starting from the Euclidean space $(\mathbb{R}^n_+, \delta_{ij})$, find the Hessian metric produced by the Shannon entropy function

$$f : \mathbb{R}^n_+ \to R,\ f(x^1, \cdots, x^n) = \frac{1}{k^2}\sum_{i=1}^{n} \ln(k^2 x^i).$$

**3.20.** Show that the inequality (3.9.20) becomes identity for the uniform distribution $p : [0, e] \to [0, \infty)$, $p(x) = 1/e$, and this is the only distribution with this property.

**3.21.** (*a*) Let $a_n(x) = \frac{\xi^n \ln(n!)}{n!}$. Show that $\lim_{n\to\infty} \left|\frac{a_{n+1}(x)}{a_n(x)}\right| = 0$ for any $x$;

(*b*) Show that the series $\sum_{n\geq 0} \frac{\xi^n \ln(n!)}{n!}$ has an infinite radius of convergence;

(*c*) Deduce that the entropy for the Poisson distribution is finite.

**3.22.** Show that the entropy of the beta distribution

$$p_{a,b}(x) = \frac{1}{B(a,b)}\, x^{a-1}(1-x)^{b-1}, \quad 0 \leq x \leq 1$$

is always non-positive, $H(\alpha, \beta) \leq 0$, for any $a, b > 0$. For which values of $a$ and $b$ does the entropy vanish?

# Chapter 4

# Kullback–Leibler Relative Entropy

Even if the entropy of a finite, discrete density is always positive, in the case of continuous density the entropy is not always positive. This drawback can be corrected by introducing another concept, which measures the relative entropy between two given densities. This chapter studies the Kullback–Leibler relative entropy (known also as the Kullback–Leibler divergence) between two probability densities in both discrete and continuous cases. The relations with the Fisher information, entropy, cross entropy, $\nabla^{(1)}$-connection are emphasized and several worked out examples are presented. The chapter ends with the study of some variational properties.

Let $p, q : \mathcal{X} \to (0, \infty)$ be two probability densities on the same statistical model $\mathcal{S}$, in the same family (e.g., exponential, mixture, etc.) or not.

## 4.1 Definition and Basic Properties

The *Kullback–Leibler relative entropy* is a non-commutative measure of the difference between two probability densities $p$ and $q$ on the same statistical manifold, and it is defined by

O. Calin and C. Udrişte, *Geometric Modeling in Probability and Statistics*,
DOI 10.1007/978-3-319-07779-6_4,

$$D_{KL}(p||q) = E_p\Big[\ln\frac{p}{q}\Big] = \begin{cases} \sum_{x^i\in\mathcal{X}} p(x^i)\ln\frac{p(x^i)}{q(x^i)}, & \text{if } \mathcal{X} \text{ is discrete;} \\ \int_{\mathcal{X}} p(x)\,\ln\frac{p(x)}{q(x)}\,dx, & \text{if } \mathcal{X} \text{ is continuous.} \end{cases}$$

In information theory the density $p$ is considered to be the true density determined from observations, while $q$ is the theoretical model density. The Kullback–Leibler relative entropy can be used to find a goodness of fit of these two densities given by the expected value of the extra-information required for coding using $q$ rather than using $p$.

Sometimes, the Kullback–Leibler relative entropy is regarded as a measure of inefficiency of assuming data distributed according to $q$, when actually it is distributed as $p$.

The following inequalities will be useful.

**Lemma 4.1.1** *(i) If $p = \{p_1, \dots, p_n, \dots\}$ and $q = \{q_1, \dots, q_n, \dots\}$ are two strictly positive discrete densities on $\mathcal{X}$, then*

$$\sum_{i\geq 1} p_i \ln p_i \geq \sum_{i\geq 1} p_i \ln q_i.$$

*(ii) If $p$ and $q$ are two strictly positive continuous densities on $\mathcal{X}$, then*

$$\int_{\mathcal{X}} p(x)\ln p(x)\,dx \geq \int_{\mathcal{X}} p(x)\ln q(x)\,dx.$$

*The previous inequalities become equalities when the densities are equal.*

*Proof:*

(*i*) Using the inequality $\ln x \leq x - 1,\ x > 0$, we find

$$\begin{aligned} \sum_i p_i \ln q_i - \sum_i p_i \ln p_i &= \sum_i p_i \ln\frac{q_i}{p_i} \leq \sum_i p_i\Big(\frac{q_i}{p_i} - 1\Big) \\ &= \sum_i q_i - \sum_i p_i = 0. \end{aligned}$$

The equality is reached for $q_i/p_i = 1$, i.e., the case of equal densities.

(*ii*) It is similar to (*i*) since an integral mimics the properties of a sum. ∎

**Proposition 4.1.2** *Let $\mathcal{S}$ be a statistical manifold.*

(*i*) *The relative entropy $D_{KL}(\cdot\,||\,\cdot)$ is positive and non-degenerate:*

$D_{KL}(p||q) \geq 0$, $\forall p, q \in \mathcal{S}$, *with* $D_{KL}(p||q) = 0$ *if and only if* $p = q$.

(*ii*) *The relative entropy is symmetric, i.e.,* $D_{KL}(p||q) = D_{KL}(q||p)$ *if and only if*

$$\int_{\mathcal{X}} (p(x) + q(x)) \ln \frac{p(x)}{q(x)}\, dx = 0.$$

(*iii*) *The relative entropy satisfies the triangle inequality*

$$D_{KL}(p||q) + D_{KL}(q||r) \geq D_{KL}(p||r)$$

*if and only if*

$$\int_{\mathcal{X}} (p(x) - q(x)) \ln \frac{q(x)}{r(x)}\, dx \leq 0.$$

*Proof:* (*i*) Applying Lemma 4.1.1, we obtain

$$\begin{aligned} D_{KL}(p||q) &= \int_{\mathcal{X}} p(x) \ln \frac{p(x)}{q(x)}\, dx \\ &= \int_{\mathcal{X}} p(x) \ln p(x)\, dx - \int_{\mathcal{X}} p(x) \ln q(x)\, dx \geq 0, \end{aligned}$$

with equality for $p = q$.
(*ii*) and (*iii*) are obtained by direct computations.

Examples 4.2.1 and 4.2.2 provide non-symmetrical Kullback–Leibler relative entropies, which do not satisfy the triangle inequality. ∎

The previous proposition shows that the Kullback–Leibler relative entropy does not satisfy all the axioms of a metric on the manifold $\mathcal{S}$. It is worth noting that the non-symmetry can be removed by defining a symmetric version of the relative entropy, $\mathcal{D}(p, q) = \frac{1}{2}\big(D_{KL}(p||q) + D_{KL}(q||p)\big)$, called quasi-metric. However, in general, the triangle inequality cannot be fixed.

## 4.2 Explicit Computations

First, we shall compute the Kullback–Leibler relative entropy for pairs of densities in the same class.

**Example 4.2.1 (Exponential Distributions)** Consider two exponential densities

$$p_1(x) = \xi^1 e^{-\xi^1 x}, \qquad p_2(x) = \xi^2 e^{-\xi^2 x}.$$

Since

$$\ln \frac{p_1(x)}{p_2(x)} = \ln \frac{\xi^1}{\xi^2} + (\xi^2 - \xi^1)x,$$

we find

$$\begin{aligned} D_{KL}(p_1||p_2) &= \int_0^\infty p_1(x) \ln \frac{p_1(x)}{p_2(x)}\, dx \\ &= \int_0^\infty \ln \frac{\xi^1}{\xi^2} p_1(x)\, dx + (\xi^2 - \xi^1) \int_0^\infty x p_1(x)\, dx \\ &= \ln \frac{\xi^1}{\xi^2} + (\xi^2 - \xi^1)\frac{1}{\xi^1} \\ &= \frac{\xi^2}{\xi^1} - \ln \frac{\xi^2}{\xi^1} - 1. \end{aligned}$$

Hence

$$D_{KL}(p_1||p_2) = f\Big(\frac{\xi^2}{\xi^1}\Big),$$

with $f(x) = x - \ln x - 1 \geq 0$. This yields $D_{KL}(p_1||p_2) \geq 0$, the equality being reached if and only if $\xi^1 = \xi^2$ i.e. if $p_1 = p_2$.

We also have $D_{KL}(p_1||p_2) = f\Big(\frac{\xi^2}{\xi^1}\Big) \neq f\Big(\frac{\xi^1}{\xi^2}\Big) = D_{KL}(p_2||p_1)$. The condition $(iii)$ can be written as

$$f\Big(\frac{\xi^2}{\xi^1}\Big) + f\Big(\frac{\xi^3}{\xi^2}\Big) \ngeq f\Big(\frac{\xi^3}{\xi^1}\Big)$$

which becomes after cancelations

$$\frac{\xi^2}{\xi^1} + \frac{\xi^3}{\xi^2} \ngeq \frac{\xi^3}{\xi^1}.$$

One can see that this relation does not hold for any $\xi^1, \xi^2, \xi^3 > 0$. Hence Proposition 4.1.2 is verified on this particular case.

**Example 4.2.2 (Normal Distributions)** Let

$$p_1(x) = \frac{1}{\sqrt{2\pi}\sigma_1} e^{-\frac{(x-\mu_1)^2}{2\sigma_1^2}}, \qquad p_2(x) = \frac{1}{\sqrt{2\pi}\sigma_2} e^{-\frac{(x-\mu_2)^2}{2\sigma_2^2}}$$

be two normal densities. Since

$$\ln \frac{p_1(x)}{p_2(x)} = \ln \frac{\sigma_2}{\sigma_1} - \frac{(x-\mu_1)^2}{2\sigma_1^2} + \frac{(x-\mu_2)^2}{2\sigma_2^2},$$

then

$$\begin{aligned} D_{KL}(p_1||p_2) &= \int_0^\infty p_1(x) \ln \frac{p_1(x)}{p_2(x)}\, dx \qquad (4.2.1) \\ &= \ln \frac{\sigma_2}{\sigma_1} - \frac{1}{2\sigma_1^2} \int (x-\mu_1)^2 p_1(x)\, dx \\ &\quad + \frac{1}{2\sigma_2^2} \int_0^\infty (x-\mu_2)^2 p_1(x)\, dx. \end{aligned}$$

The second integral term can be computed as

$$\begin{aligned} \frac{1}{2\sigma_1^2} \int_0^\infty (x-\mu_1)^2 p_1(x)\, dx &= \frac{1}{2\sigma_1^2} \frac{1}{\sqrt{2\pi}\sigma_1} \int_0^\infty (x-\mu_1)^2 e^{-\frac{(x-\mu_1)^2}{2\sigma_1^2}} \\ &= \frac{1}{2\sigma_1^2} \frac{1}{\sqrt{2\pi}\sigma_1} \int_0^\infty y^2 e^{-\frac{y^2}{2\sigma_1^2}}\, dy = \frac{1}{2}, \end{aligned}$$

see also formula (6.5.26). Let $\Delta\mu = \mu_1 - \mu_2$. Then the third integral term is computed as

$$\begin{aligned} &\frac{1}{2\sigma_2^2} \int_0^\infty (x-\mu_2)^2 p_1(x)\, dx \\ &= \frac{1}{2\sigma_2^2} \int_0^\infty (x-\mu_1+\Delta\mu)^2 \frac{1}{\sqrt{2\pi}\sigma_1} e^{-\frac{(x-\mu_1)^2}{2\sigma_1^2}}\, dx \\ &= \frac{1}{2\sigma_2^2} \int_0^\infty (y+\Delta\mu)^2 \frac{1}{\sqrt{2\pi}\sigma_1} e^{-\frac{y^2}{2\sigma_1^2}}\, dx \\ &= \frac{1}{2\sqrt{2\pi}\sigma_1\sigma_2^2} \Bigg[ \int_0^\infty y^2 e^{-\frac{y^2}{2\sigma_1^2}} dy + 2\Delta\mu \int_0^\infty y e^{-\frac{y^2}{2\sigma_1^2}} dy \\ &\qquad + (\Delta\mu)^2 \int_0^\infty e^{-\frac{y^2}{2\sigma_1^2}} dy \Bigg] \\ &= \frac{1}{2\sqrt{2\pi}\sigma_1\sigma_2^2} \Big[ \sqrt{2\pi}\sigma_1^3 + 0 + (\Delta\mu)^2 \sqrt{2\pi}\sigma_1 \Big] \\ &= \frac{1}{2\sigma_2^2} \big(\sigma_1^2 + \Delta\mu^2\big). \end{aligned}$$

Substituting in (4.2.1), we obtain

$$\begin{aligned} D_{KL}(p_1||p_2) &= \ln\frac{\sigma_2}{\sigma_1} - \frac{1}{2} + \frac{1}{2}\Big(\frac{\sigma_1}{\sigma_2}\Big)^2 + \frac{\Delta\mu^2}{2\sigma_2^2} \\ &= \frac{1}{2}\Big[\Big(\frac{\sigma_1}{\sigma_2}\Big)^2 - \ln\Big(\frac{\sigma_1}{\sigma_2}\Big)^2 - 1\Big] + \frac{(\mu_1-\mu_2)^2}{2\sigma_2^2}. \end{aligned}$$

Using the inequality $\ln x \le x - 1$, we find $D_{KL}(p_1||p_2) \ge \dfrac{(\mu_1-\mu_2)^2}{2\sigma_2^2}$, the equality being reached for $\sigma_1 = \sigma_2$.

**Example 4.2.3 (Poisson Distributions)** Consider the Poisson distributions

$$p(n,\xi) = e^{-\xi}\frac{\xi^n}{n!},\ p(n,\xi_0) = e^{-\xi_0}\frac{\xi_0^n}{n!}, \qquad n = 0,1,2,\ldots.$$

Since

$$\ln\frac{p(n,\xi)}{p(n,\xi_0)} = \xi_0 - \xi + n\ln\frac{\xi}{\xi_0},$$

the Kullback–Leibler relative entropy is

$$\begin{aligned} D_{KL}(p_\xi||p_{\xi_0}) &= \sum_{n\ge 0} p(n,\xi)\ln\frac{p(n,\xi)}{p(n,\xi_0)} \\ &= (\xi_0 - \xi)\sum_{n\ge 0} p(n,\xi) + \ln\frac{\xi}{\xi_0}\sum_{n\ge 0} np(n,\xi) \\ &= \xi_0 - \xi + \ln\frac{\xi}{\xi_0}\xi e^{-\xi}\sum_{k\ge 0}\frac{\xi^k}{k!} \\ &= \xi_0 - \xi + \xi\ln\frac{\xi}{\xi_0}. \end{aligned}$$

We shall consider next a pair of densities in different classes and compute their Kullback–Leibler relative entropy.

**Example 4.2.4** To simplify, we consider a pair consisting in a Poisson distribution

$$q(n,\xi_0) = e^{-\xi_0}\frac{\xi_0^n}{n!},\ n = 0,1,2,\ldots$$

and a geometric distribution

$$p(n,\xi) = \xi(1-\xi)^{n-1},\ n = 1,2,\ldots$$

The pairing can be done for $n = 1, 2, \dots$. We assume for convenience that $p(0, \xi) = 0$. Then, we obtain

$$\ln \frac{p(n,\xi)}{q(n,\xi_0)} = \ln \xi + (n-1)\ln(1-\xi) - (-\xi_0 + n \ln \xi_0 - \ln n!).$$

It follows

$$\begin{aligned}
D_{KL}(p_\xi || q_{\xi_0}) &= \sum_{n\geq 0} p(n,\xi) \ln \frac{p(n,\xi)}{q(n,\xi_0)} \\
&= \xi \ln \xi \sum_{n\geq 1}(1-\xi)^{n-1} + \xi \ln(1-\xi) \sum_{n\geq 1}(n-1)(1-\xi)^{n-1} \\
&+ \xi_0 \xi \sum_{n\geq 1}(1-\xi)^{n-1} - \xi \ln \xi_0 \sum_{n\geq 1} n(1-\xi)^{n-1} \\
&+ \xi \sum_{n\geq 1}(\ln n!)(1-\xi)^{n-1} \\
&= \ln \xi - \ln(1-\xi) + \xi_0 - \frac{1}{\xi}\ln \xi_0 + \xi \sum_{n\geq 1}(\ln n!)(1-\xi)^{n-1}.
\end{aligned}$$

The last series is convergent for $0 < \xi \leq 1$.

## 4.3 Cross Entropy

Consider two densities $p$ and $q$ on the sample space $\mathcal{X}$ and denote by $E_p[\,\cdot\,]$ the expectation with respect to $p$. Let $\ell_q = \ln q$ be the log-likelihood function with respect to $q$. The *cross entropy* of $p$ with respect to $q$ is defined as $S(p,q) = -E_p[\ell_q]$. More precisely,

$$S(p,q) = \begin{cases} -\displaystyle\int_{\mathcal{X}} p(x) \ln q(x)\, dx, & \text{if } \mathcal{X} \text{ is continuous} \\[2ex] -\displaystyle\sum_x p(x) \ln q(x), & \text{if } \mathcal{X} \text{ is discrete.} \end{cases}$$

The cross entropy is an information measure that can be regarded as an error metric between two given distributions.

Let $H(p)$ be the entropy of $p$. Then we have the following result:

**Proposition 4.3.1** *The relative entropy $D_{KL}(p||q)$, the entropy $H(p)$ and the cross entropy $S(p,q)$ are related by*

$$S(p,q) = D_{KL}(p||q) + H(p).$$

*Proof:* From the definition of the Kullback–Leibler relative entropy it follows that

$$\begin{aligned} D_{KL}(p||q) &= \int_{\mathcal{X}} p(x) \ln \frac{p(x)}{q(x)}\, dx = \int_{\mathcal{X}} p(x) \ln p(x)\, dx \\ &\quad - \int_{\mathcal{X}} p(x) \ln q(x)\, dx = -H(p) - E_p[\ell_q] \\ &= -H(p) + S(p,q). \end{aligned}$$

■

The following result shows that the cross entropy is minimum when the two distributions are identical.

**Corollary 4.3.2** *The entropy $H(p)$ and the cross entropy $S(p,q)$ satisfy the inequality*

$$H(p) \leq S(p,q),$$

*with equality if and only if $p = q$.*

*Proof:* It follows from $D_{KL}(p||q) \geq 0$ and previous proposition. We can also state the result as $\min\limits_{q} S(p,q) = H(p)$. ■

It is worth noting that the Kullback–Leibler relative entropy can be also written as a difference of two log-likelihood functions

$$D_{KL}(p||q) = E_p[\ell_p] - E_p[\ell_q].$$

## 4.4 Relation with Fisher Metric

We shall start with an example. The Kullback–Leibler relative entropy of two exponential densities $p_{\xi_0}$ and $p_\xi$ is

$$D_{KL}(p_{\xi_0}||p_\xi) = \frac{\xi}{\xi_0} - \ln \frac{\xi}{\xi_0} - 1, \qquad \xi_0, \xi > 0,$$

see Example 4.2.1. The first two derivatives with respect to $\xi$ are

$$\partial_\xi D_{KL}(p_{\xi_0}||p_\xi) = \frac{1}{\xi_0} - \frac{1}{\xi}, \quad \partial_\xi^2 D_{KL}(p_{\xi_0}||p_\xi) = \frac{1}{(\xi)^2}.$$

We note that the diagonal parts of these partial derivatives, obtained for $\xi = \xi_0$, are

$$\partial_\xi D_{KL}(p_{\xi_0}||p_\xi)_{|\xi=\xi_0} = 0, \quad \partial_\xi^2 D_{KL}(p_{\xi_0}||p_\xi)_{\xi=\xi_0} = g_{11}(\xi_0),$$

where $g_{11}$ is the Fisher metric. These two relations are not a coincidence, as the following results will show.

**Proposition 4.4.1** *The diagonal part of the first variation of the Kullback–Leibler relative entropy is zero,*

$$\partial_i D_{KL}(p_{\xi_0}||p_\xi)_{|\xi=\xi_0} = 0.$$

*Proof:* Differentiating in the definition of the Kullback–Leibler relative entropy yields

$$\begin{aligned}\partial_i D_{KL}(p_{\xi_0}||p_\xi) &= \partial_i \int_{\mathcal{X}} p_{\xi_0}(x) \ln p_{\xi_0}(x)\, dx - \partial_i \int_{\mathcal{X}} p_{\xi_0}(x) \ln p_\xi(x)\, dx \\ &= -\int_{\mathcal{X}} p_{\xi_0}(x) \partial_i \ln p_\xi(x)\, dx = -\int_{\mathcal{X}} p_{\xi_0}(x) \partial_i \ell_x(\xi)\, dx,\end{aligned}$$

and hence

$$\partial_i D_{KL}(p_{\xi_0}||p_\xi)_{|\xi=\xi_0} = -\int_{\mathcal{X}} p_{\xi_0}(x) \partial_i \ell_x(\xi_0)\, dx = -E_{\xi_0}[\partial_i \ell_x(\xi_0)] = 0,$$

by Proposition 1.3.2. ■

**Proposition 4.4.2** *The diagonal part of the Hessian of the Kullback–Leibler relative entropy is the Fisher metric*

$$\partial_i \partial_j D_{KL}(p_{\xi_0}||p_\xi)_{|\xi=\xi_0} = g_{ij}(\xi_0).$$

*Proof:* Differentiating in the definition of the Kullback–Leibler relative entropy implies

$$\begin{aligned}\partial_i \partial_j D_{KL}(p_{\xi_0}||p_\xi) &= \partial_i \partial_j \int_{\mathcal{X}} p_{\xi_0} \ln p_{\xi_0}\, dx - \partial_i \partial_j \int_{\mathcal{X}} p_{\xi_0} \ln p_\xi\, dx \\ &= -\int_{\mathcal{X}} p_{\xi_0} \partial_i \partial_j \ln p_\xi\, dx = -\int_{\mathcal{X}} p_{\xi_0} \partial_i \partial_j \ell_x(\xi)\, dx.\end{aligned} \tag{4.4.2}$$

Taking the diagonal value, at $\xi = \xi_0$, yields

$$\begin{aligned}\partial_i \partial_j D_{KL}(p_{\xi_0}||p_\xi)_{|\xi=\xi_0} &= -\int_{\mathcal{X}} p_{\xi_0}(x) \partial_i \partial_j \ell_x(\xi_0)\, dx \\ &= -E_{\xi_0}[\partial_i \partial_j \ell_x(\xi_0)] = g_{ij}(\xi_0),\end{aligned}$$

by Proposition 1.6.3. ■

Proposition 4.4.1 states that $p_{\xi_0}$ is a critical point for the mapping $p_\xi \to D_{KL}(p_{\xi_0}||p_\xi)$. Using Proposition 4.4.2 leads to the following result, which is specific to distance functions:

**Proposition 4.4.3** *The density* $p_{\xi_0}$ *is a minimum point for the functional* $p_\xi \to D_{KL}(p_{\xi_0}||p_\xi)$.

*Proof:* Since $g_{ij}$ is positive definite everywhere, the point $p_{\xi_0}$ is a minimum. From the non-negativity of the Kullback–Leibler divergence it follows that $p_{\xi_0}$ is in fact a global minimum. ■

The next result deals with the quadratic approximation of the Kullback–Leibler relative entropy in terms of the Fisher metric.

**Theorem 4.4.4** *Let* $\Delta\xi^i = \xi^i - \xi^i_0$. *Then*

$$D_{KL}(p_{\xi_0}||p_\xi) = \frac{1}{2}\sum_{i,j} g_{ij}(\xi_0)\Delta\xi^i\Delta\xi^j + o(\|\Delta\xi\|^2), \tag{4.4.3}$$

*where* $o(\|\Delta\xi\|^2)$ *denotes a quantity that tends to zero faster than* $\|\Delta\xi\|^2$ *as* $\Delta\xi \to 0$.

*Proof:* Let $f : \mathbb{E} \to \mathbb{R}$ be a function given by $f(\xi) = D_{KL}(p_{\xi_0}||p_\xi)$, and consider its quadratic approximation

$$f(\xi) = f(\xi_0) + \sum_i \frac{\partial f}{\partial \xi^i}(\xi_0)\Delta\xi^i + \frac{1}{2}\sum_{i,j}\frac{\partial^2 f}{\partial\xi^i\partial\xi^j}(\xi_0)\Delta\xi^i\Delta\xi^j + o(\|\Delta\xi\|^2). \tag{4.4.4}$$

Using Propositions 4.1.2, 4.4.1, and 4.4.2, we have

$$\begin{aligned} f(\xi_0) &= D_{KL}(p_{\xi_0}||p_{\xi_0}) = 0 \\ \frac{\partial f}{\partial \xi^i}(\xi_0) &= \partial_{\xi^i} D(p_{\xi_0}||p_\xi)_{|\xi=\xi_0} = 0 \\ \frac{\partial^2 f}{\partial\xi^i\partial\xi^j}(\xi_0) &= \partial_{\xi^i}\partial_{\xi^j} D(p_{\xi_0}||p_\xi)_{|\xi=\xi_0} = g_{ij}(\xi_0). \end{aligned}$$

Substituting into (4.4.4) yields (4.6.8). ■

Let $p, q \in \mathcal{S}$ be two points on the statistical model $\mathcal{S}$. The Fisher distance, $\text{dist}(p,q)$, represents the information distance between densities $p$ and $q$. It is defined as the length of the shortest curve on $\mathcal{S}$ between $p$ and $q$, i.e., the length of the geodesic curve joining $p$ and $q$. Next we shall investigate the relation between the Kullback–Leibler relative entropy $D_{KL}(p||q)$ and the Fisher distance $\text{dist}(p,q)$. We shall start with an example.

From Examples 1.6.2 and 4.2.1, the Fisher distance and the Kullback–Leibler relative entropy between two exponential densities are

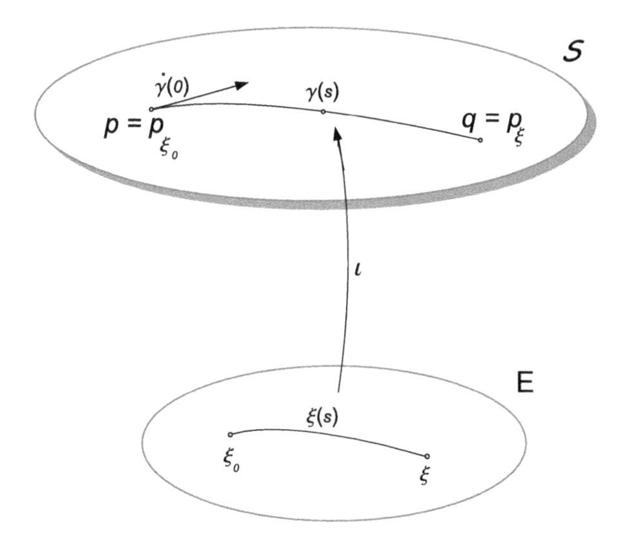


Figure 4.1: The geodesic $\gamma(s)$ between densities $p$ and $q$

$$\begin{aligned} \operatorname{dist}(p_{\xi_0}, p_\xi) &= \ln\left(\frac{\xi}{\xi_0}\right) \\ D_{KL}(p_{\xi_0}||p_\xi) &= \frac{\xi}{\xi_0} - \ln\left(\frac{\xi}{\xi_0}\right) - 1, \end{aligned}$$

where $0 < \xi_0 < \xi$. We have

$$\begin{aligned} \lim_{\xi \searrow \xi_0} \frac{D_{KL}(p_{\xi_0}||p_\xi)}{\frac{1}{2}\operatorname{dist}(p_{\xi_0}, p_\xi)^2} &= \lim_{\xi \searrow \xi_0} \frac{\frac{\xi}{\xi_0} - \ln\left(\frac{\xi}{\xi_0}\right) - 1}{\frac{1}{2}\left(\ln\left(\frac{\xi}{\xi_0}\right)\right)^2} = \lim_{x \searrow 1} \frac{x - \ln x - 1}{\frac{1}{2}(\ln x)^2} \\ &= \lim_{u \searrow 0} \frac{e^u - u - 1}{\frac{1}{2}u^2} = 1, \end{aligned}$$

by l'Hôspital's rule. Hence the asymptotics of $D_{KL}(p_{\xi_0}||p_\xi)$ as $\xi \to \xi_0$ is $\frac{1}{2}\operatorname{dist}(p_{\xi_0}, p_\xi)^2$. This result will hold true in a more general framework. The next result is a variant of Theorem 4.4.4.

**Theorem 4.4.5** *Let* $d = \operatorname{dist}(p, q)$ *denote the Fisher distance between the densities* $p$ *and* $q$ *and* $D_{KL}(p||q)$ *be the Kullback–Leibler relative entropy. Then*

$$D_{KL}(p||q) = \frac{1}{2}d^2(p, q) + o\big(d^2(p, q)\big) \tag{4.4.5}$$

*Proof:* Consider a geodesic $\gamma(s)$ on the statistical model $\mathcal{S}$ joining densities $p$ and $q$; this satisfies $\gamma(s) = \iota(\xi(s)) = p_{\xi(s)}$, with $\gamma(0) = p_{\xi_0} = p$ and $\gamma(t) = p_\xi = q$. Since the arc length along the geodesic is the Riemannian distance, we have $t = \text{dist}(p, q)$. The curve $\xi(s)$ belongs to the parameter space and has the endpoints $\xi(0) = \xi_0$ and $\xi(t) = \xi$, see Fig. 4.1.

Consider the function $\varphi(s) = f(\xi(s))$, with $f(\xi) = D_{KL}(p_{\xi_0}||p_\xi)$. A second order expansion of $\varphi$ about $t = 0$ yields

$$\varphi(t) = \varphi(0) + t\varphi'(0) + \frac{t^2}{2}\varphi''(0) + o(t^2). \tag{4.4.6}$$

Using Propositions 4.1.2, 4.4.1, and 4.4.2, we have

$$\begin{aligned}
\varphi(0) &= f(\xi(0)) = D_{KL}(p_{\xi_0}||p_{\xi_0}) = 0 \\
\varphi'(0) &= \sum_i \frac{\partial f}{\partial \xi^i}(\xi_0)\dot{\xi}(0) = 0 \\
\varphi''(0) &= \sum_{i,j} \frac{\partial^2 f}{\partial \xi^i \partial \xi^j}(\xi_0)\dot{\xi}^i(0)\dot{\xi}^j(0) + \sum_i \frac{\partial f}{\partial \xi^i}(\xi_0)\ddot{\xi}^i(0) \\
&= \sum_{i,j} g_{ij}(\xi_0)\dot{\xi}^i(0)\dot{\xi}^j(0).
\end{aligned}$$

Substituting in (4.4.6) yields

$$\varphi(t) = \frac{t^2}{2}\sum_{i,j} g_{ij}(\xi_0)\dot{\xi}^i(0)\dot{\xi}^j(0) + o(t^2) = \frac{t^2}{2}g(\dot{\gamma}(0), \dot{\gamma}(0)) + o(t^2) = \frac{t^2}{2} + o(t^2) = \frac{1}{2}d^2 + o(t^2),$$

since geodesics parameterized by the arc length are unit speed curves. Expressing the left side as $\varphi(t) = f(\xi(t)) = D_{KL}(p||q)$ leads to the desired result. ∎

**Corollary 4.4.6** *Let $d = \text{dist}(p, q)$ denote the Fisher distance between the densities $p$ and $q$ and $\mathcal{D}(p, q)$ be the Kullback–Leibler quasi-metric*

$$\mathcal{D}(p, q) = D_{KL}(p||q) + D_{KL}(q||p).$$

*Then*

$$\mathcal{D}(p, q) = d^2(p, q) + o(d^2(p, q)). \tag{4.4.7}$$

## 4.5 Relation with $\nabla^{(1)}$-Connection

The Kullback–Leibler relative entropy induces the linear connection $\nabla^{(1)}$. Its components, $\Gamma^{(1)}_{ij,k} = g(\nabla^{(1)}_{\partial_i}\partial_j, \partial_j)$, are given in terms of the Kullback–Leibler relative entropy as in the following.

**Proposition 4.5.1** *The diagonal part of the third mixed derivatives of the Kullback–Leibler relative entropy is the negative of the Christoffel symbol*

$$-\partial_{\xi^i}\partial_{\xi^j}\partial_{\xi_0^k} D_{KL}(p_{\xi_0}||p_\xi)_{|\xi=\xi_0} = \Gamma^{(1)}_{ij,k}(\xi_0).$$

*Proof:* The second derivatives in the argument $\xi$ are given by (4.4.2)

$$\partial_{\xi^i}\partial_{\xi^j} D_{KL}(p_{\xi_0}||p_\xi) = -\int_{\mathcal{X}} p_{\xi_0}(x)\partial_{\xi^i}\partial_{\xi^j}\ell_x(\xi)\, dx,$$

and differentiating in $\xi_0^k$ yields

$$\begin{aligned} -\partial_{\xi_0^k}\partial_{\xi^i}\partial_{\xi^j} D_{KL}(p_{\xi_0}||p_\xi) &= \partial_{\xi_0^k}\int_{\mathcal{X}} p_{\xi_0}(x)\partial_{\xi^i}\partial_{\xi^j}\ell_x(\xi)\, dx \\ &= \int_{\mathcal{X}} p_{\xi_0}(x)\partial_{\xi_0^k}\ell_x(\xi_0)\partial_{\xi^j}\ell_x(\xi)\, dx. \end{aligned}$$

Then considering the diagonal part

$$\begin{aligned} -\partial_{\xi_0^k}\partial_{\xi^i}\partial_{\xi^j} D_{KL}(p_{\xi_0}||p_\xi)_{|\xi=\xi_0} &= E_{\xi_0}[\partial_i\partial_j\ell(\xi)\,\partial_k\ell(\xi)] \\ &= \Gamma^{(1)}_{ij,k}(\xi_0), \end{aligned}$$

where we considered $\alpha = 1$ in formula (1.11.34). ∎

## 4.6 Third Order Approximation

This section contains a refinement of the result given by Theorem 4.4.4. This deals with the cubic approximation of the Kullback–Leibler relative entropy.

**Theorem 4.6.1** *Let $\Delta\xi^i = \xi^i - \xi_0^i$. Then*

$$D_{KL}(p_{\xi_0}||p_\xi) = \frac{1}{2}g_{ij}(\xi_0)\Delta\xi^i\Delta\xi^j + \frac{1}{6}h_{ijk}(\xi_0)\Delta\xi^i\Delta\xi^j\Delta\xi^k + o(\|\Delta\xi\|^3), \tag{4.6.8}$$

*where* $g_{ij}(\xi_0)$ *is the Fisher–Riemann metric and*

$$h_{ijk} = \partial_k g_{ij} + \Gamma^{(1)}_{ij,k},$$

*and* $o(\|\Delta\xi\|^3)$ *denotes a quantity that vanishes faster than* $\|\Delta\xi\|^3$ *as* $\Delta\xi \to 0$.

*Proof:* Following the line of the proof of Theorem 4.4.4, consider the function $f(\xi) = D_{KL}(p_{\xi_0}||p_\xi)$ and write its cubic approximation

$$\begin{aligned} f(\xi) &= \underbrace{f(\xi_0)}_{=0} + \underbrace{\frac{\partial f}{\partial \xi^i}(\xi_0)}_{=0}\Delta\xi^i + \frac{1}{2}\underbrace{\frac{\partial^2 f}{\partial\xi^i\partial\xi^j}(\xi_0)}_{=g_{ij}(\xi_0)}\Delta\xi^i\Delta\xi^j \\ &\quad + \frac{1}{6}h_{ijk}(\xi_0)\Delta\xi^i\Delta\xi^j\Delta\xi^k + o(\|\Delta\xi\|^3). \end{aligned}$$

The third coefficient can be computed as

$$\begin{aligned} h_{ijk}(\xi_0) &= \frac{\partial^3 f}{\partial\xi^i\partial\xi^j\partial\xi^k}(\xi_0) = \partial_{\xi^i}\partial_{\xi^j}\partial_{\xi^k} D_{KL}(p_{\xi_0}||p_\xi)_{|\xi=\xi_0} \\ &= -\partial_{\xi^i}\partial_{\xi^j}\partial_{\xi^k}\int_{\mathcal{X}} p_{\xi_0}(x)\ln p_\xi(x)\,dx_{|\xi=\xi_0} \\ &= -\partial_{\xi^i}\partial_{\xi^j}\partial_{\xi^k}\int_{\mathcal{X}} p_{\xi_0}(x)\ell_x(\xi)\,dx_{|\xi=\xi_0} \\ &= -E_{\xi_0}[\partial_i\partial_j\partial_k\ell] = \partial_k g_{ij}(\xi_0) + E_{\xi_0}[(\partial_i\partial_j\ell)(\partial_k\ell)] \\ &= \partial_k g_{ij}(\xi_0) + \Gamma^{(1)}_{ij,k}(\xi_0), \end{aligned}$$

where we used Proposition 1.7.1, part $(i)$, and let $\alpha = 1$ in formula (1.11.34). ■

## 4.7 Variational Properties

This section deals with inequalities and variational properties of the Kullback–Leibler relative entropy.

**Proposition 4.7.1** *For any two continuous density functions* $p$ *and* $q$*, we have*

$(i)$

$$\int_{\mathcal{X}} \frac{p^2(x)}{q(x)}\,dx \geq 1;$$

$(ii)$

$$\int_{\mathcal{X}} (p(x) - q(x))\ln\frac{p(x)}{q(x)}\,dx \geq 0.$$

*Proof:*

(*i*) Using the non-negativity of the Kullback–Leibler relative entropy and properties of logarithmic function we find

$$\begin{aligned} 0 \leq D_{KL}(p||q) &= \int_{\mathcal{X}} p(x) \ln \frac{p(x)}{q(x)}\, dx \leq \int_{\mathcal{X}} p(x) \Big( \frac{p(x)}{q(x)} - 1 \Big)\, dx \\ &= \int_{\mathcal{X}} \frac{p^2(x)}{q(x)}\, dx - \int_{\mathcal{X}} p(x)\, dx = \int_{\mathcal{X}} \frac{p^2(x)}{q(x)}\, dx - 1, \end{aligned}$$

which implies the desired inequality.

(*ii*) Using the properties of the Kullback–Leibler relative entropy, we have

$$\begin{aligned} 0 &\leq D_{KL}(p||q) + D_{KL}(q||p) = \int_{\mathcal{X}} p \ln \frac{p}{q}\, dx + \int_{\mathcal{X}} q \ln \frac{q}{p}\, dx \\ &= \int_{\mathcal{X}} (p - q) \ln p\, dx - \int_{\mathcal{X}} (p - q) \ln q\, dx = \int_{\mathcal{X}} (p - q) \ln \frac{p}{q}\, dx. \end{aligned}$$

■

**Example 4.7.1** We check (*i*) of Proposition 4.7.1 in the case of exponential densities $p = \xi_1 e^{-\xi_1 x}$ and $q(x) = \xi_2 e^{-\xi_2 x}$, $x > 0$. We have

$$\begin{aligned} \int_0^\infty \frac{p^2(x)}{q(x)}\, dx &= \int_0^\infty \frac{\xi_1^2 e^{-2\xi_1 x}}{\xi_2 e^{-\xi_2 x}}\, dx = \frac{\xi_1^2}{\xi_2} \int_0^\infty e^{-(2\xi_1 - \xi_2)}\, dx \\ &= \frac{\xi_1^2}{\xi_2} \cdot \frac{1}{2\xi_1 - \xi_2} \geq \frac{2\xi_1\xi_2 - \xi_2^2}{2\xi_1\xi_2 - \xi_2^2} = 1, \end{aligned}$$

where we used

$$\xi_1^2 \geq 2\xi_1\xi_2 - \xi_2^2 \Longleftrightarrow (\xi_1 - \xi_2)^2 \geq 0.$$

In the following $\mathcal{S}$ will denote the manifold of all continuous densities on $\mathcal{X}$. One way of defining the distance between two densities $p, q \in \mathcal{S}$ is using the $L_2$-norm

$$d(p, q) = \|p - q\|_2 = \sqrt{\int_{\mathcal{X}} |p(x) - q(x)|^2\, dx},$$

provided the $L_2$-norm is finite.

*Given two densities $p, r \in \mathcal{S}$, we shall ask the question of finding a density $q \in \mathcal{S}$ such that the sum $F(q) = d(p,q) + d(q,r)$ is minimum.*

This occurs when the density $q$ realizes the equality in the triangle inequality $d(p,q) + d(q,r) \geq d(p,r)$. Since Minkowski's integral inequality becomes equality for proportional integrands, it follows that the density $q$ is a convex combination

$$q(x) = \frac{\lambda}{1+\lambda} r(x) + \frac{1}{1+\lambda} p(x),$$

with $\lambda \in [0,1]$. Hence, the minimum of the sum $F(q)$ is realized for all the densities between $p$ and $r$.

We obtain a different result in the case when the distance $d(p,q)$ is replaced by the Kullback–Leibler relative entropy $D_{KL}(p||q)$, or by the *Hellinger distance*

$$d_H(p,q) = 2 \int_{\mathcal{X}} \left( \sqrt{p(x)} - \sqrt{q(x)} \right)^2 dx.$$

In the following we deal with several similar variational problems.

**Problem:** *Given two distinct densities $p, r \in \mathcal{S}$, find all densities $q \in \mathcal{S}$ for which the sum $G(q) = D_{KL}(p||q) + D_{KL}(q||r)$ is minimum.*

We shall employ the method of Lagrange multipliers considering the functional with constraints

$$q \longmapsto \int_{\mathcal{X}} \mathcal{L}(q)\, dx = G(q) + \lambda \Big( \int_{\mathcal{X}} q\, dx - 1 \Big),$$

where the Lagrangian is

$$\mathcal{L}(q) = p \ln \frac{p}{q} + q \ln \frac{q}{r} + \lambda q.$$

The density $q$, which realizes the minimum of $G(q) \geq 0$, satisfies the Euler–Lagrange equation

$$\begin{aligned}
\frac{\partial \mathcal{L}}{\partial q} &= 0 \Longleftrightarrow \\
\ln q - \frac{p}{q} &= \ln r - \lambda - 1 \Longleftrightarrow \\
\ln q - \ln p - \frac{p}{q} &= \ln r - \ln p - \lambda - 1 \Longleftrightarrow \\
\ln \frac{p}{q} + \frac{p}{q} &= 1 + \lambda + \ln \frac{p}{r}.
\end{aligned}$$

Exponentiating yields

$$\frac{p}{q}e^{p/q} = \frac{p}{r}e^{1+\lambda},$$

which after making the substitution $u = p/q$ becomes the Lambert equation

$$ue^u = \frac{p}{r}e^{1+\lambda},$$

with the solution

$$u = W\left(\frac{p}{r}e^{1+\lambda}\right),$$

where $W(\cdot)$ is the Lambert function.[1] Hence the minimum for the functional $G(q)$ is reached for the density

$$q = \frac{p}{W\left(\frac{p}{r}e^{1+\lambda}\right)}, \tag{4.7.9}$$

with the constant $\lambda$ determined from the unitary integral constraint $\int_{\mathcal{X}} q(x)\,dx = 1$. The value of $\lambda$ is unique since $W(x)$ is increasing with $W(0+) = 0$, and $W(\infty) = \infty$ and

$$\begin{aligned}
\lim_{\lambda\to-\infty}\int_{\mathcal{X}} \frac{p(x)}{W\left(\frac{p(x)}{r(x)}e^{1+\lambda}\right)}\,dx &= \frac{\int_{\mathcal{X}} p\,dx}{W(0+)} = +\infty;\\
\lim_{\lambda\to\infty}\int_{\mathcal{X}} \frac{p(x)}{W\left(\frac{p(x)}{r(x)}e^{1+\lambda}\right)}\,dx &= \frac{\int_{\mathcal{X}} p\,dx}{W(\infty)} = 0.
\end{aligned}$$

Hence there is only one density $q$ that minimizes the sum $D_{KL}(p||q) + D_{KL}(q||r)$, and it is given by formula (4.7.9).

In the case $p = r$, we have the following result.

**Proposition 4.7.2** *Let p be a fixed density. Then the symmetric relative entropy*

$$D_{KL}(p||q) + D_{KL}(q||p)$$

*achieves its minimum for $q = p$, and the minimum is equal to zero.*

*Proof:* Since we have $D_{KL}(p||q) \geq 0$ and $D_{KL}(q||p) \geq 0$, with identity achieved for $p = q$, it follows that $D_{KL}(p||q) + D_{KL}(q||p) \geq 0$ with identity for $p = q$. ■

Curious enough, swapping the arguments of the relative entropy, we obtain a different variational problem with a much nicer solution.

[1] Named after Johann Heinrich Lambert; it is also called the Omega function.

**Problem:** *Given two densities $p, r \in \mathcal{S}$, find all distributions $q \in \mathcal{S}$ for which the sum $F(q) = D_{KL}(p||q) + D_{KL}(r||q)$ is minimum.* Using the method of Lagrange multipliers we consider the functional with constraints

$$q \longmapsto \int_{\mathcal{X}} \mathcal{L}(q)\, dx = F(q) + \lambda\Big(\int_{\mathcal{X}} q\, dx - 1\Big),$$

with the Lagrangian

$$\mathcal{L}(q) = p \ln\frac{p}{q} + r \ln\frac{r}{q} + \lambda q.$$

The density $q$, which realizes the minimum of $F(q) \geq 0$, satisfies the Euler–Lagrange equation

$$\begin{aligned}
\frac{\partial \mathcal{L}}{\partial q} &= 0 \Longleftrightarrow \\
-\frac{p}{q} - \frac{r}{q} + \lambda &= 0 \Longleftrightarrow \\
\frac{p+r}{q} &= \lambda \Longleftrightarrow \\
q &= \frac{1}{\lambda}(p+r).
\end{aligned}$$

The multiplier $\lambda$ is determined from the integral constraint

$$1 = \int_{\mathcal{X}} q\, dx = \frac{1}{\lambda}\int_{\mathcal{X}} (q+r)\, dx = \frac{2}{\lambda} \Longrightarrow \lambda = 2.$$

Hence, the density that minimizes the sum $D_{KL}(p||q) + D_{KL}(r||q)$ is

$$q(x) = \frac{1}{2}\big(p(x) + r(x)\big).$$

Using the same idea of proof, one can easily generalize the previous result:

*Given $n$ distinct densities $p_1, \ldots, p_n \in \mathcal{S}$, the density which minimizes the functional*

$$q \longmapsto \sum_{k=1}^{n} D_{KL}(p_k||q)$$

*is the average of the densities*

$$q(x) = \frac{p_1(x) + \cdots + p_n(x)}{n}.$$

## 4.8 Problems

**4.1.** (*a*) Let $q$ be a fixed density on $(-\infty, \infty)$. Show that for any $\alpha \in [0,1]\backslash\{1/2\}$, there is a unique number $a$ such that

$$\int_{-\infty}^{a} q(x)\,dx = \alpha, \qquad \int_{a}^{\infty} q(x)\,dx = 1 - \alpha.$$

(*b*) Define

$$p(x) = \begin{cases} \dfrac{1}{2\alpha} q(x), & \text{if } x < a \\ \\ \dfrac{1}{2(1-\alpha)} q(x), & \text{if } x \geq a. \end{cases}$$

Verify that $p(x)$ is a probability density on $\mathbb{R}$.

(*c*) Consider the symmetric difference of Kullback–Leibler relative entropies

$$\Delta(\alpha) = D_{KL}(p||q) - D_{KL}(q||p).$$

(*d*) Show that

$$\Delta(\alpha) = \Big(\alpha + \frac{1}{2}\Big) \ln \frac{1}{2\alpha} + \Big(\frac{3}{2} - \alpha\Big) \ln \frac{1}{2(1-\alpha)}.$$

(*e*) Prove that $\Delta(\alpha) \neq 0$ for $\alpha \neq 1/2$, and deduct that $D_{KL}(p||q) \neq D_{KL}(q||p)$.

**4.2.** Let $\alpha \in \mathbb{R}$ and $p, p_i, q, q_i$ density functions on $\mathcal{X}$. Prove the following algebraic properties of the cross entropy function:

$$\begin{aligned} &(a) \quad S(p_1 + p_2, q) = S(p_1, q) + S(p_2, q) \\ &(b) \quad S(\alpha p, q) = \alpha S(p, q) \\ &(c) \quad S(p, q_1 q_2) = S(p, q_1) + S(p, q_2) \\ &(d) \quad S(p, q^\alpha) = S(\alpha p, q). \end{aligned}$$

**4.3.** Let $q$ be a fixed density on $\mathbb{R}$, and consider the set of all densities which have a constant cross entropy with respect to $q$,

$$M_q(k) = \{p; S(p,q) = k\}.$$

Show that the maximum of the entropy function, $H(p)$, on the set $M_k(q)$ is realized for $p = q$, and in this case $\max\limits_{p} H(p) = k$.

**4.4.** Use the inequality $\ln x \le x - 1$, $\forall x > 0$, to show the following inequality satisfied by the cross entropy:

$$S(p, q) \ge 1 - \int_{\mathcal{X}} p(x)q(x)\, dx.$$

**4.5.** Consider the exponential distributions $p(x) = \xi e^{-\xi x}$ and $q(x) = \theta e^{-\theta x}$, $x \ge 0$, $\xi, \theta > 0$. Compute the cross entropy $S(p, q)$.

**4.6.** Consider the Poisson distribution $p(n, \xi) = e^{-\xi}\frac{\xi^n}{n!}$, with $n = 0, 1, 2, \dots$.

(*a*) Show that

$$D_{KL}(p_\xi || p_{\xi_0}) + D_{KL}(p_{\xi_0} || p_{\xi_1}) - D_{KL}(p_\xi || p_{\xi_1}) = (\xi_0 - \xi) \ln \frac{\xi_0}{\xi_1}.$$

(*b*) Which conditions must be satisfied by $\xi, \xi_0, \xi_1$ such that the triangle inequality for the Kullback–Leibler holds?

(*c*) When does the triangle inequality for the Kullback–Leibler become identity?

**4.7.** Use Proposition 4.4.2 to find the Fisher information metric starting from the Kullback–Leibler relative entropy for the following statistical models:

(*a*) Poisson distribution.

(*b*) exponential distribution.

(*c*) normal distribution.

(*d*) gamma distribution.

**4.8.** Use Proposition 4.5.1 to find the Christoffel symbol $\Gamma^{(1)}_{ij,k}$ starting from the Kullback–Leibler relative entropy for the following statistical models:

(*a*) Poisson distribution.

(*b*) exponential distribution.

(*c*) normal distribution.

(*d*) gamma distribution.

**4.9.** Find the dual contrast function of the Kullback–Leibler relative entropy (see Sect. 11.4) and then compute the coefficient $\Gamma_{ij,k}^{(-1)}$ using Proposition 4.5.1 for the following distributions:

(*a*) Poisson distribution.

(*b*) exponential distribution.

(*c*) normal distribution.

(*d*) gamma distribution.

**4.10.** Let $p_{a,b}(x)$ and $p_{a',b'}(x)$ be two gamma distributions. Show that the Kullback–Leibler relative entropy is

$$\begin{aligned} D_{KL}(p_{a,b}, p_{a',b'}) =& (a - a')\psi(a) - \ln\Gamma(a) + \ln\Gamma(a') \\ & + a'\ln(b/b') + a(b' - b)/b, \end{aligned}$$

where $\psi(x)$ denotes the digamma function.

**4.11.** Let $p_{a,b}(x)$ and $p_{a',b'}(x)$ be two beta distributions.

(*a*) Show that the Kullback–Leibler relative entropy is

$$\begin{aligned} D_{KL}(p_{a,b}, p_{a',b'}) =& \ln\frac{B(a',b')}{B(a,b)} + (a - a')\psi(a) + (b - b')\psi(b) \\ & + (a' - a + b' - b)\psi(a + b). \end{aligned}$$

(*b*) Show that the cross entropy is given by

$$\begin{aligned} S(p_{a,b}, p_{a',b'}) =& \ln B(a',b') - (a' - 1)\psi(a) - (b' - 1)\psi(b) \\ & + (a' + b' - 2)\psi(a + b). \end{aligned}$$

# Chapter 5

# Informational Energy

The informational energy is a concept inspired from the kinetic energy expression of Classical Mechanics. From the information theory point of view, the *informational energy* is a measure of uncertainty or randomness of a probability system, and was introduced and studied for the first time by Onicescu [67, 68] in the mid-1960s.

The informational energy and entropy are both measures of randomness, but they describe distinct features. This chapter deals with the informational energy in the framework of statistical models. The chapter contains the main properties of informational energy, its first and second variation, relation with entropy, and numerous worked-out examples.

## 5.1 Definitions and Examples

Let $\mathcal{S} = \{p_\xi = p(x;\xi) | \xi = (\xi^1, \dots, \xi^n) \in \mathbb{E}\}$ be a statistical model. The *informational energy* on $\mathcal{S}$ is a function $I : \mathbb{E} \to \mathbb{R}$ defined by

$$I(\xi) = \int_{\mathcal{X}} p^2(x, \xi)\, dx. \tag{5.1.1}$$

Observe that the energy is convex and invariant under measure preserving transformations, properties similar to those of entropy.

O. Calin and C. Udrişte, *Geometric Modeling in Probability and Statistics*,
DOI 10.1007/978-3-319-07779-6_5,

In the finite discrete case, when $\mathcal{X} = \{x^1, \dots, x^n\}$, formula (5.1.1) is replaced by

$$I(\xi) = \sum_{k=1}^{n} p^2(x^k, \xi). \tag{5.1.2}$$

While (5.1.2) is obviously finite, we need to require the integral (5.1.1) to be finite. However, if $\mathcal{X} = \mathbb{R}$, we have the following result.

**Proposition 5.1.1** *Let $p(x)$ be a probability density on $\mathbb{R}$ satisfying:*

*(i) $p(x)$ is continuous*

*(ii) $p(x) \to 0$ as $x \to \pm\infty$.*

*Then the informational energy of $p$ is finite, i.e., the following integral is convergent*

$$I(p) = \int_{-\infty}^{\infty} p^2(x)\, dx < \infty.$$

*Proof:* Let $0 < a < 1$. Then there is number $A > 0$ such that $p(x) < a$ for $|x| > A$. This follows from the fact that $p(x) \searrow 0$ as $|x| \to \infty$, see Fig. 5.1. Writing

$$I(p) = \int_{-\infty}^{-A} p^2(x)\, dx + \int_{-A}^{A} p^2(x)\, dx + \int_{A}^{\infty} p^2(x)\, dx,$$

we note that

$$\int_{-\infty}^{-A} p^2(x)\, dx < a \int_{-\infty}^{-A} p(x)\, dx = aF(a) < a$$

$$\int_{A}^{\infty} p^2(x)\, dx < a \int_{A}^{\infty} p(x)\, dx = a(1 - F(a)) < a,$$

where $F(x)$ denotes the distribution function of $p(x)$. Since the function $p(x)$ is continuous, it reaches its maximum on the interval $[-A, A]$, denoted by $M$. Then we have the estimation

$$\int_{-A}^{A} p^2(x)\, dx < M \int_{-A}^{A} p(x)\, dx = M\big(F(A) - F(-A)\big) < 2M.$$

It follows that $I(p) \le 2a + 2M < \infty$, which ends the proof. ■

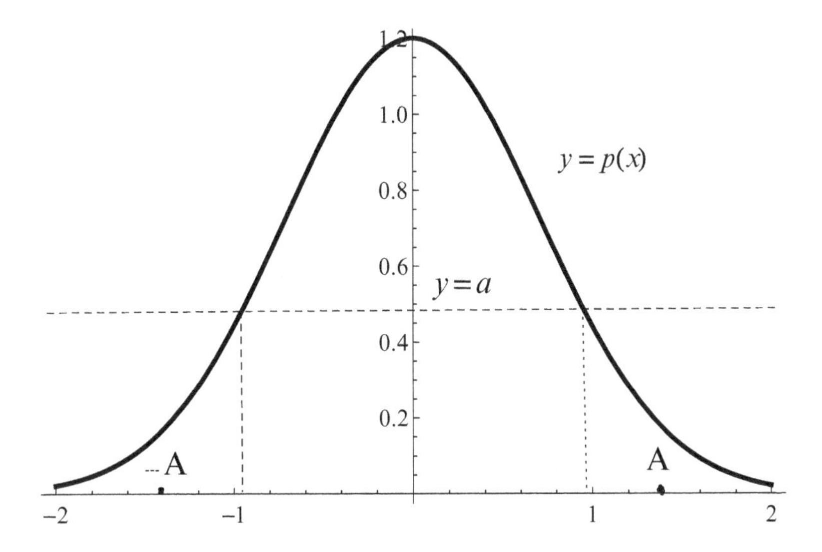


Figure 5.1: The graph of $y = p(x)$; $p(x) < a < 1$ for $|x| > A$

We make the remark that the continuity of $p(x)$ is essential. For instance

$$p(x) = \begin{cases} \frac{1}{2\sqrt{x}}, & \text{if } 0 < x < 1 \\ 0, & \text{otherwise.} \end{cases}$$

is discontinuous and has an infinite informational energy.

Observe that condition $(ii)$ is independent of condition $(i)$. For instance, the following infinite mixture of Gaussians

$$p(x) = \frac{6}{\pi^2} \sum_{k=1}^{\infty} \frac{1}{k^2} \cdot \frac{k^3}{\sqrt{2\pi}} exp\Big( - \frac{(x-k)^2 k^6}{2} \Big)$$

is a continuous density which does not satisfy condition $(ii)$, since $p(k) \geq \frac{6k}{\sqrt{2\pi}\pi^2}$.

**Example 5.1.1 (Discrete Finite Distribution)** *Consider an experiment with the following probability distribution table*

| *event* | $x^1$ | $x^2$ | $\dots$ | $x^n$ |
|---|---|---|---|---|
| *probability* | $p_1$ | $p_2$ | $\dots$ | $p_n$ |

*where the probabilities $p_1, \dots, p_n$ sum up to 1. The first $n-1$ probabilities can be taken as parameters $\xi^1, \dots, \xi^{n-1}$. Then the informational energy is given by*

$$\begin{aligned} I &= \sum_{i=1}^{n} p_i^2 = (\xi^1)^2 + \ldots + (\xi^{n-1})^2 + (1 - \xi^1 - \ldots - \xi^{n-1})^2 \\ &= 1 + 2\Big[\sum_j (\xi^j)^2 - \sum_j \xi^j - \sum_{i \neq j} \xi^i \xi^j\Big]. \end{aligned} \tag{5.1.3}$$

**Example 5.1.2 (Constant Discrete Finite Distribution)** If an experiment has $n$ outcomes which are equiprobable, i.e., they have the same probability $p = 1/n$, then the informational energy is $I = 1/n$. This follows easily from

$$I = \sum_{i=1}^{n} p_i^2 = \frac{1}{n^2} + \ldots + \frac{1}{n^2} = \frac{n}{n^2} = \frac{1}{n}.$$

We note that the informational energy is bounded from below and above by 0 and 1, respectively. The informational energy tends to have an opposite variation to entropy, fact known in Thermodynamics as the third principle. If the entropy of a system decreases, its informational energy increases, and vice versa. We shall deal with a more general case in the next section. We shall sketch the idea in the case of a discrete distribution in the next example.

In the infinite discrete case, when $\mathcal{X} = \{x^1, \dots, x^n, \dots\}$, formula (5.1.1) is replaced by

$$I(\xi) = \sum_{k=1}^{\infty} p^2(x^k, \xi). \tag{5.1.4}$$

only if the series $\sum_{k=1}^{\infty} p^2(x^k, \xi)$ is convergent for any $x \in \mathcal{X}$ and $\xi \in \mathbb{E}$.

In the next example the concept of *randomness* refers to the lack of predictability of the outcomes of a tossed coin.

**Example 5.1.3** The informational energy increases when the randomness decreases.

We shall give the explanation on a very particular case. A similar procedure can be carried over in the general case. Consider an experiment with two random outcomes $x_1$ and $x_2$. The maximum randomness is achieved when none of the outcomes is more likely to occur. This corresponds to equal probabilities $P(x_1) = p_1 = \frac{1}{2}$ and $P(x_2) = p_2 = \frac{1}{2}$. This can be thought of as flipping a fair coin. If the coin is weighted, then the randomness decreases, because the side with less weight will be more likely to land up. In the case of a weighted coin consider the new probabilities $p_1' = p_1 - x$ and $p_2' = p_2 + x$ where $x$ is a number between 0 and $\frac{1}{2}$. The probability distribution table in both situations is

| | T | H |
|---|---|---|
| $p$ | $\frac{1}{2}$ | $\frac{1}{2}$ |
| $p'$ | $\frac{1}{2} - x$ | $\frac{1}{2} + x$ |

The informational energy for the fair coin is $I = p_1^2 + p_2^2 = \frac{1}{2}$. The informational energy for the weighted coin can be written as

$$\begin{aligned} I' &= p_1'^2 + p_2'^2 = \Big(\frac{1}{2} - x\Big)^2 + \Big(\frac{1}{2} + x\Big)^2 \\ &= \frac{1}{4} + x^2 - x + \frac{1}{4} + x^2 + x \\ &= \frac{1}{2} + 2x^2 > \frac{1}{2} = I. \end{aligned}$$

Therefore $I' > I$, i.e., the informational energy for the weighted coin is larger than for the fair coin.

Both informational energy and entropy are measures of randomness. However, it is worthy to note that the two notions of randomness captured by the entropy and informational energy are distinct. Otherwise, it would suffice to study only one of the concepts. It is easy to construct examples of distributions with the same entropy and distinct informational energy, and vice versa. Therefore, studying both energy and entropy provides a more complete picture of the randomness of a distribution.

The following property deals with the bounds of the informational energy in the discrete finite case.

**Proposition 5.1.2** *The informational energy of a system with $n$ elementary outcomes is bounded above by 1 and below by $1/n$, i.e.,*

$$\frac{1}{n} \leq I \leq 1.$$

*The minimum of the informational energy is reached in the case when all the outcomes have the same probability. This minimum is $1/n$.*

*Proof:* The upper bound comes from the estimation

$$1 = (p_1 + \ldots + p_n)^2 = \underbrace{\sum_{i=1}^{n} p_i^2}_{=I} + \sum_{i \neq j}^{n} p_i p_j \geq I.$$

For the lower bound part, we shall assume that $x^1, \ldots, x^n$ are the outcomes of an experiment and $p_1 = P(x^1), \ldots, p_n = P(x^n)$ are the associated probabilities. Since $p_i$ is not necessarily equal to $1/n$, it makes sense to consider the differences

$$x^i = p_i - \frac{1}{n}, \qquad i = 1, \ldots, n.$$

Since $\sum_{i=1}^{n} p_i = 1$ we obtain

$$\sum_{i=1}^{n} x^i = \underbrace{(p_1 + \ldots + p_n)}_{=1} - \underbrace{(\frac{1}{n} + \ldots \frac{1}{n})}_{n \text{ times}} = 1 - 1 = 0.$$

Using the definition of the informational energy yields

$$\begin{aligned}
I &= p_1^2 + \ldots + p_n^2 \\
&= \Big(\frac{1}{n} + x_1\Big)^2 + \ldots + \Big(\frac{1}{n} + x_1\Big)^2 \\
&= \Big(\frac{1}{n^2} + 2\frac{1}{n}x_1 + x_1^2\Big)^2 + \ldots + \Big(\frac{1}{n^2} + 2\frac{1}{n}x_n + x_n^2\Big)^2 \\
&= \underbrace{\Big(\frac{1}{n^2} + \ldots + \frac{1}{n^2}\Big)}_{=1/n} + \frac{2}{n}\Big(\underbrace{x_1 + \ldots + x_n}_{=0}\Big) + \underbrace{(x_1^2 + \ldots + x_n^2)} \\
&= \frac{1}{n} + \underbrace{(x_1^2 + \ldots + x_n^2)}_{\geq 0} \geq \frac{1}{n}.
\end{aligned}$$

The equality in the above inequality is reached when all $x_i = 0$. In this case $p_i = \dfrac{1}{n} + x_i = \dfrac{1}{n}$ and the informational energy reaches the minimum value $I = \frac{1}{n}$. Here we also notice that while for $p_i = \frac{1}{n}$ the informational energy reaches the minimum, the entropy $H$ reaches its maximum. ■

In the following we shall present an alternate proof of the fact that the minimum of the informational energy is realized for the uniform distribution. Let the distribution $q = \{q_i\}$, $q_i = p_i + s_i$, $i = 1, \ldots, n$

be the perturbed distribution of $p = \{p_i\}$. Since $\sum_{i=1}^{n} s_i = \sum_{i=1}^{n} q_i - \sum_{i=1}^{n} p_i = 0$, then $s_n = -\sum_{i=1}^{n-1} s_i$, and hence the informational energies

$$\begin{aligned} I(p) &= \sum_{i=1}^{n-1} p_i^2 + p_n^2, \\ I(q) &= \sum_{i=1}^{n-1} (p_i + s_i)^2 + (p_n + s_n)^2 \\ &= \sum_{i=1}^{n-1} (p_i + s_i)^2 + (p_n - \sum_{i=1}^{n-1} s_i)^2 \end{aligned}$$

are functions of the $n-1$ variables $s_1, \ldots, s_{n-1}$. The distribution $p$ is a minimum point for the informational energy if

$$\frac{\partial I}{\partial s_i}\Big|_{s_1=\ldots=s_{n-1}=0} = 0, \qquad \forall i = 1, \ldots, n-1,$$

and the Hessian $\frac{\partial^2 I}{\partial s_i \partial s_j}\Big|_{s_1=\ldots=s_{n-1}=0}$ is positive definite. We have

$$\begin{aligned} \frac{\partial I}{\partial s_i} &= 2(p_i + s_i) - \Big(p_n - \sum_{k=1}^{n-1} s_k\Big) \Longrightarrow \\ \frac{\partial I}{\partial s_i}\Big|_{s_1=\ldots=s_{n-1}=0} &= 0 \Longleftrightarrow p_i = p_n, \qquad i = 1, \ldots, n-1. \end{aligned}$$

Hence the uniform distribution is a critical point for the informational energy. The Hessian is given by the following $(n-1) \times (n-1)$ matrix

$$H_n = \begin{pmatrix} 4 & 2 & \ldots & 2 \\ 2 & 4 & \ldots & 2 \\ \vdots & \vdots & \ddots & \vdots \\ 2 & 2 & \ldots & 4 \end{pmatrix} = 2 \begin{pmatrix} 2 & 1 & \ldots & 1 \\ 1 & 2 & \ldots & 1 \\ \vdots & \vdots & \ddots & \vdots \\ 1 & 1 & \ldots & 2 \end{pmatrix} = 2D_n.$$

To compute $\det D_n$, we add all the rows to the first row, we extract the factor $n$, and then subtract the first row from all the other rows to obtain an upper diagonal determinant with the entries on the main diagonal equal to 1. We have explicitly

$$\det D_n = \det \begin{pmatrix} n & n & \dots & n \\ 1 & 2 & \dots & 1 \\ \vdots & \vdots & \ddots & \vdots \\ 1 & 1 & \dots & 2 \end{pmatrix} = n \det \begin{pmatrix} 1 & 1 & \dots & 1 \\ 1 & 2 & \dots & 1 \\ \vdots & \vdots & \ddots & \vdots \\ 1 & 1 & \dots & 2 \end{pmatrix}$$

$$= n \det \begin{pmatrix} 1 & 1 & \dots & 1 \\ 0 & 1 & \dots & 0 \\ \vdots & \vdots & \ddots & \vdots \\ 0 & 0 & \dots & 1 \end{pmatrix} = n, \qquad \forall n \geq 2.$$

Consequently, the Hessian matrix $H_n$ is non-degenerate. Moreover, the Hessian is positive definite. It follows that the uniform distribution realizes the minimum for the informational energy.

The next properties deal with bounds for the informational energy functional in the case of continuous distributions $p : [a, b] \to [0, \infty)$.

**Proposition 5.1.3** *The informational energy functional, defined on continuous distributions on* $[a, b]$, *satisfies the inequality*

$$\frac{1}{b-a} \leq I(p).$$

*The minimum of the informational energy functional is reached in the case of the uniform distribution* $p(x) = 1/(b-a)$.

*Proof:* If we let $q = 1$ in the following Cauchy's integral inequality

$$\int_a^b |p(x)q(x)|\, dx \leq \Big(\int_a^b p^2(x)\, dx\Big)^{1/2} \Big(\int_a^b q^2(x)\, dx\Big)^{1/2},$$

we find

$$1 = \int_a^b p(x)\, dx \leq \Big(\int_a^b p^2(x)\, dx\Big)^{1/2} (b-a)^{1/2},$$

which leads to the desired inequality after dividing by $(b-a)^{1/2}$ and taking the square. The equality is reached when the functions $p(x)$ and $q(x) = 1$ are proportional, i.e., when $p(x)$ is constant. This corresponds to the case of a uniform distribution.

As an alternate proof, we can apply Jensen integral inequality for the convex function $g(u) = u^2$:

$$\begin{aligned} g\Big(\frac{1}{b-a}\int_a^b p(x)\,dx\Big) &\le \frac{1}{b-a}\int_a^b g\big(p(x)\big)\,dx \Longleftrightarrow \\ \frac{1}{(b-a)^2} &\le \frac{1}{b-a}\int_a^b p^2(x)\,dx \Longleftrightarrow \\ \frac{1}{b-a} &\le \int_a^b p^2(x)\,dx \Longleftrightarrow \\ \frac{1}{b-a} &\le I(p). \end{aligned}$$

Again, the equality is reached for a constant function $p$, which corresponds to the uniform distribution. ■

Among all distributions defined on $[a, b]$, the uniform distribution is the one with the smallest informational energy. Hence the informational energy provides a measure of closeness of an arbitrary distribution to the uniform one.

We shall compute the informational energy for a few particular cases.

**Example 5.1.4 (Poisson Distribution)** Since in the case of the Poisson distribution the statistical manifold is one-dimensional, the informational energy depends only on the variable $\xi$:

$$I(\xi) = \sum_{n\ge 0} p^2(n,\xi) = e^{-2\xi}\sum_{n\ge 0}\frac{\xi^{2n}}{(n!)^2} = e^{-2\xi} I_0(2\xi),$$

where

$$I_0(z) = \sum_{n\ge 0}\frac{(z/2)^{2n}}{(n!)^2}$$

is the modified Bessel function of order 0. We note the informational energy decreases to zero as $\xi \to \infty$, and $I(\xi) < I(0) = 1$, for any $\xi > 0$, see Fig. 5.2.

**Example 5.1.5 (Normal Distribution)** In this case the informational energy can be computed explicitly

$$\begin{aligned} I(\mu,\sigma) &= \int_{\mathbb{R}} p(x;\mu,\sigma)^2\,dx = \frac{1}{2\pi\sigma^2}\int_{\mathbb{R}} e^{-\frac{(x-\mu)^2}{\sigma^2}}\,dx \\ &= \frac{1}{2\pi\sigma^2}\cdot\sigma\sqrt{\pi} = \frac{1}{2\sigma\sqrt{\pi}}. \end{aligned}$$

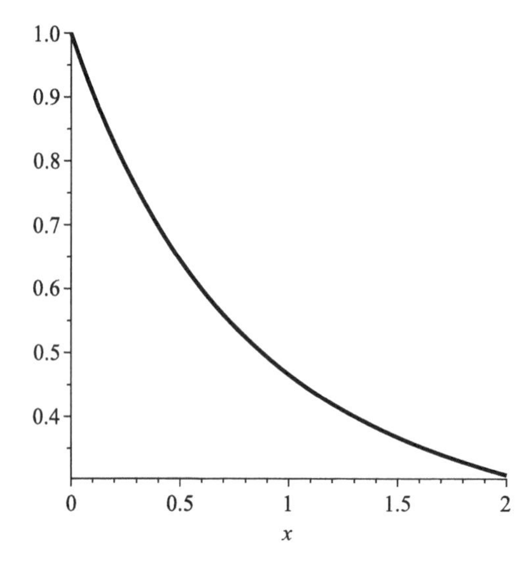


Figure 5.2: The graph of $x \to e^{-x} I_0(x)$

The function $I(\mu, \sigma)$ does not depend on the mean $\mu$, and it is a decreasing function of $\sigma$.

**Example 5.1.6 (Exponential Distribution)** This is another case when the informational energy can be worked out explicitly

$$I(\xi) = \int_0^\infty p(x,\xi)^2\,dx = \int_0^\infty \xi^2 e^{-2\xi x}\,dx = \frac{\xi}{2},$$

which is increasing in terms of $\xi$.

**Example 5.1.7 (Gamma Distribution)** The model is defined by the family of distributions

$$p_\xi(x) = p_{\alpha,\beta}(x) = \frac{1}{\beta^\alpha \Gamma(\alpha)} x^{\alpha-1} e^{-x/\beta},$$

with parameters $\xi = (\xi^1, \xi^2) = (\alpha, \beta) \in (0,\infty) \times (0,\infty)$ and sample space $\mathcal{X} = (0,\infty)$. Assuming $\alpha > 1/2$, with the substitution $a = 2\alpha - 1$ and $b = \beta/2$, the informational energy becomes

$$\begin{aligned} I(\xi) &= \int_0^\infty p^2_{\alpha,\beta}(x)\,dx = \int_0^\infty \frac{1}{\beta^{2\alpha}\Gamma(\alpha)^2}\, x^{2\alpha-2}\, e^{-2x/\beta}\,dx \\ &= \frac{1}{\beta^{2\alpha}\Gamma(\alpha)^2}\, b^a \Gamma(a) \int_0^\infty \frac{1}{b^a \Gamma(a)} x^{a-1} e^{-\frac{x}{b}}\,dx \end{aligned}$$

$$
\begin{aligned}
&= \frac{1}{\beta^{2\alpha}\Gamma(\alpha)^2} b^a \Gamma(a) \int_0^\infty p_{a,b}(x)\,dx \\
&= \frac{1}{\beta^{2\alpha}\Gamma(\alpha)^2} \cdot \frac{\beta^{2\alpha-1}}{2^{2\alpha-1}} \Gamma(2\alpha-1) \\
&= \frac{1}{\beta\, 2^{2\alpha-1}} \cdot \frac{\Gamma(2\alpha)}{2\alpha-1} \cdot \frac{1}{\Gamma(\alpha)^2}.
\end{aligned}
$$

The case $\alpha \leq 1/2$ is eliminated by the divergence of the improper integral. Using the Legendre's duplication formula

$$
\Gamma(2\alpha) = \frac{2^{2\alpha-1}}{\sqrt{\pi}} \Gamma(\alpha)\Gamma(\alpha+1/2), \tag{5.1.5}
$$

the computation can be continued as

$$
\begin{aligned}
I(\xi) &= \frac{1}{\beta(2\alpha-1)\sqrt{\pi}} \cdot \frac{\Gamma(\alpha+1/2)}{\Gamma(\alpha)} = \frac{1}{\beta(2\alpha-1)} \cdot \frac{\Gamma(\alpha+1/2)}{\Gamma(1/2)\Gamma(\alpha)} \\
&= \frac{1}{\beta(2\alpha-1)} \frac{1}{B(\alpha, 1/2)},
\end{aligned}
$$

where

$$
B(x,y) = \int_0^1 t^{x-1}(1-t)^{y-1}\,dt
$$

is the *beta function*, and we used that $\Gamma(1/2) = \sqrt{\pi}$. Hence

$$
I(\alpha,\beta) = \frac{1}{\beta(2\alpha-1)B(\alpha,1/2)}.
$$

**Example 5.1.8 (Beta Distribution)** The density of a beta distribution on the sample space $\mathcal{X} = [0,1]$ is

$$
p_{a,b}(x) = \frac{1}{B(a,b)} x^{a-1}(1-x)^{b-1},
$$

with $a, b > 0$. Let $\alpha = 2a - 1 > 0$ and $\beta = 2b - 1 > 0$. Then the informational energy is

$$
\begin{aligned}
I(a,b) &= \int_0^1 p_{a,b}^2(x)\,dx = \frac{1}{B^2(a,b)} \int_0^1 x^{2a-1}(1-x)^{2b-2}\,dx \\
&= \frac{1}{B^2(a,b)} \int_0^1 x^{\alpha-1}(1-x)^{\beta-1}\,dx \\
&= \frac{B(\alpha,\beta)}{B^2(a,b)} = \frac{B(2a-1,2b-1)}{B^2(a,b)}.
\end{aligned}
$$

Using the expression of beta function in terms of gamma functions and the Legendre's duplication formula (5.1.5), the energy can be also written as

$$\begin{aligned} I(a,b) &= \frac{1}{(2a-1)(2b-1)} \cdot \frac{\Gamma(2a)\Gamma(2b)}{\Gamma(2a+2b-2)} \cdot \frac{\Gamma(a+b)^2}{\Gamma(a)^2\Gamma(b)^2} \\ &= \frac{(a+b-1/2)(a+b-1)}{(a-1/2)(b-1/2)} \cdot \frac{\Gamma(2a)\Gamma(2b)}{\Gamma(2a+2b)} \cdot \frac{\Gamma(a+b)^2}{\Gamma(a)^2\Gamma(b)^2} \\ &= \frac{(a+b-1/2)(a+b-1)}{(a-1/2)(b-1/2)} \cdot \frac{\Gamma(a+1/2)\Gamma(b+1/2)\Gamma(a+b)}{2\sqrt{\pi}\Gamma(a+b+1/2)}. \end{aligned}$$

**Example 5.1.9 (Lognormal Distribution)** Consider the distribution

$$p_{\mu,\sigma}(x) = \frac{1}{\sqrt{2\pi}\,\sigma x} e^{-\frac{(\ln x-\mu)^2}{2\sigma^2}}$$

with sample space $\mathcal{X} = (0, \infty)$ and positive parameters $\mu$ and $\sigma$. Using the substitution $y = \ln x - \mu$ yields

$$\begin{aligned} I(\mu,\sigma) &= \int_0^\infty p^2_{\mu,\sigma}(x)\,dx \\ &= \frac{1}{2\pi\sigma^2} \int_0^\infty \frac{1}{x^2} e^{-\frac{(\ln x-\mu)^2}{\sigma^2}}\,dx \\ &= \frac{1}{2\pi\sigma^2} \int_{-\infty}^\infty e^{-\frac{y^2}{\sigma^2}-y-\mu}\,dy \\ &= \frac{1}{2\sigma\sqrt{\pi}} e^{\frac{\sigma^2}{4}-\mu}, \end{aligned}$$

where we used that

$$\int_{\mathbb{R}} e^{-ay^2+by+c}\,dy = \sqrt{\frac{\pi}{a}}\, e^{\frac{b^2}{4a}+c}.$$

**Example 5.1.10 (Dirac Distribution)** Consider the family of density functions

$$\varphi_\epsilon(x) = \begin{cases} 1/\epsilon, & \text{if } x_0 - \epsilon/2 < x < x_0 + \epsilon/2 \\ 0, & \text{otherwise.} \end{cases}$$

Since the Dirac distribution centered at $x_0$ can be written as the limit, see Example 3.2.7

$$\delta(x - x_0) = \lim_{\epsilon \searrow 0} \varphi_\epsilon(x), \quad a < x < b,$$

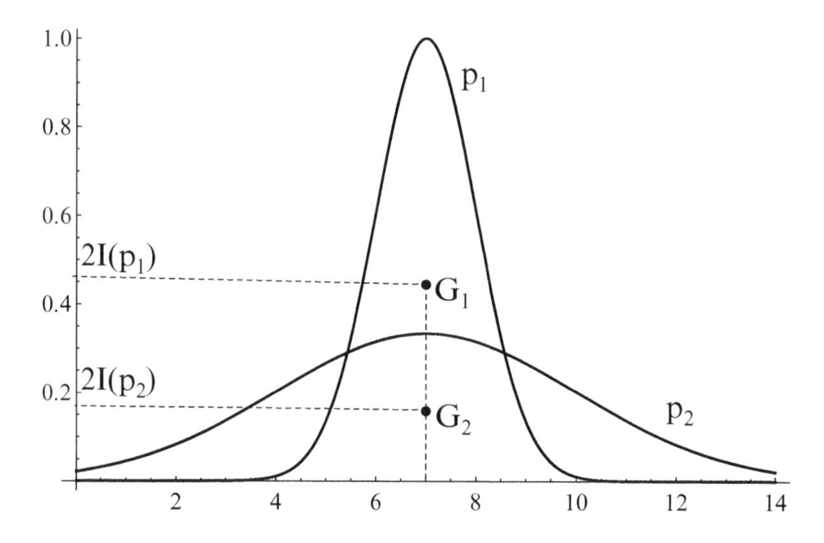


Figure 5.3: The distribution with the lower centroid has the smaller informational energy

then by the Dominated Convergence Theorem

$$I(\delta(x - x_0)) = \lim_{\epsilon \searrow 0} I(\varphi_\epsilon) = \lim_{\epsilon \searrow 0} \int_{x_0-\epsilon/2}^{x_0+\epsilon/2} \frac{1}{\epsilon^2}\, dx = \lim_{\epsilon \searrow 0} \frac{1}{\epsilon} = +\infty.$$

Hence the Dirac distribution $\delta(x-x_0)$ has infinite information energy.

## 5.2 Informational Energy and Constraints

Consider a density function $p(x)$ defined on the interval $[a, b]$. The coordinates of the center of mass, $G$, of the subgraph region

$$\{(x, y); y \leq p(x), a \leq x \leq b\}$$

are given by the well-known formulas

$$x_G = \int_a^b xp(x)\, dx,\ y_G = \frac{1}{2}\int_a^b p^2(x)\, dx.$$

It follows that $x_G = \mu$ and $y_G = \frac{1}{2}I(p)$. Hence the informational energy measures the height of the center of mass associated with the subgraph of the density function. On the other side, the energy is more or less insensitive to the changes of the mean.

Since the informational energy measures the $y$-coordinate of the centroid, it follows that the distribution of a statistical model $\mathcal{S} = \{p_\xi\}$ with the smallest informational energy corresponds to the most "stable" subgraph (here "stable" has a gravitational connotation). In Fig. 5.3 there are represented two densities, $p_1$ and $p_2$. Since the centroid $G_2$ is lower than $G_1$, it follows that the energy of $p_2$ is smaller than the energy $p_1$.

The next result deals with the case when we consider a constraint on the mean. If the mean is fixed, we are looking for the distribution with the subgraph centroid, $G$, having a fixed $x$-coordinate and the lowest possible $y$-coordinate.

**Proposition 5.2.1** *Let $\mu \in (a, b)$. Among all probability densities defined on $(a, b)$ with given mean $\mu$, there is only one distribution with the smallest informational energy. This has the form*

$$p(x) = \lambda_1 x + \lambda_2,$$

*with*

$$\lambda_1 = \frac{12\Big(\mu - \frac{a+b}{2}\Big)}{(b-a)^3} \tag{5.2.6}$$

$$\lambda_2 = \frac{12\Big(\mu(a+b) - (a^2 - ab + b^2)/3\Big)}{(b-a)^3}. \tag{5.2.7}$$

*Proof:* Following a constrained optimization problem, we need to minimize the objective functional

$$\frac{1}{2} I(p) = \frac{1}{2} \int_a^b p^2(x) dx,$$

subject to constraints

$$\int_a^b x p(x) dx = \mu, \qquad \int_a^b p(x)\, dx = 1. \tag{5.2.8}$$

This leads to the following optimization problem with constraints

$$p \to \int_a^b p^2(x) dx - 2\lambda_1 \bigg( \int_a^b x p(x) dx - \mu \bigg) - 2\lambda_2 \bigg( \int_a^b p(x) dx - 1 \bigg),$$

where $\lambda_i$ denote the Lagrange multipliers. The associated Lagrangian is

$$L(p) = p^2(x) - 2\lambda_1 x p - 2\lambda_2 p.$$

The critical point condition $\frac{\partial L}{\partial p} = 0$ provides the linear density

$$p(x) = \lambda_1 x + \lambda_2.$$

The parameters $\lambda_1, \lambda_2$ are determined from the constraints (5.2.8) and are given by formulas (5.2.6)–(5.2.7). ∎

## 5.3 Product of Statistical Models

Let $\mathcal{S} \times \mathcal{U}$ be a product of statistical manifolds, see Example 1.3.9, and consider $f \in \mathcal{S} \times \mathcal{U}$, with $f(x, y) = p(x)q(y)$, $p \in \mathcal{S}$, $q \in \mathcal{U}$. Then

$$\begin{aligned} I_{\mathcal{S}\times\mathcal{U}}(f) &= \iint_{\mathcal{X}\times\mathcal{Y}} p^2(x)q^2(y)\,dxdy \\ &= \int_{\mathcal{X}} p^2(x)\,dx \int_{\mathcal{Y}} p^2(y)\,dy \\ &= I_{\mathcal{S}}(p)I_{\mathcal{P}}(q), \end{aligned}$$

i.e., the informational energy of an element of $\mathcal{S} \times \mathcal{U}$ is the product of the informational energies of the projections on $\mathcal{S}$ and $\mathcal{U}$.

## 5.4 Onicescu's Correlation Coefficient

Given two distributions $p, q \in \mathcal{S}$, the correlation coefficient introduced by Onicescu is

$$\mathcal{R}(p, q) = \frac{\int_{\mathcal{X}} p(x)q(x)\,dx}{\sqrt{I(p)I(q)}},$$

if the distributions are continuous, and

$$\mathcal{R}(p, q) = \frac{\sum_k p(x_k)q(x_k)}{\sqrt{I(p)I(q)}},$$

if the distributions are discrete.

**Proposition 5.4.1** *The correlation coefficient has the following properties:*

(*i*) $\mathcal{R}(p, q) = \mathcal{R}(q, p)$;

(*ii*) $\mathcal{R}(p, q) \leq 1$, *with identity if* $p = q$.

*Proof:*

(*i*) It follows from the symmetry of the definition relations.

(*ii*) For the continuous case we use the Cauchy's integral inequality

$$\Big(\int p(x)q(x)\,dx\Big)^2 \leq \int p^2(x)\,dx \int q^2(x)\,dx,$$

while the discrete distributions use the inequality

$$\Big(\sum_{k=1}^{n} p(x_k)q(x_k)\Big)^2 \leq \sum_{k=1}^{n} p^2(x_k) \sum_{k=1}^{n} q^2(x_k).$$

The identity in both inequalities is reached when the distributions are proportional, i.e., $p(x) = \lambda q(x)$. This easily implies $\lambda = 1$. ∎

## 5.5 First and Second Variation

The following result deals with the first variation of the informational energy on a statistical manifold $\mathcal{S}=\{p_\xi=p(x;\xi)|\xi=(\xi_1,\ldots,\xi_n)\in\mathbb{E}\}$.

**Proposition 5.5.1** *A point $\xi = (\xi^1,\ldots,\xi^n)$ is a critical point for the integral function $I$ if and only if*

$$\int_{\mathcal{X}} p(x,\xi)\,\partial_{\xi^i} p(x,\xi)\,dx = 0. \tag{5.5.9}$$

*In the discrete case this becomes*

$$\sum_k p(x_k,\xi)\,\partial_{\xi^i} p(x_k,\xi) = 0.$$

*Proof:* The critical points $\xi \in \mathbb{E}$ for the energy $I$ satisfy the equation $\partial_{\xi_i} I(\xi) = 0$. Then from

$$\partial_{\xi^i} I(\xi) = \partial_{\xi^i} \int p^2(x,\xi)\,dx = 2\int p(x,\xi)\,\partial_{\xi^i} p(x,\xi)\,dx$$

we obtain the desired conclusion. ∎

We note that the first variation formula (5.5.9) can be also written as an expectation

$$E_\xi[\partial_{\xi^i} p] = 0, \qquad \forall i = 1,\ldots,n. \tag{5.5.10}$$

Since $\{\partial_{\xi^1}p, \dots, \partial_{\xi^n}p\}$ is a basis of the tangent space $T_p\mathcal{S}$, then any vector $X \in T_p\mathcal{S}$ can be written as a linear combination

$$X = \alpha^i\, \partial_{\xi^i}p, \quad \alpha_i \in \mathbb{R}.$$

Using (5.5.10) yields

$$E_\xi[X] = E_\xi\big[\alpha^i \partial_{\xi^i}p\big] = \alpha^i E_\xi\big[\partial_{\xi^i}p\big] = 0.$$

We arrive at the following reformulation of the above result:

**Proposition 5.5.2** *A critical point for the informational energy is a point $\xi$ such that*

$$E_\xi[X] = 0, \qquad \forall X \in T_{p_\xi}\mathcal{S}.$$

This means that at critical points of $I(\xi)$ the expectation vanishes in all directions.

The Hessian coefficients of the informational energy are given by

$$\begin{aligned}\partial_{\xi^j\xi^i}I(\xi) &= 2\partial_{\xi^j}\int_{\mathcal{X}} p(x,\xi)\,\partial_{\xi^i}p(x,\xi)\,dx\\ &= 2\int_{\mathcal{X}}\Big(\partial_{\xi^j}p(x,\xi)\,\partial_{\xi^i}p(x,\xi) + p(x,\xi)\,\partial_{\xi^j\xi^i}p\Big)\,dx,\end{aligned}$$

which in the discrete case takes the following form

$$\partial_{\xi^i\xi^i}I(\xi) = 2\sum_k\Big(\partial_{\xi^j}p(x_k)\,\partial_{\xi^i}p(x^k) + p(x^k)\,\partial_{\xi^i\xi^j}p(x_k)\Big).$$

In the following we shall apply the first and second variations to the discrete finite distribution, see Example 5.1.1. The informational energy is given by formula (5.1.3). We have

$$\begin{aligned}\frac{\partial I}{\partial \xi^j}(\xi) &= 2\xi^j + \frac{\partial}{\partial \xi^j}(1-\xi^1-\ldots-\xi^{n-1})^2\\ &= 2\xi^j - 2(1-\xi^1-\ldots-\xi^{n-1})\\ &= 2(\xi^j-\xi^n).\end{aligned}$$

Hence $\dfrac{\partial I}{\partial \xi_j}(\xi) = 0$ if and only if $\xi^j = \xi^n$, for all $j = 1, \dots, n$. The Hessian coefficients are given by

$$\begin{aligned}\partial_{\xi_i\xi_j}I(\xi) &= 2\sum_{k=1}^{n-1}\Big(\partial_{\xi_j}(\xi_k)\partial_{\xi_i}(\xi_k) + \xi_k\,\underbrace{\partial_{\xi_i\xi_j}(\xi_k)}_{=0}\Big)\\ &= 2\sum_{k=1}^{n-1}(\delta_{jk}\delta_{ik}) = 2\mathbb{I}_{n-1},\end{aligned}$$

which is positive definite everywhere. Hence

$$\xi_1 = \ldots = \xi_{n-1} = \xi_n = \frac{1}{n}$$

is a minimum point for the informational energy $I(\xi)$. We thus recovered part of Proposition 5.1.2.

## 5.6 Informational Energy Minimizing Curves

Given two distributions $p_0$ and $p_1$ on a statistical manifold $\mathcal{S}$ with finite dimensional parameter space, we are interested in finding a smooth curve on $\mathcal{S}$ joining the distributions and having the smallest cumulative informational energy along the curve. More precisely, we are looking for a curve $\gamma : [0,1] \longrightarrow \mathcal{S}$ with $\gamma(0) = p_0$ and $\gamma(1) = p_1$, which minimizes the action integral

$$\xi \to \int_0^1 I(\xi(u))\,ds = \int_0^1 I(\xi(u))\sqrt{g_{ij}(\xi(u))\dot{\xi}^i(u)\dot{\xi}^j(u)}\,du$$

with the square of arc-element $ds^2 = g_{ij}(\xi)d\xi^i d\xi^j$, where the Fisher–Riemann metric is denoted by $g_{ij}(\xi)$. There are a couple of examples where we can describe the curves of minimum information energy explicitly.

**Example 5.6.1 (Exponential Distribution)** From Examples 1.6.2 and 5.1.6 the information energy and the Fisher information metric are

$$I(\xi) = \frac{\xi}{2}, \qquad g_{11}(\xi) = \frac{1}{\xi^2}.$$

The action becomes

$$\begin{aligned}\xi \to \int_0^1 I(\xi(u))\,ds &= \int_0^1 \frac{1}{2}\xi(u)\sqrt{\frac{1}{\xi^2}\dot{\xi}^2(u)}\,du \\ &= \frac{1}{2}\int_0^1 |\dot{\xi}(u)|\,du,\end{aligned}$$

which is half the length of the curve $\xi(u)$. The minimum is reached for the linear function $\xi(u) = au + b$, with $a$, $b$ constants.

**Example 5.6.2 (Normal Distribution)** Partial explicit computations can be carried out in the case of normal distribution. In this case $\xi^1 = \mu \in \mathbb{R}$, $\xi^2 = \sigma > 0$. Since the informational energy is given by $I(\mu, \sigma) = \dfrac{1}{2\sigma\sqrt{\pi}}$, see Example 5.1.5, and the Fisher information matrix is

$$g_{ij} = \begin{pmatrix} \frac{1}{\sigma^2} & 0 \\ 0 & \frac{2}{\sigma^2} \end{pmatrix},$$

the action that needs to be minimized becomes

$$\begin{aligned} (\mu, \sigma) \quad &\to \quad \int_0^1 \frac{1}{2\sigma^2(u)\sqrt{\pi}} \sqrt{\dot{\mu}^2(u) + 2\dot{\sigma}^2(u)}\, du \\ &= \quad \int_0^1 \frac{1}{\sqrt{\pi}\nu^2(u)} \sqrt{\dot{\mu}^2(u) + \dot{\nu}^2(u)}\, du, \end{aligned}$$

where we substituted $\nu = \sigma\sqrt{2}$. The minimum informational curves satisfy the Euler–Lagrange equations with the Lagrangian

$$L(\mu, \dot{\mu}, \nu, \dot{\nu}) = \frac{1}{\nu^2}\sqrt{\dot{\mu}^2 + \dot{\nu}^2}.$$

Since $\dfrac{\partial L}{\partial \mu} = 0$, there is a constant $C$ such that $\dfrac{\partial L}{\partial \dot{\mu}} = C$, or

$$C\nu^2\sqrt{\dot{\mu}^2 + \dot{\mu}^2} = \dot{\mu}.$$

Using this relation, we obtain the following relations

$$\begin{aligned} \frac{\partial L}{\partial \dot{\nu}} &= \frac{C\dot{\nu}}{\dot{\mu}} \\ \frac{\partial L}{\partial \nu} &= -\frac{2}{\nu^3}\sqrt{\dot{\mu}^2 + \dot{\nu}^2} = -\frac{2\dot{\mu}}{C\nu^5}. \end{aligned}$$

Hence, the Euler–Lagrange equation $\dfrac{d}{du}\Big(\dfrac{\partial L}{\partial \dot{\nu}}\Big) = \dfrac{\partial L}{\partial \nu}$ becomes

$$C^2\Big(\frac{\dot{\nu}}{\dot{\mu}}\Big)^{\cdot} + \frac{2\dot{\mu}}{\nu^5} = 0.$$

In the case $C = 0$ we obtain $\mu =$ constant, which corresponds to vertical lines.

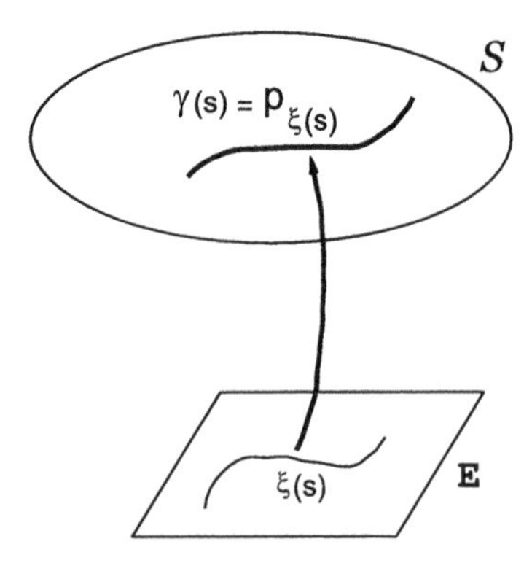


Figure 5.4: The curve $\gamma(s) = p_{\xi(s)}$ on the statistical manifold $\mathcal{S}$

## 5.7 The Laws of Thermodynamics

This section deals with an informational version of the second and third law of Thermodynamics.

**Thermodynamic Processes.** We consider a process that is described at each instance of time by a probability distribution. This state is specific, for instance, to the quantum particles that are characterized by a wave function, which is a probability density. The time evolution of the state is a curve, called a *Thermodynamic process.* In the case of statistical manifolds a Thermodynamic process is a regular curve on the statistical model $\mathcal{S} = \{p_\xi; \xi \in \mathbb{E}\}$. This is a differentiable mapping of an interval $(a, b)$ into $\mathcal{S}$

$$(a, b) \ni s \longrightarrow p_{\xi(s)} \in \mathcal{S},$$

where $(a, b) \ni s \longrightarrow \xi(s)$ is a smooth curve in the parameters space $\mathbb{E}$, see Fig. 5.4.

Let $\gamma(s) = p_{\xi(s)}$ be a process. For each $s \in (a, b)$, the curve $\gamma(s)$ is a probability distribution on $\mathcal{X}$. The velocity along $\gamma(s)$ is

$$\dot{\gamma}(s) = \frac{d}{ds} p_{\xi(s)} = \sum_i \partial_{\xi^i} p_{\xi(s)}\, \dot{\xi}^i(s). \qquad (5.7.11)$$

We note that

$$\int_{\mathcal{X}} \dot{\gamma}(s)\, dx = \int_{\mathcal{X}} \frac{d}{ds} p_{\xi(s)}(x)\, dx = \frac{d}{ds} \int_{\mathcal{X}} p_{\xi(s)}(x)\, dx = 0. \qquad (5.7.12)$$

The parameter $s$ can be regarded as time, flowing from lower to larger values. Let $p_{\xi_0} \in \mathcal{S}$ be fixed, and consider a curve $\gamma$ starting at $p_{\xi_0}$, i.e., $\gamma(s) = p_{\xi(s)}$, $\gamma(0) = p_{\xi(0)} = p_{\xi_0}$.

**The Second Law of Thermodynamics.** Let $H(p_{\xi_0})$ be the entropy at $p_{\xi_0}$. The *second law of Thermodynamics* states that the entropy of an isolated system tends to increase over time. In our case, the entropy tends to increase along the curve $\gamma$ if

$$H(p_{\xi_0}) < H(p_{\xi(s)})$$

for $s > 0$. This means the derivative to the right is positive

$$\lim_{s\searrow 0} \frac{H(p_{\xi(s)}) - H(p_{\xi_0})}{s - 0} > 0.$$

This condition can be stated in a couple of ways.

**Proposition 5.7.1** *The entropy increases along the curve* $\gamma(s) = p_{\xi(s)}$ *if either one of the following conditions is satisfied:*

(*i*) $\int_{\mathcal{X}} \dot{\gamma}(s)(x) \ln \gamma(s)(x)\, dx < 0.$

(*ii*) $D_{KL}\big(\gamma(s)||\gamma(s+\epsilon)\big) \geq \int_{\mathcal{X}} (\gamma(s+\epsilon) - \gamma(s)) \ln \gamma(s+\epsilon)(x)\, dx,$ *for* $\epsilon > 0$ *sufficiently small, where* $D_{KL}$ *stands for the Kullback–Leibler relative entropy.*

*Proof:*

(*i*) Differentiating in the formula of entropy, and using (5.7.12), we have

$$\begin{aligned} 0 < \frac{d}{ds} H(p_{\xi(s)}) &= -\frac{d}{ds} \int_{\mathcal{X}} p_{\xi(s)}(x) \ln p_{\xi(s)}(x)\, dx \\ &= -\int_{\mathcal{X}} \Big( \frac{d}{ds} p_{\xi(s)}(x) \ln p_{\xi(s)}(x) + \frac{d}{ds} p_{\xi(s)}(x) \Big)\, dx \\ &= -\int_{\mathcal{X}} \frac{d}{ds} p_{\xi(s)}(x) \ln p_{\xi(s)}(x)\, dx \\ &= -\int_{\mathcal{X}} \frac{d}{ds} p_{\xi(s)}(x) \ln p_{\xi(s)}(x)\, dx \\ &= -\int_{\mathcal{X}} \dot{\gamma}(s)(x) \ln \gamma(s)(x)\, dx, \end{aligned}$$

which leads to the desired inequality.

(*ii*) We have the following sequence of equivalences:

$$\begin{aligned} H(p_{\xi(s)}) &\leq H(p_{\xi(s+\epsilon)}) \\ \Longleftrightarrow \int p_{\xi(s)} \ln p_{\xi(s)} &\geq \int p_{\xi(s+\epsilon)} \ln p_{\xi(s+\epsilon)} \end{aligned}$$

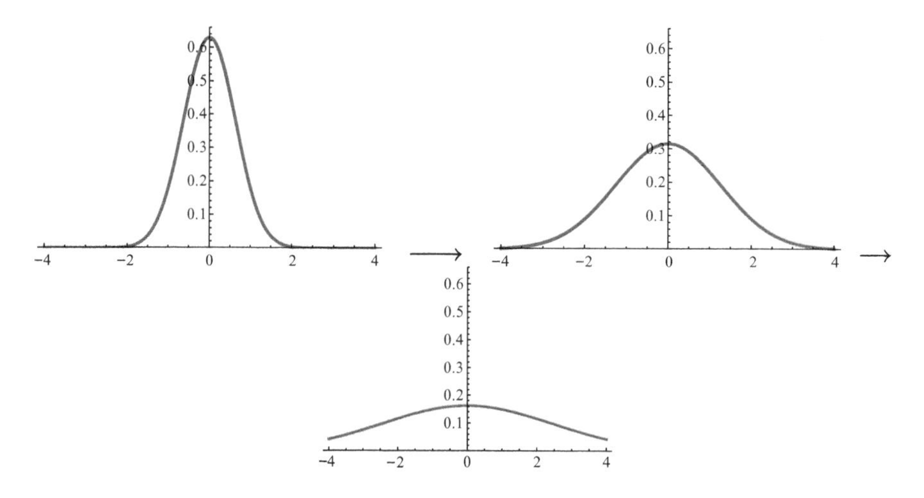


Figure 5.5: The change in the shape of the probability density of a normal distribution as the entropy increases

$$\begin{aligned}
\Longleftrightarrow D_{KL}\big(p_{\xi(s)}||p_{\xi(s+\epsilon)}\big) &\geq \int \Big(p_{\xi(s+\epsilon)} - p_{\xi(s)}\Big) \ln p_{\xi(s+\epsilon)} \\
\Longleftrightarrow D_{KL}\big(\gamma(s)||\gamma(s+\epsilon)\big) &\geq \int_{\mathcal{X}} (\gamma(s+\epsilon) - \gamma(s)) \\
&\quad \ln \gamma(s+\epsilon)(x)\, dx.
\end{aligned}$$

■

**Example 5.7.1 (The Normal Distribution)** The entropy of a normal distribution increases over time as the standard deviation $\sigma$ increases, see Fig. 5.5. When the times gets large, then $\sigma \to \infty$, and hence, the probability density tends to zero.

**Example 5.7.2 (The Exponential Distribution)** The entropy of an exponential distribution, $1 - \ln \xi$, tends to infinity as $\xi \searrow 0$, i.e., when the mean $\frac{1}{\xi} \to \infty$. In this case the density function tends to zero, see Fig. 5.6.

**The Third Law of Thermodynamics.** This law says that the entropy and the kinetic energy of an isolated Thermodynamical system have opposite variations. More precisely, this means that if the entropy of a system increases during a Thermodynamic process, then its kinetic energy tends to decrease during the same process. Here the role of kinetic energy is played by the informational energy.

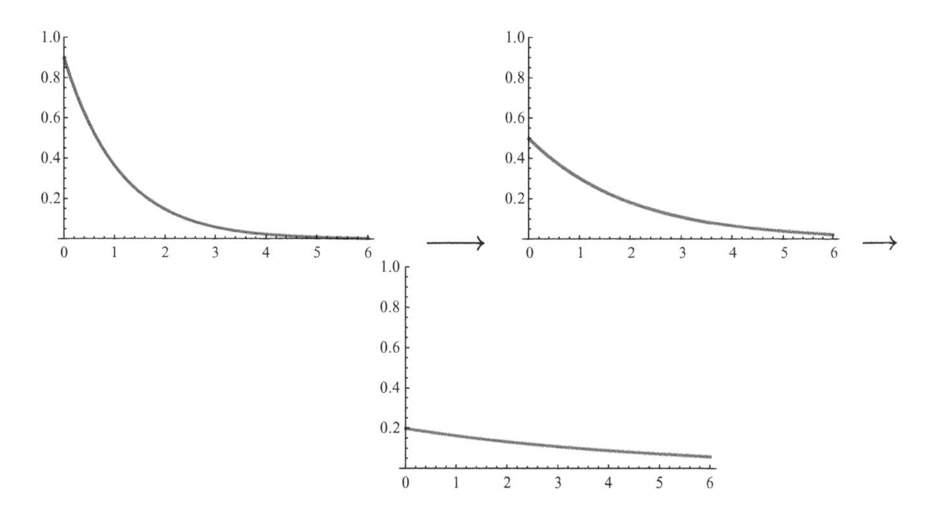


Figure 5.6: The shape of the probability density of an exponential distribution as the entropy increases

The opposite variations of the entropy and informational energy along the curve $\gamma(s) = p_{\xi(s)}$ can be written in our case as

$$\Big(H(p_{\xi(s+\epsilon)}) - H(p_{\xi(s)})\Big)\Big(I(p_{\xi(s+\epsilon)}) - I(p_{\xi(s)})\Big) < 0,$$

for $\epsilon > 0$. The instantaneous relation

$$\frac{d}{ds}H(p_{\xi(s)}) \cdot \frac{d}{ds}I(p_{\xi(s)}) < 0 \tag{5.7.13}$$

can be written as

$$\int \frac{d}{ds}p_{\xi(s)}(x)\,\ln p_{\xi(s)}(x)\,dx \cdot \int \frac{d}{ds}p_{\xi(s)}(x)\,p_{\xi(s)}(x)\,dx > 0. \tag{5.7.14}$$

**Definition 5.7.2** *A statistical manifold $\mathcal{S} = \{p_\xi; \xi \in \mathbb{E}\}$ satisfies the third law of Thermodynamics if the inequality (5.7.14) is satisfied for all curves on the model.*

Next we shall encounter a few examples of statistical manifolds that satisfy the third law of Thermodynamics.

**Example 5.7.3 (Exponential Distribution)** *In this case the statistical manifold $\mathcal{S} = \{p_\xi; \xi \in \mathbb{E}\}$ is given by $p_\xi = \xi e^{-\xi x}$, $\mathbb{E} = (0, \infty)$, and $\mathcal{X} = (0, \infty)$. Using Examples 3.2.3 and 5.1.6 we have*

$$\frac{d}{ds}H(p_{\xi(s)}) \cdot \frac{d}{ds}I(p_{\xi(s)}) = \frac{d}{ds}(1 - \ln \xi(s)) \frac{d}{ds}\frac{\xi(s)}{2} = -\frac{\dot{\xi}(s)^2}{2\xi(s)} < 0,$$

*which is the condition (5.7.13).*

**Example 5.7.4 (Normal Distribution)** The statistical manifold is parametrized by $p_\xi = \frac{1}{\sqrt{2\pi}\sigma}e^{-\frac{(x-\mu)^2}{2\sigma^2}}$, with $\xi = (\mu, \sigma) \in \mathbb{R} \times (0, \infty)$. Using Examples 3.2.1 and 5.1.5, we have

$$\frac{d}{ds}H(p_{\xi(s)}) \cdot \frac{d}{ds}I(p_{\xi(s)}) = \frac{d}{ds}\ln\left(\sigma(s)\sqrt{2\pi e}\right) \frac{d}{ds}\frac{1}{2\sigma(s)\sqrt{\pi}} = -\frac{\dot{\sigma}(s)^2}{\sigma(s)^2} < 0,$$

which recovers condition (5.7.13).

The previous two examples show that any process curve $\gamma(s)$ satisfies the third law of Thermodynamics. In general, on an arbitrary statistical manifold, there might be some processes for which this does not hold. The processes for which the third law of Thermodynamics holds are some distinguished curves on the statistical model, corresponding to some *natural* evolution processes. It is interesting to note that in the case of exponential and normal distributions, any curve is *natural.*

## 5.8 Uncertainty Relations

In Quantum Mechanics there is a tradeoff between the accuracy of measuring the position and the velocity of a particle. This section deals with a similar tradeoff between the entropy and the informational energy of a system. This will be shown by considering lower bounds for the sum between the aforementioned measures.

**Lemma 5.8.1** *For any number $p > 0$ we have*

$$1 + \ln p \leq p,$$

*with equality if and only if $p = 1$.*

*Proof:* Let $f(p) = 1 + \ln p - p$. Then $f(0+) = -\infty$, $f(\infty) = -\infty$, and $f(1) = 0$. Since $f'(p) = \frac{1-p}{p}$, then $f$ is increasing on $(0, 1)$ and decreasing on $(1, \infty)$. Then $p = 1$ is a maximum point and hence $f(p) \leq f(1) = 0$, for all $p > 0$, which is equivalent with the desired inequality. ■

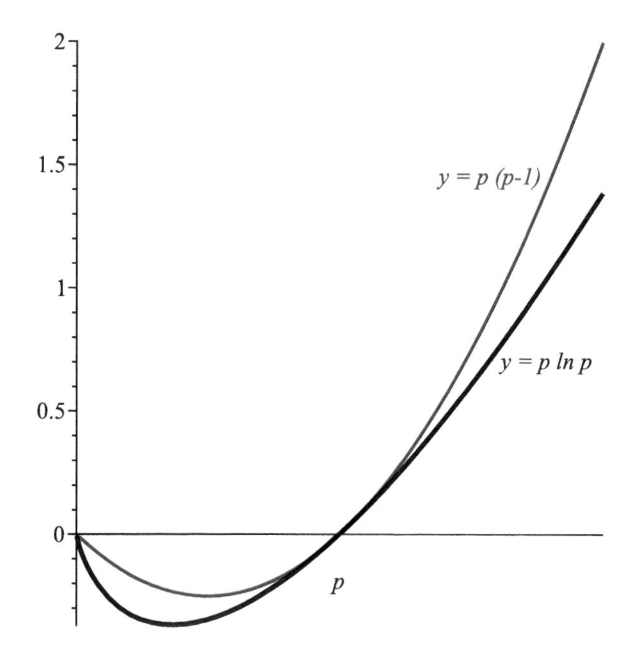


Figure 5.7: The inequality $p \ln p \leq p^2 - p$ for $p > 0$

**Theorem 5.8.2** *Let $p$ be a discrete or continuous distribution. Then*

$$H(p) + I(p) \geq 1. \tag{5.8.15}$$

*Proof:* We shall do the proof in the following three distinct cases:

*The discrete case:* Let $\mathcal{X} = \{x^1, \dots, x^n\}$, with $n$ finite or infinite, and $p_i = P(x^i)$. From Lemma 5.8.1 we have $1 + \ln p_i \leq p_i$, and hence $p_i \ln p_i \leq p_i^2 - p_i$, for $i = 1, \dots, n$. Summing over $i$ yields

$$\begin{aligned} -H(p) = \sum_{i=1}^{n} p_i \ln p_i &\leq \sum_{i=1}^{n} p_i^2 - \sum_{i=1}^{n} p_i \\ &= I(p) - 1, \end{aligned}$$

which leads to (5.8.15).

*The continuous case:* From Lemma 5.8.1, we have $1 + \ln p(x) \leq p(x)$, for any $x \in \mathcal{X}$. Then $p(x) \ln p(x) \leq p^2(x) - p(x)$, see Fig. 5.7.

Integrating yields

$$\begin{aligned} -H(p) &= \int_{\mathcal{X}} p(x) \ln p(x)\, dx \leq \int_{\mathcal{X}} p^2(x)\, dx - \int_{\mathcal{X}} p(x)\, dx \\ &= I(p) - 1, \end{aligned}$$

which yields relation (5.8.15). ∎

**Corollary 5.8.3** *In the case of a discrete distribution; the entropy is always non-negative.*

*Proof:* Let $p$ be a discrete probability distribution. From Proposition 5.1.2 we have $I(p) \leq 1$. Using (5.8.15), we get

$$H(p) \geq 1 - I(p) \geq 0.$$

∎

The next consequence states that the entropy is positive if the informational energy is small.

**Corollary 5.8.4** *If $p$ is a continuous distribution with an informational energy $I(p) \in (0, 1)$, then $H(p) > 0$.*

Next we shall verify relation (5.8.15) in two particular cases.

**Example 5.8.1** Consider the case of the exponential distribution $p(x, \xi) = \xi e^{-\xi x}$, with $x \in \mathcal{X} = (0, \infty)$ and $\xi \in \mathbb{E} = (0, \infty)$. By Examples 3.2.3 and 5.1.6, the entropy and the informational energy are, respectively, given by

$$H = 1 - \ln \xi, \; I = \frac{\xi}{2}.$$

Making use of the well-known inequality $\frac{\ln x}{x} \leq \frac{1}{e}$ for $x > 0$, and using $e > 2$, yields

$$\ln x \leq \frac{x}{e} < \frac{x}{2}.$$

This implies

$$H + I = 1 - \ln \xi + \frac{\xi}{2} \geq 1.$$

**Example 5.8.2** Consider the case of the normal distribution with mean $\mu$ and variance $\sigma^2$. From Examples 3.2.1 and 5.1.5 the entropy and the informational energy are

$$H = \ln(\sigma\sqrt{2\pi e}), \qquad I = \frac{1}{2\sigma\sqrt{\pi}}.$$

Denoting by $u = \dfrac{1}{2\sigma\sqrt{\pi}}$, we have

$$\begin{aligned} H + I &= \ln\Big(\frac{1}{u}\sqrt{\frac{e}{2}}\Big) + u \\ &= u - \ln u + \frac{1}{2}\ln\frac{e}{2} > 1 + \frac{1}{2}\ln\frac{e}{2} > 1, \end{aligned}$$

since $e > 2$ and $u \geq 1 + \ln u$ for $u > 0$.

The next concept is an analog of the kinetic energy. Define the *adjusted informational energy* of a distribution by

$$J(p) = \frac{1}{2}I(p) = \frac{1}{2}\int_{\mathcal{X}} p^2(x)\,dx.$$

The next result shows that the sum between the entropy and the adjusted informational energy has a positive lower bound.

**Proposition 5.8.5** *For any distribution $p$, we find*

$$H(p) + J(p) \geq 1 - \ln 2.$$

*Proof:* Using that the function $f(u) = \frac{1}{2}u - \ln u$ has a minimum at $u = 2$ and the minimum value is $f(2) = 1 - \ln 2$, we have the estimation

$$\begin{aligned} H(p) + J(p) &= -\int p(x)\ln p(x)\,dx + \frac{1}{2}\int p^2(x)\,dx \\ &= \int p(x)\big(\frac{1}{2}p(x) - \ln p(x)\big)\,dx \\ &\geq (1 - \ln 2)\int p(x)\,dx = 1 - \ln 2. \end{aligned}$$

■

**Theorem 5.8.6** *In the case of a discrete distribution, we have*

$$e^{-H(p_1,\dots,p_n)} \leq I(p_1,\dots,p_n).$$

*Proof:* If in the Jensen's inequality, with $f$ convex,

$$f\Big(\sum_{i=1}^{n} p_i x_i\Big) \leq \sum_{i=1}^{n} p_i f(x_i),$$

we let $f(u) = e^u$, $x_i = \ln p_i$, then we have

$$e^{-H} = e^{\Sigma p_i \ln p_i} < \sum_i p_i e^{x_i} = \sum_i p_i^2 = I.$$

■

**Remark 5.8.7** Theorem 5.8.2 provides a lower bound for the entropy in terms of informational energy as

$$H(p_1, \ldots, p_n) > 1 - I(p_1, \ldots, p_n) > 0.$$

However, Theorem 5.8.6 provides the better lower bound

$$H(p_1, \ldots, p_n) > -\ln I(p_1, \ldots, p_n) > 1 - I(p_1, \ldots, p_n) > 0.$$

## 5.9 A Functional Extremum

We end this chapter with a result regarding the minimum of an integral action depending on a probability density. The next result generalizes several inequalities regarding the entropy and informational energy.

**Proposition 5.9.1** *Let $F$ be a smooth convex function ($F'' > 0$) on the interval $[a, b]$. Then the functional*

$$p \longmapsto \int_a^b F\big(p(x)\big)\, dx$$

*has a minimum that is reached only when $p$ is the uniform distribution on $[a, b]$, and this minimum is equal to $(b-a)F\Big(\frac{1}{b-a}\Big)$.*

*Proof:* Since $F$ is convex, Jensen's integral inequality can be written as

$$F\Big(\frac{1}{b-a}\int_a^b p(x)\, dx\Big) \leq \frac{1}{b-a}\int_a^b F\big(p(x)\big)\, dx,$$

which after using that $\int_a^b p(x)dx = 1$ becomes

$$(b-a)F\Big(\frac{1}{b-a}\Big) \leq \int_a^b F\big(p(x)\big)\, dx.$$

The expression in the left side does not depend on the distribution $p(x)$. Since the Jensen inequality becomes identity only in the case when $p(x) =$ constant, it follows that the right side of the aforementioned inequality reaches its minimum for the uniform distribution on $[a, b]$.

*Variant of proof:* We can also show the above result by using the method of Lagrange multipliers. Adding the constraint $\int_a^b p(x)\,dx - 1 = 0$, the variational problem will apply now to the functional

$$p \longmapsto \int_a^b F\big(x, p(x)\big)\,dx - \lambda\Big(\int_a^b p(x)\,dx - 1\Big) = \int_a^b \mathcal{L}\big(p(x)\big)\,dx,$$

where $\lambda$ is a Lagrange multiplier and

$$\mathcal{L}\big(p(x)\big) = F\big(p(x)\big) - \lambda\Big(p(x) - \frac{1}{b-a}\Big)$$

is the Lagrangian. In order to find the critical points of the aforementioned functional, we consider a smooth variation $p_\epsilon$ of the function $p$, with $p_0 = p$. Differentiating with respect to $\epsilon$ and equating to 0, yields

$$\begin{aligned} 0 &= \frac{d}{d\epsilon}\int_a^b \mathcal{L}\big(p(x)\big)\,dx\Big|_{\epsilon=0} = \int_a^b \frac{d\mathcal{L}}{dp}\frac{dp_\epsilon}{d\epsilon}(x)\,dx\Big|_{\epsilon=0} \\ &= \int_a^b \frac{d\mathcal{L}}{dp}(p(x))\,\eta(x)\,dx, \end{aligned}$$

for any variation $\eta(x) = \dfrac{dp_\epsilon(x)}{d\epsilon}\Big|_{\epsilon=0}$. Hence $p$ satisfies the Euler–Lagrange equation

$$\frac{d\mathcal{L}}{dp}(p(x)) = 0,$$

which becomes

$$F'(p) = \lambda.$$

Since $G(u) = F'(u)$ is increasing, the equation $G(p) = \lambda$ has the unique solution $p(x) = G^{-1}(\lambda)$, which is a constant. Using the constraint $\displaystyle\int_a^b p(x)dx = 1$, it follows that $p(x) = \dfrac{1}{b-a}$, which is the uniform distribution. The minimality follows from the sign of the second derivative,

$$\frac{d^2}{dp^2}\int_a^b \mathcal{L}(p(x))\,dx = \int_a^b F''(p) > 0.$$

■

**Corollary 5.9.2** *The informational energy functional, on smooth distributions over* $[a, b]$*, satisfies*

$$\frac{1}{b-a} \le I(p),$$

*with the equality reached for the uniform distribution.*

*Proof:* Choose $F(u) = u^2$ and apply Proposition 5.9.1. ■

**Corollary 5.9.3** *The entropy of a smooth distribution on $[a, b]$ satisfies*

$$H(p) \leq \ln(b - a),$$

*with the identity reached for the uniform distribution.*

*Proof:* Choose $F(u) = u \ln u$. Since $F''(u) = 1/u > 0$ on $(0, \infty)$, the function $F$ is convex. Applying Proposition 5.9.1, we have

$$\begin{aligned} (b-a)F\Big(\frac{1}{b-a}\Big) &\leq \int_a^b p(x) \ln p(x)\, dx \Longleftrightarrow \\ \ln \frac{1}{b-a} &\leq -H(p), \end{aligned}$$

which is equivalent with the desired inequality. ■

## 5.10 Problems

**5.1.** Show that the informational energy is convex, i.e., for any two densities $p, q : \mathcal{X} \to \mathbb{R}$ we have

$$I(\alpha p + \beta q) \leq \alpha I(p) + \beta I(q),$$

$\forall \alpha, \beta \in [0, 1]$, with $\alpha + \beta = 1$.

**5.2.** Find the Onicescu correlation coefficient $\mathcal{R}(p_\xi, p_\theta)$, where $p_\xi(x) = \xi e^{-\xi x}$ and $p_\theta(x) = \theta e^{-\theta x}$, $x \geq 0$, $\xi, \theta > 0$ are two exponential distributions.

**5.3.** Consider two discrete distributions $p, q : \{x_1, x_2\} \to [0, 1]$, with $p(x_1) = p(x_2) = 1/2$, $q(x_1) = \xi^1$, and $q(x_2) = \xi^2$, with $\xi^1 + \xi^2 = 1$, $\xi^i \geq 0$. Show that the Onicescu correlation coefficient is $\mathcal{R}(p, q) = \dfrac{1}{\sqrt{2((\xi^1)^2 + (\xi^2)^2)}}$.

**5.4.** Show that the density

$$p(x) = \begin{cases} \frac{1}{2\sqrt{x}}, & \text{if } 0 < x < 1 \\ 0, & \text{otherwise.} \end{cases}$$

is discontinuous and has an infinite informational energy.

**5.5.** Show that the following infinite mixture of Gaussians

$$p(x) = \frac{6}{\pi^2} \sum_{k=1}^{\infty} \frac{1}{k^2} \cdot \frac{k^3}{\sqrt{2\pi}} exp\Big( - \frac{(x-k)^2 k^6}{2} \Big)$$

is a continuous density satisfying $p(k) \geq \frac{6k}{\sqrt{2\pi}\pi^2}$.

**5.6.** Construct two distributions such that:

(a) they have the same entropy and distinct informational energies.

(b) they have the same informational energy and distinct entropies.

**5.7.** Can you find two distinct distributions with the same entropy and informational energy?

**5.8.** Show the following inequality between the cross entropy, $S(p,q)$, of two densities and their informational energies

$$S(p,q) \geq 1 - I(p)^{1/2} I(q)^{1/2}.$$

# Chapter 6

# Maximum Entropy Distributions

This chapter is dedicated to the study of entropy maximization under moment constraints. We present results of entropy maximization under constraints of mean, variance, or any $N$ moments. The solution of these variational problems belongs to the exponential family. However, explicit solutions exist only in a few particular cases. A distinguished role is played by the study of the Maxwell–Boltzmann distribution.

## 6.1 Moment Constraints

Sometimes just looking at data we can infer the underlying distribution. This may occur in simple distributions cases, like normal, exponential, lognormal, etc. However, if the underlying distribution is complicated, we cannot guess it, especially if it is not one of the usual distributions. In this case we need to approximate the distribution by one of the well-known distributions, which matches well the data and is the most unbiased. This problem can be formalized as in the following:

*Given the data $x^1, x^2, \dots, x^N$, which is the most natural distribution that fits the data?*

O. Calin and C. Udrişte, *Geometric Modeling in Probability and Statistics*,
DOI 10.1007/978-3-319-07779-6_6,

If $\hat{m}_1, \hat{m}_2, \ldots, \hat{m}_k$ are the estimations of the first $k$ moments from data, we shall consider the distribution that matches the first $k$ moments to the data

$$m_1 = \hat{m}_1, \quad m_2 = \hat{m}_2, \ldots, m_k = \hat{m}_k,$$

and has a maximum ignorance for the rest of the moments. We shall assume that the maximum ignorance distribution is the one with the maximum entropy. The problem will be treated under its both existence and uniqueness aspects, which usually depend on the sample space structure.

We shall start the study with the general case of matching $N$ moments. The sample space $\mathcal{X}$ is considered to be either a finite or an infinite interval.

## 6.2 Matching the First $N$ Moments

This section deals with the general problem of finding the density $p$ of maximum entropy subject to the first $N$ moment constraints. Given the numbers $m_1, m_2, \ldots, m_N$, we are interested in finding the distribution $p$ that maximizes the following entropy functional with Lagrange multipliers

$$J(p) = -\int_{\mathcal{X}} p(x) \ln p(x)\, dx + \sum_{j=0}^{N} \lambda_j \Big( \int_{\mathcal{X}} x^j p(x)\, dx - m_j \Big),$$

where we choose for convenience $m_0 = 1$. Taking the functional derivative with respect to $p(x)$ and equating to zero yields

$$-\ln p(x) - 1 + \sum_{j=0}^{N} \lambda_j x^j = 0.$$

This implies that the maximum entropy distribution belongs to the following exponential family

$$p(x) = e^{-1+\lambda_0+\lambda_1 x+\lambda_2 x^2+\ldots+\lambda_N x^N} = Ce^{\lambda_1 x+\lambda_2 x^2+\ldots+\lambda_N x^N}, \qquad (6.2.1)$$

with the normalization constant given by $C = e^{-1+\lambda_0}$, and Lagrange multipliers $\lambda_j$ determined from the moment constraints. However, solving the constraint system for $\lambda_j$ is a non-trivial job for $N \geq 2$. The explicit computation of the Lagrange multipliers in the cases $N = 0, 1, 2$ will be covered in the next sections.

Even if the uniqueness is a complicated problem for general $N$, the existence can be always proved for any number of constraints. We shall deal with this problem in the following.

It suffices to show that among all distributions $q(x)$ that satisfy the moment constraints

$$\int_{\mathcal{X}} x^j q(x)\,dx = m_j, \qquad j = 0, 1, \ldots, N,$$

with $m_0 = 1$, the maximum entropy is realized for the distribution $p(x)$ given by (6.2.1). The following computation uses the non-negativity and non-degeneracy of the Kullback–Leibler relative entropy, see Proposition 4.1.2, part (i)

$$D_{KL}(q||p) \geq 0$$

with $D_{KL}(q||p) = 0$ if and only if $q = p$. Then we have

$$\begin{aligned}
H(q) &= -\int_{\mathcal{X}} q \ln q = -\int_{\mathcal{X}} q \ln\left(\frac{q}{p}p\right) = -\int_{\mathcal{X}} q \ln \frac{q}{p} - \int_{\mathcal{X}} q \ln p \\
&= -D_{KL}(q||p) - \int q \ln p \leq -\int_{\mathcal{X}} q \ln p \\
&= -\int_{\mathcal{X}} q(x)(-1 + \lambda_0 + \lambda_1 x + \ldots + \lambda_N x^N)\,dx \\
&= -(-1 + \lambda_0 + \lambda_1 m_1 + \ldots + \lambda_N m_N) \\
&= -\int_{\mathcal{X}} p(x)(-1 + \lambda_0 + \lambda_1 x + \ldots + \lambda_N x^N)\,dx \\
&= -\int_{\mathcal{X}} p \ln p = H(p).
\end{aligned}$$

Therefore, for any density $q$ satisfying the moment constraints, we have $H(q) \leq H(p)$, with equality when $q = p$. Hence the density $p$ achieves the maximum entropy. This does not exclude the existence of another maximum entropy distribution, different than $p(x)$.

The fact that the maximizing entropy distribution is unique will be proved in the next sections for the particular case $N \leq 2$. This will be done by showing the uniqueness of the Lagrange multipliers $\lambda_j$ satisfying the given constraints.

## 6.3 Case $N = 0$: Constraint-Free Distribution

Consider a random variable for which we have absolutely no information on its probability distribution. Then the best one can do in this case is to consider the underlying distribution to be the uniform distribution. The reason for this choice is that this distribution achieves the maximum entropy. The result in this case is given in the following:

**Proposition 6.3.1** *Among all distributions defined on the finite interval $(a, b)$, the one with the largest entropy is the uniform distribution.*

*Proof:* Since in this case there is only one constraint

$$\int_a^b p(x)\, dx = m_0 (= 1), \tag{6.3.2}$$

formula (6.2.1) becomes

$$p(x) = e^{-1+\lambda_0} = C,$$

constant. Using (6.3.2) yields the unique value of the Lagrange multiplier $\lambda_0 = 1 - \ln(b - a)$. Hence, the uniform distribution is the unique distribution with maximum entropy. ∎

We note that this result was proved in a different way in Chap. 4.

## 6.4 Case $N = 1$: Matching the Mean

Consider some data $x^1, x^2, \ldots, x^N$, which take values in the finite interval $(a, b)$. One can estimate the data mean by

$$m_1 = \frac{x^1 + x^2 + \ldots + x^N}{N},$$

and look for a continuous probability distribution $p : (a, b) \to [0, +\infty)$ with mean

$$\mu = \int_a^b x p(x)\, dx \tag{6.4.3}$$

such that

(1) $\mu = m_1$;

(2) the probability distribution $p$ has the maximum entropy subject to the constraint (1).

In order to solve the problem we set up a variational problem with constraints like in Sect. 6.2. The maximum entropy density function given by (6.2.1) can be written in the simple form

$$p(x) = e^{\lambda_1 x + \lambda_0 - 1}. \tag{6.4.4}$$

The constants $\lambda_1$ and $\lambda_0$ can be determined from the condition constraints

$$\begin{cases} \displaystyle\int_a^b x p(x)\, dx = m_1 \\ \\ \displaystyle\int_a^b p(x)\, dx = 1 \end{cases} \Leftrightarrow \begin{cases} \displaystyle e^{\lambda_0 - 1} \int_a^b x e^{\lambda_1 x}\, dx = m_1 \\ \\ \displaystyle e^{\lambda_0 - 1} \int_a^b e^{\lambda_1 x}\, dx = 1 \end{cases} \Leftrightarrow$$

$$\begin{cases} \displaystyle e^{\lambda_0 - 1} \left[ \frac{b e^{\lambda_1 b} - a e^{\lambda_1 a}}{\lambda_1} - \frac{e^{b\lambda_1} - e^{a\lambda_1}}{\lambda_1^2} \right] = m_1 \\ \\ \displaystyle e^{\lambda_0 - 1} \frac{e^{\lambda_1 b} - e^{\lambda_1 a}}{\lambda_1} = 1. \end{cases}$$

Dividing the equations, we eliminate $\lambda_0$,

$$\frac{b e^{\lambda_1 b} - a e^{\lambda_1 a}}{e^{\lambda_1 b} - e^{\lambda_1 a}} - \frac{1}{\lambda_1} = m_1.$$

Dividing by $e^{\lambda_1 b}$, after some algebraic manipulations we obtain the following equation satisfied by $\lambda_1$

$$a + \frac{b - a}{1 - e^{\lambda_1 (a - b)}} = m_1 + \frac{1}{\lambda_1}. \tag{6.4.5}$$

We shall prove that the Eq. (6.4.5) has a unique solution $\lambda_1$. First, consider the functions

$$f(x) = a + \frac{b - a}{1 - e^{(a-b)x}} \tag{6.4.6}$$

$$g(x) = m_1 + \frac{1}{x}, \tag{6.4.7}$$

and study the solutions of the equation $f(x) = g(x)$.

**Lemma 6.4.1** *The equation $f(x) = g(x)$ has a unique solution $x^*$.*

*(a) If $m_1 > \frac{a+b}{2}$, the solution is $x^* > 0$;*

*(b) If $m_1 < \frac{a+b}{2}$, the solution is $x^* < 0$;*

*(c) If $m_1 = \frac{a+b}{2}$, the solution is $x^* = 0$.*

*Proof:* The asymptotic behavior of $f(x)$ as $x \to 0$ is given by

$$\begin{aligned} f(x) &= \frac{1}{x} + \frac{a+b}{2} + \frac{1}{12}(b-a)^2 x + O(x^2) \\ &= g(x) + \Big(\frac{a+b}{2} - m_1\Big) + \frac{1}{12}(b-a)^2 x + O(x^2). \end{aligned}$$

$(a)$ If $m_1 > \frac{a+b}{2}$, by continuity reasons, there is an $\epsilon > 0$ such that

$$\Big(\frac{a+b}{2} - m_1\Big) + \frac{1}{12}(b-a)^2 x + O(x^2) < 0, \qquad \forall\, 0 < x < \epsilon.$$

Hence $f(x) < g(x)$ for $0 < x < \epsilon$.

On the other side, since $\lim_{x\to\infty} f(x) = b > \lim_{x\to\infty} g(x) = m_1$, we have $f(x) > g(x)$, for $x$ large. Since $f(x)$ and $g(x)$ are continuous, by the Intermediate Value Theorem, there is an $x^* > 0$ such that $f(x^*) = g(x^*)$.

$(b)$ If $m_1 < \frac{a+b}{2}$, in a similar way, there is an $\epsilon > 0$ such that

$$f(x) - g(x) = \Big(\frac{a+b}{2} - m_1\Big) + \frac{1}{12}(b-a)^2 x + O(x^2) > 0, \quad \forall\, -\epsilon < x < 0.$$

Hence $f(x) > g(x)$ for $-\epsilon < x < 0$.

Since $\lim_{x\to-\infty} f(x) = a < \lim_{x\to-\infty} g(x) = m_1$, there is an $x^* < 0$ for which $f(x^*) = g(x^*)$.

$(c)$ If $m_1 = \frac{a+b}{2}$, then

$$f(x) - g(x) = \frac{1}{12}(b-a)^2 x + O(x^2)$$

and hence $f(0) = g(0)$. ■

We conclude with the following existence and uniqueness result:

**Theorem 6.4.2** *Given $m_1 \in (a, b)$, there is only one probability distribution $p$ on $[a, b]$ with maximum entropy and having the mean equal to $m_1$.*

*If $m_1 \neq \frac{a+b}{2}$, then the distribution is an exponential distribution;*

*If $m_1 = \frac{a+b}{2}$, then the distribution is an uniform distribution.*

*Proof:* The proof follows from the fact that there are unique solutions $\lambda_1, \lambda_0$ satisfying the constraints. The first of the Lagrange multipliers is $\lambda_1 = x^*$, where $x^*$ is the unique solution of the equation $f(x) = g(x)$. The other multiplier, $\lambda_0$, depends on $\lambda_1$ in the unique way

$$e^{\lambda_0 - 1} = \frac{\lambda_1}{e^{\lambda_1 b} - e^{\lambda_1 a}}.$$

The resulting probability density $p$ is given by expression (6.4.4)

$$p(x) = e^{\lambda_1 x + \lambda_0 - 1}.$$

In the case $\lambda_1 = 0$ (i.e., for $m_1 = \frac{a+b}{2}$), the distribution becomes the uniform distribution $p(x) = \frac{a+b}{2}$.

Using that $p'(x) = \lambda_1 p(x)$, the following can be inferred about the shape of the distribution:

(a) if $m_1 < \frac{a+b}{2}$, then $x^* = \lambda_1 > 0$ and hence $p(x)$ is increasing, and skewed to the left.

(b) if $m_1 > \frac{a+b}{2}$, then $x^* = \lambda_1 < 0$ and hence $p(x)$ is decreasing, and skewed to the right.

(c) if $m_1 = \frac{a+b}{2}$, then $x^* = \lambda_1 = 0$ and hence $p(x)$ is the uniform distribution, $p(x) = \frac{a+b}{2}$. ■

All probability distributions with the same mean on a statistical manifold form a Boltzman–Gibbs submanifold, see Sect. 3.10. The previous result states the existence and uniqueness of the maximum entropy distribution on a Boltzman–Gibbs submanifold, where the state space is the compact interval $[a, b]$.

However, solving for $\lambda_1$ and $\lambda_0$ in a direct way is in general difficult, and not always possible. We shall end this section by providing a non-constructive variant of proof for Theorem 6.4.2, which is based on the Inverse Function Theorem. Setting

$$F_1(\lambda_0, \lambda_1) = \int_a^b x e^{\lambda_1 x} e^{\lambda_0 - 1}\, dx, \quad F_2(\lambda_0, \lambda_1) = \int_a^b e^{\lambda_1 x} e^{\lambda_0 - 1}\, dx,$$

we need to show that the system of constraints

$$F_1(\lambda_0, \lambda_1) = m_1, \qquad F_2(\lambda_0, \lambda_1) = 1$$

has a unique solution $(\lambda_0, \lambda_1)$. Since

$$\begin{aligned}
\frac{\partial F_1}{\partial \lambda_1} &= \int_a^b x^2 p(x)\, dx = m_2, \\
\frac{\partial F_1}{\partial \lambda_0} &= F_1(\lambda_1, \lambda_0) = m_1, \\
\frac{\partial F_2}{\partial \lambda_1} &= \int_a^b x p(x)\, dx = m_1, \\
\frac{\partial F_2}{\partial \lambda_0} &= F(\lambda_1, \lambda_0) = 1,
\end{aligned}$$

then we can evaluate the determinant of partial derivatives as in the following

$$\Delta = \begin{vmatrix} \frac{\partial F_1}{\partial \lambda_1} & \frac{\partial F_1}{\partial \lambda_0} \\ \frac{\partial F_2}{\partial \lambda_1} & \frac{\partial F_2}{\partial \lambda_0} \end{vmatrix} = \begin{vmatrix} m_2 & m_1 \\ m_1 & 1 \end{vmatrix} = m_2 - (m_1)^2 = Var(p) > 0,$$

Applying the Inverse Function Theorem, the aforementioned system of constraints has a unique solution for any value of $m_1$. Hence, there are unique smooth functions $G_1$ and $G_0$ defined on the interval $(a, b)$ such that $\lambda_1 = G_1(m_1)$ and $\lambda_0 = G_0(m_1)$.

It is worth noting that $\Delta = 0$ if and only if the variance of $p$ vanishes. This occurs when $p$ is a Dirac distribution on $(a, b)$, see Example 3.2.7.

In the following we treat the case when $a = 0$ and $b = \infty$.

**Theorem 6.4.3** *Among all distributions on $(0, \infty)$, with given positive mean $\mu$, the one with the largest entropy is the exponential distribution $p(x) = \frac{1}{\mu} e^{-\frac{x}{\mu}}$.*

*Proof:* The functional with constraints which needs to be maximized in this case is

$$\begin{aligned} J(p) &= -\int_0^\infty p(x) \ln p(x)\, dx + \lambda_1 \Big[ \int_0^\infty x p(x)\, dx - \mu \Big] \\ &\quad + \lambda_0 \Big[ \int_0^\infty p(x)\, dx - 1 \Big], \end{aligned}$$

where $\lambda_1, \lambda_0$ are Lagrange multipliers. The maximum entropy density given by (6.2.1) in this case takes the form

$$p(x) = C e^{\lambda_1 x}, \qquad C = e^{\lambda_0 - 1},$$

with the constants $C$ and $\lambda_1$ to be determined from the constraints

$$\int_0^\infty p(x)\, dx = 1, \qquad \int_0^\infty x p(x)\, dx = \mu.$$

We obtain $C = \dfrac{1}{\mu}$, $\lambda_1 = -\dfrac{1}{\mu}$ and hence the distribution $p(x)$ becomes the exponential distribution. The maximum value of the entropy is reached for the aforementioned distribution and it is equal to $1 + \ln \mu$, see Example 3.2.3. ■

## 6.5 $N = 2$: Matching Mean and Variance

Sometimes we need to match the first two moments of a distribution to a sample data and express maximum of ignorance for the higher moments. Consider the data $x^1, x^2, \ldots, x^N$ that take values in the interval $(a, b)$. The estimators for the mean and variance are given by

$$\begin{aligned} \hat{\mu} &= \frac{x^1 + x^2 + \ldots + x^N}{N} \\ \hat{\sigma}^2 &= \frac{1}{N-1} \sum_{j=1}^{N} (x^j - m_1)^2. \end{aligned}$$

We are interested in the probability distribution $p : (a, b) \to [0, +\infty)$ of maximum entropy that has the mean and variance equal to the mean and variance of the previous sample, i.e.,

$$\mu = \hat{\mu}, \qquad \sigma^2 = \hat{\sigma}^2. \tag{6.5.8}$$

This problem is equivalent with the one of finding the probability density $p(x)$ of maximum entropy which has prescribed values for the first two moments, $m_1$ and $m_2$, as described in Sect. 6.2. The maximum entropy density, which is given by (6.2.1), in this case becomes

$$p(x) = e^{\lambda_2 x^2 + \lambda_1 x + \lambda_0 - 1}, \qquad a < x < b. \tag{6.5.9}$$

The constants $\lambda_0$, $\lambda_1$, and $\lambda_2$ are the solutions of the following system of constraints

$$\int_a^b e^{\lambda_2 x^2 + \lambda_1 x + \lambda_0 - 1}\, dx = 1 \tag{6.5.10}$$

$$\int_a^b x e^{\lambda_2 x^2 + \lambda_1 x + \lambda_0 - 1}\, dx = m_1 \tag{6.5.11}$$

$$\int_a^b x^2 e^{\lambda_2 x^2 + \lambda_1 x + \lambda_0 - 1}\, dx = m_2. \tag{6.5.12}$$

We shall treat the problem in the following two distinct cases: (i) $(a, b) = (-\infty, +\infty)$ and (ii) $(a, b)$ finite.

### The Case $a = -\infty$, $b = +\infty$

In this case the integrals on the left side of the system (6.5.10)–(6.5.12) can be computed explicitly. The computation uses the following well-known evaluation of improper integrals

$$\int_{-\infty}^{\infty} e^{-\alpha x^2 + \beta x}\, dx = \sqrt{\frac{\pi}{\alpha}} e^{\frac{\beta^2}{4\alpha}} \tag{6.5.13}$$

$$\int_{-\infty}^{\infty} x e^{-\alpha x^2 + \beta x}\, dx = \sqrt{\frac{\pi}{\alpha}} \Big(\frac{\beta}{2\alpha}\Big) e^{\frac{\beta^2}{4\alpha}} \tag{6.5.14}$$

$$\int_{-\infty}^{\infty} x^2 e^{-\alpha x^2 + \beta x}\, dx = \sqrt{\frac{\pi}{\alpha}} \frac{1}{2\alpha} \Big(1 + \frac{\beta^2}{2\alpha}\Big) e^{\frac{\beta^2}{4\alpha}}, \tag{6.5.15}$$

where $\alpha > 0$. We note that if $\alpha < 0$, the integrals diverge.

Substituting $\lambda_2 = -\alpha$ and $\lambda_1 = \beta$ we obtain

$$\int_{-\infty}^{\infty} e^{\lambda_2 x^2 + \lambda_1 x + \lambda_0 - 1}\, dx = \sqrt{\frac{\pi}{-\lambda_2}} e^{\frac{\lambda_1^2}{-4\lambda_2}} e^{\lambda_0 - 1}$$

$$\int_{-\infty}^{\infty} x e^{\lambda_2 x^2 + \lambda_1 x + \lambda_0 - 1}\, dx = \sqrt{\frac{\pi}{-\lambda_2}} \Big(\frac{\lambda_1}{-2\lambda_2}\Big) e^{\frac{\lambda_1^2}{-4\lambda_2}} e^{\lambda_0 - 1}$$

$$\int_{-\infty}^{\infty} x^2 e^{\lambda_2 x^2 + \lambda_1 x + \lambda_0 - 1}\, dx = \sqrt{\frac{\pi}{-\lambda_2}} \Big(\frac{1}{-2\lambda_2}\Big) \Big(1 - \frac{\lambda_1^2}{2\lambda_2}\Big) e^{\frac{\lambda_1^2}{-4\lambda_2}} e^{\lambda_0 - 1},$$

where we supposed $\lambda_2 < 0$. Then the system of constraints (6.5.10)–(6.5.12) becomes

$$\begin{aligned} 1 &= \sqrt{\frac{-\lambda_2}{\pi}} e^{\frac{\lambda_1^2}{4\lambda_2}} e^{1-\lambda_0} && (6.5.16) \\ -\frac{\lambda_1}{2\lambda_2} &= m_1 \sqrt{\frac{-\lambda_2}{\pi}} e^{\frac{\lambda_1^2}{4\lambda_2}} e^{1-\lambda_0} && (6.5.17) \\ -\frac{1}{2\lambda_2}\Big(1 - \frac{\lambda_1^2}{2\lambda_2}\Big) &= m_2 \sqrt{\frac{-\lambda_2}{\pi}} e^{\frac{\lambda_1^2}{4\lambda_2}} e^{1-\lambda_0}. && (6.5.18) \end{aligned}$$

Dividing (6.5.16) and (6.5.17) yields the following relationship between multipliers

$$\lambda_1 = -2\lambda_2 m_1. \tag{6.5.19}$$

Dividing (6.5.18) to (6.5.17), we obtain

$$\frac{\lambda_1}{1 - \frac{\lambda_1^2}{2\lambda_2}} = \frac{m_1}{m_2},$$

and using (6.5.19) yields the following equation in $\lambda_1$

$$\frac{\lambda_1 m_1}{1 + m_1\lambda_1} = \frac{m_1^2}{m_2},$$

with the solution

$$\lambda_1 = \frac{m_1}{m_2 - m_1^2}. \tag{6.5.20}$$

Substituting back in (6.5.19), and assuming $m_1 \neq 0$, we find

$$\lambda_2 = \frac{-1}{2(m_2 - m_1^2)}. \tag{6.5.21}$$

We note that the denominator of the previous expression, being a variance, is always positive, and hence $\lambda_2 < 0$.

The multiplier $\lambda_0$ can be found from (6.5.16). Substituting the values of $\lambda_2$ and $\lambda_1$ yields

$$\begin{aligned} \lambda_0 &= \frac{\lambda_1^2}{4\lambda_2} + 1 = \frac{1}{2}\ln\Big(\frac{\pi}{-\lambda_2}\Big) \\ &= \frac{-m_1^2}{2(m_2 - m_1^2)} - 1 - \frac{1}{2}\ln\Big(2\pi(m_2 - m_1^2)\Big). \end{aligned}$$

We shall treat next the case $m_1 = 0$. Then (6.5.19) implies $\lambda_1 = 0$ and the other multipliers are given by $\lambda_2 = -\frac{1}{2m_2}$ and $\lambda_0 = 1 - \frac{1}{2}\ln(\pi m_2)$.

We note that $m_2 \neq 0$, because otherwise the variance vanishes, which is an excluded case.

Since the values of the multipliers are unique, then the exponential distribution (6.5.9) is also unique. The following result shows that this is actually a normal distribution.

**Theorem 6.5.1** *Among all distributions on $\mathbb{R}$, with given mean $\mu$ and variance $\sigma^2$, the one with the largest entropy is the normal distribution $p(x) = \dfrac{1}{\sigma\sqrt{2\pi}} e^{-\frac{(x-\mu)^2}{2\sigma^2}}$.*

*Proof:* It suffices to minimize the following action with constraints

$$
\begin{aligned}
p \;\longmapsto\; & \int_{-\infty}^{\infty} \mathcal{L}\big(p(x), x\big)\, dx \\
= \; & -\int_{-\infty}^{\infty} p(x) \ln p(x)\, dx - \gamma\Big[\int_{-\infty}^{\infty} (x-\mu)^2 p(x)\, dx - \sigma^2\Big] \\
& -\beta\Big[\int_{-\infty}^{\infty} x p(x)\, dx - \mu\Big] + \alpha\Big[\int_{-\infty}^{\infty} p(x)\, dx - 1\Big],
\end{aligned}
$$

with the Lagrange multipliers $\alpha, \beta, \gamma$. Collecting the Lagrangian, we get

$$
\mathcal{L} = -p \ln p - \gamma(x-\mu)^2 p - \beta(x-\mu)p + \alpha p + \gamma\sigma^2 - \alpha.
$$

The Euler–Lagrange equation $\dfrac{\partial \mathcal{L}}{\partial p} = 0$ can be written as

$$
\ln p = -\gamma(x-\mu)^2 - \beta(x-\mu) + c, \qquad c = \alpha - 1,
$$

and hence the distribution takes the form

$$
p(x) = e^c\, e^{-\gamma(x-\mu)^2 - \beta(x-\mu)}. \tag{6.5.22}
$$

The constants $c, \beta, \gamma$ will be determined from the constraints

$$
\int_{-\infty}^{\infty} p(x)\, dx = 1, \quad \int_{-\infty}^{\infty} x p(x)\, dx = \mu, \quad \int_{-\infty}^{\infty} (x-\mu)^2 p(x)\, dx = \sigma^2. \tag{6.5.23}
$$

The following integral formulas will be useful in the next computation. They are obtained from (6.5.13) and (6.5.15) substituting $\beta = 0$:

$$\int_{-\infty}^{\infty} e^{-ay^2}\,dy = \sqrt{\frac{\pi}{a}} \tag{6.5.24}$$

$$\int_{-\infty}^{\infty} ye^{-ay^2+by}\,dy = \sqrt{\frac{\pi}{a}}\left(\frac{b}{4a}\right)^{\frac{b^2}{4a}} \tag{6.5.25}$$

$$\int_{-\infty}^{\infty} y^2e^{-ay^2}\,dy = \sqrt{\frac{\pi}{a}}\frac{1}{2a}. \tag{6.5.26}$$

Using the second constraint of (6.5.23), formula (6.5.25), and the expression (6.5.22) yields

$$\begin{aligned} 0 &= \int_{-\infty}^{\infty} xp(x)\,dx - \mu = \int_{-\infty}^{\infty} (x-\mu)p(x)\,dx \\ &= e^c \int_{-\infty}^{\infty} (x-\mu)\,e^{-\gamma(x-\mu)^2-\beta(x-\mu)}\,dx \\ &= e^c \int_{-\infty}^{\infty} ye^{-\gamma y^2-\beta y}\,dy = e^c\sqrt{\frac{\pi}{\gamma}}\left(\frac{-\beta}{2\gamma}\right)^{\frac{\beta^2}{4\gamma}}, \end{aligned}$$

so $\beta = 0$ and substituting in (6.5.22) we get

$$p(x) = e^c e^{-\gamma(x-\mu)^2}. \tag{6.5.27}$$

Then using the first constraint of (6.5.23), formula (6.5.24), and (6.5.27), we have

$$\begin{aligned} 1 &= \int_{-\infty}^{\infty} p(x)\,dx = e^c \int e^{-\gamma(x-\mu)^2}\,dx \\ &= e^c \int_{-\infty}^{\infty} e^{-\gamma y^2}\,dy = e^c\sqrt{\frac{\pi}{\gamma}}, \end{aligned}$$

so

$$e^c = \sqrt{\frac{\gamma}{\pi}}. \tag{6.5.28}$$

The third constraint of (6.5.23), formulas (6.5.26), and (6.5.28) yield

$$\begin{aligned} \sigma^2 &= \int_{-\infty}^{\infty} (x-\mu)^2p(x)\,dx = e^c\int_{-\infty}^{\infty} (x-\mu)^2e^{-\gamma(x-\mu)^2}\,dx \\ &= e^c\int_{-\infty}^{\infty} y^2e^{-\gamma y^2}\,dy = e^c\sqrt{\frac{\pi}{\gamma}}\frac{1}{2\gamma} = \frac{1}{2\gamma}, \end{aligned}$$

and hence

$$\gamma = \frac{1}{2\sigma^2}. \tag{6.5.29}$$

Substituting in (6.5.28), we find

$$e^c = \frac{1}{\sigma\sqrt{2\pi}}. \tag{6.5.30}$$

Using the values of the constants provided by (6.5.29) and (6.5.30), the expression (6.5.27) becomes

$$p(x) = \frac{1}{\sigma\sqrt{2\pi}} e^{-\frac{(x-\mu)^2}{2\sigma^2}},$$

which is the normal distribution. The maximum value of the entropy in this case is $\ln(\sigma\sqrt{2\pi e})$, see Example 3.2.1. ■

**The Case $a < \infty$, $b < \infty$**

In this case the integrals on the left side of system (6.5.10)–(6.5.12) cannot be computed in terms of elementary functions, and hence explicit formulas for the multipliers $\lambda_i$ in terms of the momenta $m_i$ are not available. In this case we shall approach the existence and uniqueness of the solutions of system (6.5.10)–(6.5.12) from a qualitative point of view. Setting

$$\begin{aligned} F_1(\lambda_0, \lambda_1, \lambda_2) &= \int_a^b e^{\lambda_2 x^2 + \lambda_1 x + \lambda_0 - 1}\, dx \\ F_2(\lambda_0, \lambda_1, \lambda_2) &= \int_a^b x e^{\lambda_2 x^2 + \lambda_1 x + \lambda_0 - 1}\, dx \\ F_3(\lambda_0, \lambda_1, \lambda_2) &= \int_a^b x^2 e^{\lambda_2 x^2 + \lambda_1 x + \lambda_0 - 1}\, dx, \end{aligned}$$

the system (6.5.10)–(6.5.12) becomes

$$F_1(\lambda_0, \lambda_1, \lambda_2) = 1 \tag{6.5.31}$$

$$F_2(\lambda_0, \lambda_1, \lambda_2) = m_1 \tag{6.5.32}$$

$$F_3(\lambda_0, \lambda_1, \lambda_2) = m_2. \tag{6.5.33}$$

Assume the nonvanishing condition $\Delta = \det\left(\frac{\partial F_i}{\partial \lambda_j}\right) \neq 0$. Then, by the Inverse Function Theorem, the aforementioned system of constraints has a unique solution for any values of $m_1$ and $m_2$, i.e., there are unique smooth functions $G_0$, $G_1$ and $G_2$ defined on the interval $(a,b)$ such that $\lambda_j = G_j(m_1, m_2)$, $j \in \{0,1,2\}$.

Now we go back to the nonvanishing condition. A computation shows that

$$\Delta = \begin{vmatrix} m_2 & m_1 & 1 \\ m_3 & m_2 & m_1 \\ m_4 & m_3 & m_2 \end{vmatrix} = m_2^3 + m_1^2 m_4 + m_3^2 - (m_2 m_4 + 2m_1 m_2 m_3)$$

$$= m_2^3 + m_3^2 - m_4\sigma^2 - 2m_1 m_2 m_3,$$

where $m_k = \int_a^b x^k p(x)\,dx$ is the $k$-th moment and $\sigma^2 = m_2 - m_1^2$ is the variance. The condition $\Delta \neq 0$ is equivalent to

$$m_2^3 + m_3^2 \neq m_4\sigma^2 + 2m_1 m_2 m_3. \qquad (6.5.34)$$

Relation (6.5.34) is a necessary condition for the existence and uniqueness of the distribution with maximum entropy on the sample space $(a,b)$.

## 6.6 The Maxwell–Boltzmann Distribution

This section deals with a distribution describing the equilibrium state in statistical mechanics, see Rao [71]. Consider the $n$-dimensional phase space described by the coordinate system $x^1, \dots, x^n$. Consider the particles density function $p(x^1, \dots, x^n)$ as the limit ratio of number of particles in a small volume $\Delta v$ around the point $(x^1, \dots, x^n)$ to $\Delta v$.

Assume that the particle with coordinates $(x^1, \dots, x^n)$ has the potential energy $V(x^1, \dots, x^n)$. One restriction, which is imposed on the particle system, is that the average potential energy per particle to be constant, i.e.,

$$\int_{\mathcal{X}} V(x)p(x)\,dx = k, \qquad (6.6.35)$$

where we consider $dx = dx^1 \dots dx^n$, $\mathcal{X}$ a subdomain of $\mathbb{R}^n$, $V : \mathcal{X} \to \mathbb{R}_+$ a smooth function, and $k > 0$ a positive constant.

The particle system is said to be in equilibrium if, given the kinematic constraint (6.6.35), the distribution of particles is as close as possible to the uniform distribution. One way of approaching this problem is to look for the distribution $p(x)$ with the maximum entropy, which satisfies the restriction (6.6.35). The resulting probability density is given in the following result.

**Theorem 6.6.1** *Among all distributions defined on $\mathcal{X}$ with a given expected value $k$ of the potential $V$, the one with the largest entropy is the Maxwell–Boltzmann distribution $p(x) = ce^{\beta V(x)}$. The constants $c$ and $\beta$ are uniquely determined from the distribution constraints.*

*Proof:* In order to maximize the entropy $-\int_{\mathcal{X}} p(x) \ln p(x)\, dx$, with $dx = dx^1 \dots dx^n$, subject to the constraints

$$\int_{\mathcal{X}} p(x)\, dx = 1, \qquad \int_{\mathcal{X}} V(x)p(x)\, dx = k,$$

we shall consider the action

$$p \longrightarrow -\int_{\mathcal{X}} p(x) \ln p(x)\, dx + \beta\Big[\int_{\mathcal{X}} V(x)p(x)\, dx - k\Big] + \gamma\Big[\int_{\mathcal{X}} p(x)\, dx - 1\Big].$$

The Euler–Lagrange equation for the Lagrangian

$$\mathcal{L} = -p \ln p + \beta V(x)p + \gamma p$$

is $\ln p = \beta V(x) + \gamma - 1$, with the solution

$$p(x) = ce^{\beta V(x)}, \qquad c = e^{\gamma - 1}.$$

The constants $c$ and $\beta$ are determined from the following constraints

$$\int_{\mathcal{X}} p(x)\, dx = 1 \iff c = \left(\int_{\mathcal{X}} e^{\beta V(x)}\, dx\right)^{-1} \tag{6.6.36}$$

$$\int_{\mathcal{X}} V(x)p(x)\, dx = k \iff \int_{\mathcal{X}} V(x)e^{\beta V(x)}\, dx = \frac{k}{c}. \tag{6.6.37}$$

Relation (6.6.36) shows that $c$ depends uniquely on $\beta$. Substituting (6.6.36) into (6.6.37) yields the following equation in $\beta$

$$\int_{\mathcal{X}} e^{\beta V(x)}\, dx \int_{\mathcal{X}} V(x)e^{\beta V(x)}\, dx = k. \tag{6.6.38}$$

Consider

$$\Phi(\beta) = \int_{\mathcal{X}} e^{\beta V(x)}\, dx \int_{\mathcal{X}} V(x)e^{\beta V(x)}\, dx,$$

which is an increasing function of $\beta$. Since

$$\lim_{\beta\to+\infty}\Phi(\beta) = +\infty, \qquad \lim_{\beta\to-\infty}\Phi(\beta) = 0,$$

the continuity of $\Phi(\beta)$ implies that the equation (6.6.38) has a solution and this is unique. Hence there is a unique pair $(c,\beta)$ satisfying the problem constraints. Therefore, the Maxwell–Boltzmann distribution is unique. ■

We shall provide in the following another proof which does not use the variational principle. The next argument follows Rao [71]. Using the inequality between two density functions

$$D_{KL}(p||q) = \int_{\mathcal{X}} p(x)\ln\frac{p(x)}{q(x)}\,dx \geq 0,$$

we can write

$$\begin{aligned} -\int_{\mathcal{X}} p(x)\ln p(x)\,dx &\leq -\int_{\mathcal{X}} p(x)\ln q(x)\,dx \\ &= -\int_{\mathcal{X}} p(x)\big(\beta V(x)+\lambda\big)\,dx = -\big(\beta k+\lambda\big) \end{aligned}$$

where we choose $\ln q(x) = \beta V(x)+\lambda$, and we used the constraint (6.6.35). Then $-\big(\beta k+\lambda\big)$ is a fixed upper bound for the entropy $H(p)$, which is reached for the choice $p(x) = e^{\beta V(x)+\lambda} = ce^{\beta V(x)}$. The constant $c = e^{\lambda}$ and $\beta$ are determined from the following constraints

$$1 = \int_{\mathcal{X}} p(x)\,dx = \int_{\mathcal{X}} ce^{\beta V(x)}\,dx \tag{6.6.39}$$

$$k = \int_{\mathcal{X}} V(x)p(x)\,dx = \int_{\mathcal{X}} cV(x)e^{\beta V(x)}\,dx. \tag{6.6.40}$$

Let

$$F_1(c,\beta) = \int_{\mathcal{X}} ce^{\beta V(x)}\,dx, \qquad F_2(c,\beta) = \int_{\mathcal{X}} cV(x)e^{\beta V(x)}\,dx.$$

Then

$$\begin{aligned} \frac{\Delta(F_1,F_2)}{\Delta(c,\beta)} &= \begin{vmatrix} \dfrac{\partial F_1}{\partial c} & \dfrac{\partial F_1}{\partial \beta} \\ \dfrac{\partial F_2}{\partial c} & \dfrac{\partial F_2}{\partial \beta} \end{vmatrix} = \begin{vmatrix} \dfrac{1}{c} & k \\ \dfrac{k}{c} & \displaystyle\int_{\mathcal{X}} V^2 p \end{vmatrix} \\ &= \frac{1}{c}\Big[\int_{\mathcal{X}} V^2(x)p(x)\,dx - \Big(\int_{\mathcal{X}} V(x)p(x)\,dx\Big)^2\Big] \\ &= \frac{1}{c}Var\big(V(x)\big) \neq 0. \end{aligned}$$

By the Inverse Function Theorem, the system (6.6.39)–(6.6.40) has a unique solution.

Next we shall consider a few particular cases of Maxwell–Boltzmann distributions.

**Example 6.6.1** Let $\mathcal{X} = [0, \infty)$ and $V(x) = x$. In this case we recover the exponential distribution $p(x) = \frac{1}{k}e^{-\frac{x}{k}}$, see Theorem 6.4.3.

**Example 6.6.2** Let $\mathcal{X} = \mathbb{R}$ and $V(x) = x^2$. In this example $k > 0$ and $\beta < 0$. The constraints can be solved in this case as follows

$$1 = \int_{\mathbb{R}} ce^{\beta x^2}\,dx = c\sqrt{\frac{\pi}{-\beta}} \Longrightarrow c = \sqrt{\frac{-\beta}{\pi}} \tag{6.6.41}$$

$$\begin{aligned} k &= \int_{\mathbb{R}} ce^{\beta x^2} x^2\,dx = c\sqrt{\frac{\pi}{-\beta}}\Big(\frac{1}{-2\beta}\Big) \\ &= -\frac{1}{2\beta} \Longrightarrow \beta = -\frac{1}{2k}. \end{aligned} \tag{6.6.42}$$

Substituting back into (6.6.41) yields $c = \sqrt{\frac{1}{2\pi k}}$, and we obtain the distribution $p(x) = \frac{1}{\sqrt{2\pi k}}\, e^{-\frac{x^2}{2k}}$. Hence, the Maxwell–Boltzmann distribution associated with the quadratic potential $V(x) = x^2$ is the normal distribution with mean zero and variance $k$.

**Example 6.6.3** Let $\mathcal{X} = \mathbb{R}^2$ and consider the quadratic potential $V(x_1, x_2) = x_1^2 + x_2^2$, and denote $\alpha = -\beta$. Since we have

$$\begin{aligned} \iint_{\mathbb{R}^2} e^{-\alpha(x_1^2+x_2^2)}\,dx_1dx_2 &= \frac{\pi}{\alpha} \\ \iint_{\mathbb{R}^2} (x_1^2 + x_2^2)e^{-\alpha(x_1^2+x_2^2)}\,dx_1dx_2 &= \frac{\pi}{4\alpha^3}, \end{aligned}$$

substituting in the constraints yields

$$c\frac{\pi}{\alpha} = 1, \qquad c\frac{\pi}{4\alpha^3} = k.$$

Solving for $\alpha$ and $c$ we obtain

$$\alpha = \frac{1}{2\sqrt{k}}, \qquad c = \frac{1}{2\pi\sqrt{k}}.$$

Hence the Maxwell–Boltzmann distribution associated with the quadratic potential $V(x) = x_1^2 + x_2^2$ is the bivariate normal distribution

$$p(x) = \frac{1}{2\pi\sqrt{k}} e^{\frac{-(x_1^2+x_2^2)}{2\sqrt{k}}}.$$

**Example 6.6.4 (Maxwell's Distribution of Velocities)** In this example the coordinates of velocities of gas particles in three orthogonal directions are denoted by $x_1, x_2, x_3$. The sample space is $\mathcal{X} = \mathbb{R}^3$ and the quadratic potential is $V(x_1, x_2, x_3) = x_1^2 + x_2^2 + x_3^2$. The Maxwell–Boltzmann distribution is given by

$$p(x) = ce^{-(x_1^2+x_2^2+x_3^2)},$$

where the constant $c$ is obtained by integration using spherical coordinates

$$\begin{aligned}
\frac{1}{c} &= \iiint_{\mathbb{R}^3} e^{-(x_1^2+x_2^2+x_3^2)}\, dx_1 dx_2 dx_3 \\
&= \int_0^\infty \int_0^{2\pi} \int_0^\pi e^{-\rho^2} \rho^2 \sin\phi \, d\phi \, d\theta \, d\rho = 4\pi \int_0^\infty e^{-\rho^2} \rho^2 \, d\rho \\
&= 2\pi \int_0^\infty e^{-t} t^{\frac{3}{2}-1} \, dt = 2\pi \Gamma\Big(\frac{3}{2}\Big) = \pi^{3/2} \Longrightarrow c = \pi^{-3/2}.
\end{aligned}$$

Hence, the Maxwell distribution of velocities is given by

$$p(x) = \pi^{-3/2} e^{-(x_1^2+x_2^2+x_3^2)}. \tag{6.6.43}$$

In the following we shall compute the entropy of this probability density. We start with the integral

$$\begin{aligned}
J &= \iiint_{\mathbb{R}^3} e^{-(x_1^2+x_2^2+x_3^2)} (x_1^2 + x_2^2 + x_3^2)\, dx_1 dx_2 dx_3 \\
&= \int_0^\infty \int_0^{2\pi} \int_0^\pi e^{-\rho^2} \rho^2 \rho^2 \sin\phi \, d\phi \, d\theta \, d\rho = 4\pi \int_0^\infty e^{-\rho^2} \rho^4 \, d\rho \\
&= 2\pi \int_0^\infty e^{-t} t^{\frac{5}{2}-1} \, dt = 2\pi \Gamma\Big(\frac{5}{2}\Big) = 2\pi \frac{3}{2}\frac{1}{2} \Gamma\Big(\frac{1}{2}\Big) = \frac{3}{2}\pi^{3/2}.
\end{aligned}$$

The entropy of the distribution given by (6.6.43) becomes

$$\begin{aligned}
H(p) &= -\int_{\mathcal{X}} p(x) \ln p(x) \, dx = -\ln(\pi^{-3/2}) \int_{\mathcal{X}} p(x) \, dx + \pi^{-3/2} J \\
&= -\ln(\pi^{-3/2}) + \frac{3}{2} = \frac{3}{2}(\ln\pi + 1).
\end{aligned}$$

The next result deals with a variational property where the entropy is replaced by the relative entropy.

**Proposition 6.6.2** *Let $p$ be a given distribution on $\mathcal{X}$. Among all distributions on $\mathcal{X}$, with given positive mean $\mu$, the one with the smallest Kullback–Leibler relative entropy $D_{KL}(q||p)$ is the distribution given by $q(x) = p(x)e^{-ax-b}$, with the constants $a, b$ determined from the constraints $\int_{\mathcal{X}} q(x)\,dx = 1$, $\int_{\mathcal{X}} xq(x)\,dx = \mu$.*

*Proof:* Consider the following variational problem with constraints

$$q \longmapsto \int_{\mathcal{X}} q(x) \ln \frac{q(x)}{p(x)}\,dx + \lambda_1 \Big[\int_{\mathcal{X}} xq(x)\,dx - \mu\Big] + \lambda_2 \Big[\int_{\mathcal{X}} q(x)\,dx - 1\Big].$$

The associated Lagrangian is given by

$$\mathcal{L}(q) = q \ln q - q \ln p + \lambda_1 xq + \lambda_2 q,$$

and the solution of the Euler–Lagrange equation is

$$\frac{\partial L}{\partial q} = 0 \Longleftrightarrow q = pe^{-\lambda_1 x - \lambda_2 - 1}.$$

Making the substitutions $a = \lambda_1$, $b = \lambda_2 + 1$, we get the desired result.

The uniqueness of the constants $a$, $b$ follows from an application of the Inverse Function Theorem for the functions

$$F_1(a, b) = \int_{\mathcal{X}} p(x)e^{-ax-b}\,dx, \qquad F_2(a, b) = \int_{\mathcal{X}} p(x)xe^{-ax-b}\,dx,$$

and using the nonzero value of their Jacobian

$$\frac{\Delta(F_1, F_2)}{\Delta(a, b)} = \Big(\int_{\mathcal{X}} xq(x)\,dx\Big)^2 - \int_{\mathcal{X}} x^2 q(x)\,dx = -Var(q) \neq 0.$$

■

## 6.7 Problems

**6.1.** (*a*) Prove that among all one-dimensional densities $p(x)$ on $\mathbb{R}$ with the same standard deviation, the one with the maximum entropy is the Gaussian distribution.

(*b*) If $p(x_1, \ldots, x_n)$ is an $n$-dimensional density, define the second order moments by

$$A_{ij} = \int_{\mathbb{R}} \cdots \int_{\mathbb{R}} x_i x_j p(x_1, \ldots, x_n)\, dx_1 \ldots dx_n.$$

Find the density $p(x_1, \ldots, x_n)$ with the maximum entropy and given second order moments $A_{ij}$.

**6.2.** Let $D = \{(\mu, \sigma); \mu^2 + \sigma^2 < 1\}$. Verify if the normal density of probability

$$p(x; \mu, \sigma) = \frac{1}{\sigma\sqrt{2\pi}}\, e^{-\frac{(x-\mu)^2}{2\sigma^2}}, \ (\mu, \sigma) \in D$$

is an extremal of the functional

$$I(p(\cdot)) = \frac{1}{2} \int\!\!\int_D (p_\mu^2 + p_\sigma^2 - 48xyp)\, d\mu d\sigma,$$

with a suitable $\varphi(\mu, \sigma)$ satisfying the constraint

$$\mu\, \frac{\partial p}{\partial \mu} - \sigma\, \frac{\partial p}{\partial \sigma} = \varphi(\mu, \sigma)\, p.$$

**6.3.** Determine the densities of probability, localized in $[0, 1]^2$, which are extremals of the functional

$$I(z(\cdot)) = \frac{1}{2} \int\!\!\int_{[0,1]^2} (z_x^2 + z_y^2)\, dxdy,$$

constrained by $z_x z_y - z = 0$.

**6.4.** Determine the positive functions $g_{11}(x, y)$ şi $g_{22}(x, y)$ such that the density of probability $z(x, y) = A\left((x - y)^3 + e^{x+y}\right)$ to be an extremal of the functional

$$I(z(\cdot)) = \frac{1}{2} \int\!\!\int_{[0,1]^2} (g_{11}(x, y) z_x^2 + g_{22}(x, y) z_y^2)\, dxdy,$$

constrained by $\frac{\partial^2 z}{\partial x^2} - \frac{\partial^2 z}{\partial y^2} = 0$.

**6.5.** Let $\Omega = [0,1]^2$. Find the extremals of the functional

$$I(u(\cdot)) = \frac{1}{2}\int_{\Omega}(u_x^2 + u_y^2)\, dxdy,$$

knowing that $u_x u_y$ is a density of probability and $u(0,0) = 0$, $u(1,1) = 1$.

**6.6.** Let $\Omega = [0,1]^2$. Find the extremals of the functional

$$I(u(\cdot)) = \frac{1}{2}\int_{\Omega}(u_t^2 - u_x^2)\, dtdx,$$

knowing that $u_t + u_x$ is a density of probability and $u(0,0) = 0$, $u(1,1) = 1$.

**6.7.** Can you construct a density function for which relation (6.5.34) does not hold?

The total entropy along the curve $\gamma(t) = p_{\xi(t)}$, $t \in [0,1]$, is defined by

$$H_\gamma = \int_\gamma H(\xi)\, ds = \int_0^1 H(\xi(t))\sqrt{g_{ij}(\xi(t))\dot{\xi}^i(t)\dot{\xi}^j(t)}\, dt,$$

where $ds^2 = g_{ij}(\xi(t))\dot{\xi}^i(t)\dot{\xi}^j(t)$ is the square arc element.

**6.8.** Show that in the case of the exponential distribution the entropy along a curve depends on its end points only. More precisely, if consider the curve $\gamma(t) = \xi(t)e^{-\xi(t)x}$, $t \in [0,1]$, with $\xi(0) = \xi_0$, $\xi(1) = \xi_1$, show that

$$H_\gamma = \int_\gamma H(\xi)\, ds = \left|\ln\frac{\xi_1}{\xi_0}\right|\left(1 - \ln\sqrt{\xi_0\xi_1}\right),$$

which is independent of the curve $\xi(t)$.

**6.9.** Let $p : [a,b] \to [0,\infty)$ be a probability density and consider the transformation $\varphi : [a,b] \to [a,b]$, $\varphi(x) = a + b - x$. Denote $\hat{p} = p \circ \varphi$.

(a) Show that $\hat{p}$ is a probability distribution on $[a,b]$.

(b) Prove that $H(\hat{p}) = H(p)$.

(c) Let $m_1$, $\hat{m_1}$ denote the means of $p$ and $\hat{p}$, respectively. Show that

$$\hat{m_1} = a + b - m_1.$$

Note the means are symmetric with respect to the median $\frac{a+b}{2}$.

(d) Let $V$ and $\hat{V}$ denote the variances of $p$ and $\hat{p}$, respectively. Prove that $\hat{V} = V$.

# Part II

# Statistical Manifolds

# Chapter 7

# An Introduction to Manifolds

This chapter contains a brief introduction to the classical theory of differential geometry. The fundamental notions presented here deal with differentiable manifolds, tangent space, vector fields, differentiable maps, 1-forms, tensors, linear connections, Riemannian manifolds, and the Levi–Civita connection. The material of this chapter forms the basis for next chapters.

## 7.1 The Concept of Manifold

A manifold is a multidimensional geometric object that can be considered as a space which is locally similar to the Euclidean space. Since differentiation is a locally defined property, then the differentiation can be defined on a manifold in a similar way as it is defined on the Euclidean space. A point on a manifold can be described by several sets of parameters, which are regarded as local coordinate systems.

The advantage of working on a manifold is that one can consider and study those geometric concepts (functions, invariants, vector fields, tensor fields, connections, etc.) that make sense globally and can also be described quantitatively in local coordinate systems. This property initially made sense in Physics and Relativity Theory, where each coordinate system corresponds to a system of reference.

O. Calin and C. Udrişte, *Geometric Modeling in Probability and Statistics*,
DOI 10.1007/978-3-319-07779-6_7,

Therefore, the main objects of study in that case are velocity, acceleration, force, matter fields, momenta, etc., i.e., objects that remain invariant under a change of the system of reference. This means that while these objects make sense globally, they can be described quantitatively in terms of local coordinates.

The earth's surface is one of the most suggestive examples of manifolds. One is aware of this manifold only locally, where it resembles a piece of plane. A local observer situated on earth's surface can measure coordinates at any point by choosing an origin and a unit of measure, the result of this work being a local map of the region. Even if drawn at different scales, any two maps of overlapping regions are correlated, in the sense that one can "understand" their relationship. If these maps constitute an entire cartography[1] of the planet, then they form an *atlas*. Nowadays people are more familiar with the googlemaps system. The maps can be transformed by translation, contraction, or dilation, which move from one map to another, the transformation being smooth and assuring the correlation between maps. The local knowledge of the earth surface contained in an atlas forms the notion of manifold.

Consider now the system of artificial satellites rotating around the earth. Each satellite can cover a certain region of the earth surface. All satellites are supposed to cover the entire earth surface, with some overlap. The information retrieved from the satellites forms an atlas and the manifold notion emerges again.

Suppose now that a certain country is monitored by a grid of cellular phone towers, each tower servicing a specific region. This is an example that can be considered as a manifold again, each tower region being considered as a local chart. In general, the manifold notion emerges when we can describe only locally an entire global object. The word "local" in this case describes a region one can encompass with the eye, or an area which can be covered by the local cellular phone tower. "Global" describes either an entire country or continent or even the whole earth surface.

There are two distinct points of view when studying a manifold. One is the *intrinsic* point of view of a local observer, situated on the manifold, whose knowledge is bound to a local chart, or system of reference. For instance, digging a ditch or measuring the distance between a house and a nearby tree is an intrinsic activity. The other

[1] Cartography is the study and practice of making maps.

point of view is called *extrinsic*. The extrinsic knowledge is acquired by an observer while elevating himself above the manifold and looking at it from outside. The information about the earth surface obtained by a monkey climbing an eucalypt tree that grows on the earth surface, is extrinsic, while the point of view of a microorganism living on the ground is intrinsic. The notions of intrinsic and extrinsic can be described geometrically by considering either only the metric of the manifold, or taking into account also the normal vector to the manifold. There are some geometrical notions that can be described exclusively in an extrinsic way. Since the round shape of the earth was recently fully mapped by satellites, understanding the shape of the earth is an extrinsic feature. However, this should not be mistaken with curvature, which can be described intrinsically in terms of the local metric (Gauss' Egregium Theorem).

A useful tool used in describing some geometric objects on a manifold is the concept of *tensor*. Many physical quantities, such as force, velocity, acceleration, work, etc., can be described successfully as tensors. Their main feature of a tensor is that it can be described quantitatively in a local chart, and its coordinates transform by a matrix multiplication when changing charts. Therefore, if a tensor vanishes in one chart, then it vanishes in all charts. Since it turns out that the difference of two tensors is also a tensor, the last two features allow for a very powerful method of proving relations between tensors by checking them in a suitable local chart. The work in local coordinates used to prove global relations has been proved extremely useful and has been developed into the so-called tensorial formalism. For instance, if one needs to show that the tensors $T$ and $P$ are equal, it suffices to only show that their components are equal in a chart, $T_{ij} = P_{ij}$.

Many geometrical objects studied in differential geometry are tensors; however, they are called by distinct names, such as metric, vector field, 1-form, volume form, curvature, etc. All these are objects independent of the system of coordinates and can be defined globally but may be written locally in a local system of coordinates using local components. For example, a *vector field* is an object that may be written in local coordinates as $V = \sum V^i \frac{\partial}{\partial x^i}$, where $\left\{ \frac{\partial}{\partial x^i} \right\}_{i=1,\dots,n}$ is a basis of the local system of coordinates chosen. This means that its components measured in this system of reference are given by $V^1, \dots, V^n$. Similarly, a *1-form* is an object that can be written in local coordinates as $\omega = \sum \omega_i dx^i$, where $\{dx^i\}_{i=1,\dots,n}$ is a basis of the 1-forms of the local system of coordinates chosen. A *metric* is a tensor written as $g = \sum g_{ij} dx^i \otimes dx^j$, where $\otimes$ is an operation called *tensorial product.*

## 7.2 Manifold Definition

This section presents the precise definition of manifolds. All manifolds considered in this book are *real,* i.e., the local model is the Euclidean space $\mathbb{R}^n$.

The construction of a manifold starts with a *metric space* (the underlying structure of the manifold), i.e., a space on which is defined a distance function.

**Definition 7.2.1** *Let $M$ be a set of points. A distance function is a mapping $d : M \times M \to [0, \infty)$ with the following properties:*

*(i) non-degenerate: $d(x, y) = 0$ if and only if $x = y$;*

*(ii) symmetric: $d(x, y) = d(y, x)$, for all $x, y \in M$;*

*(iii) satisfies the triangle inequality: $d(x, z) \leq d(x, y) + d(y, z)$, for all $x, y, z \in M$.*

*The pair $(M, d)$ is called a metric space.*

**Example 7.2.2** Let $M = \mathbb{R}^n$ and consider $x, y \in M$, with $x = (x^1, \dots, x^n)$, $y = (y^1, \dots, y^n)$. Then the Euclidean distance is given by $d_E(x, y) = \Big[\sum_{k=1}^{n} (x^k - y^k)^2\Big]^{1/2}$. The metric space $(M, d_E)$ is called the Euclidean space.

**Example 7.2.3** The mapping $d_T : \mathbb{R}^n \times \mathbb{R}^n \to [0, \infty)$ given by $d_T(x, y) = \sum_{k=1}^{n} |x^k - y^k|$ is called the taxi-cab distance. It bears its name after the distance followed by a cab in a city with perpendicular and equidistant streets such as New York city.

**Definition 7.2.4** *Let $(M, d)$ be a metric space. Consider $x \in M$ and $r > 0$. The ball of radius $r$ and centered at $x$ is the set $B_r(x) = \{y \in M; d(x, y) < r\}$. A subset $U$ of $M$ is called open if for any $x \in U$, there is a $r > 0$ such that $B_r(x) \subset U$.*

The equivalence of the definitions of functions continuity in the framework of metric spaces is stated as in the following.

**Proposition 7.2.5** *Let $f : (M, d_M) \to (N, d_N)$ be a mapping between two metric spaces. The following are equivalent:*

(*a*) *For any open set $V$ in $N$, the pullback $f^{-1}(V) = \{x \in M; f(x) \in V\}$ is an open set in $M$.*

(*b*) *For any convergent sequence $x_n \to x$ in $M$, (i.e., $d_M(x_n, x) \to 0, n \to \infty$) we have $f(x_n) \to f(x)$ in $N$, (i.e., $d_N\big(f(x_n), f(x)\big) \to 0,\ n \to \infty$).*

A function $f : M \to N$ is called *continuous* if any of the foregoing parts ($a$) or ($b$) holds true. If $f$ is invertible and both $f$ and $f^{-1}$ are continuous, then $f$ is called a *homeomorphism* between $M$ and $N$.

**Definition 7.2.6** *Let $U \subset M$ be an open set. Then the pair $(U, \phi)$ is called a chart (coordinate system) on $M$, if $\phi : U \to \phi(U) \subset \mathbb{R}^n$ is a homeomorphism of the open set $U$ in $M$ onto an open set $\phi(U)$ of $\mathbb{R}^n$. The coordinate functions on $U$ are defined as $x^j : U \to \mathbb{R}$, and $\phi(p) = (x^1(p), \dots, x^n(p))$, namely $x^j = u^j \circ \phi$, where $u^j : \mathbb{R}^n \to \mathbb{R}$, $u^j(a_1, \dots, a_n) = a_j$ is the $j$th projection.*

The integer $n$ is the dimension of the coordinate system. Roughly speaking, the dimension is the number of coordinates needed to describe the position of a point in $M$.

**Definition 7.2.7** *An atlas $\mathcal{A}$ of dimension $n$ associated with the metric space $M$ is a collection of charts $\{(U_\alpha, \phi_\alpha)\}_\alpha$ such that*

*1) $U_\alpha \subset M$, $\forall \alpha$, $\bigcup_\alpha U_\alpha = M$ (i.e., $U_\alpha$ covers $M$),*

*2) if $U_\alpha \cap U_\beta \neq \emptyset$, the restriction to $\phi_\alpha(U_\alpha \cap U_\beta)$ of the map*

$$F_{\alpha\beta} = \phi_\beta \circ \phi_\alpha^{-1} : \phi_\alpha(U_\alpha \cap U_\beta) \to \phi_\beta(U_\alpha \cap U_\beta)$$

*is differentiable from $\mathbb{R}^n$ to $\mathbb{R}^n$ (i.e., the systems of coordinates overlap smoothly), see Fig. 7.1.*

There might be several atlases on a given metric space $M$. Two atlases $\mathcal{A}$ and $\mathcal{A}'$ are called *compatible* if their union is an atlas on $M$. The set of compatible atlases with a given atlas $\mathcal{A}$ can be partially ordered by inclusion. Its maximal element is called the *complete atlas* $\overline{\mathcal{A}}$. This

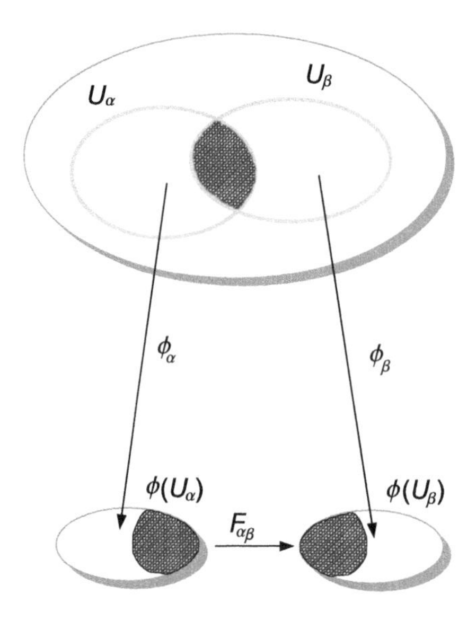


Figure 7.1: Correlated charts on a differential manifold

atlas contains all the charts that overlap smoothly with the charts of the given atlas $\mathcal{A}$. The dimension $n$ of the space $\mathbb{R}^n$, which models the manifold structure, is called the dimension of the atlas $\mathcal{A}$.

**Definition 7.2.8** *A differentiable manifold $M$ is a metric space endowed with a complete atlas. The dimension $n$ of the atlas is called the dimension of the manifold.*

We owe a remark about the completeness of an atlas. The completeness feature is required to assure for maximum chartographic information, in the sense that any considered chart is already filed in the atlas; equivalently, no new charts can be considered besides the ones that are already part of the atlas.

However, in practice it suffices to supply an arbitrary atlas (usually not the maximal one), the maximal atlas resulting from the combination of all atlases.

## 7.3 Examples of Manifolds

In this section we supply a few examples of useful manifolds.

1) The simplest differentiable manifold is the Euclidean space itself, $\mathbb{R}^n$. In this case the atlas has only one chart, the identity map, $\mathrm{Id} : \mathbb{R}^n \to \mathbb{R}^n$, $\mathrm{Id}(x) = x$.

2) Any open set $U$ of $\mathbb{R}^n$ is a differential manifold, with only one chart, $(U, \mathrm{Id})$.

3) Any non-intersecting curve $c : (a, b) \to \mathbb{R}^n$, with $\dot{c}(t) \neq 0$, is a one-dimensional manifold. In this case $M = c\big((a, b)\big)$ and the atlas consists of only one chart $(U, \phi)$, with $U = c\big((a, b)\big)$, and $\phi : U \to (a, b)$, $\phi = c^{-1}_{|U}$.

4) The sphere $\mathbb{S}^2 = \{x = (x^1, x^2, x^3) \in \mathbb{R}^3 \ ; \ (x^1)^2 + (x^2)^2 + (x^3)^2 = 1\}$ is a differentiable manifold of dimension 2. We shall supply in the following two atlases. The first atlas contains six charts, being given by $\mathcal{A} = \{U_i, \phi_i\}_{i=\overline{1,3}} \cup \{V_i, \psi_i\}_{i=\overline{1,3}}$, where

$$U_1 = \{x \ ; \ x^1 > 0\}, \quad \phi_1 : U_1 \to \mathbb{R}^2, \quad \phi_1(x) = (x^2, x^3),$$

$$V_1 = \{x \ ; \ x^1 < 0\}, \quad \psi_1 : V_1 \to \mathbb{R}^2, \quad \psi_1(x) = (x^2, x^3),$$

$$U_2 = \{x \ ; \ x^2 > 0\}, \quad \phi_2 : U_2 \to \mathbb{R}^2, \quad \phi_2(x) = (x^1, x^3),$$

$$V_2 = \{x \ ; \ x^2 < 0\}, \quad \psi_2 : V_2 \to \mathbb{R}^2, \quad \psi_2(x) = (x^1, x^3),$$

$$U_3 = \{x \ ; \ x^3 > 0\}, \quad \phi_3 : U_3 \to \mathbb{R}^2, \quad \phi_3(x) = (x^1, x^2),$$

$$V_3 = \{x \ ; \ x^3 < 0\}, \quad \psi_3 : V_3 \to \mathbb{R}^2, \quad \psi_3(x) = (x^1, x^2).$$

The second atlas is $\mathcal{A}' = \{(U, \phi_N), (V, \phi_S)\}$, where $U = \mathbb{S}^2 \backslash \{(0, 0, 1)\}$, $V = \mathbb{S}^2 \backslash \{(0, 0, -1)\}$, and the stereographic projections $\phi_N : U \to \mathbb{R}^2$, $\phi_S : V \to \mathbb{R}^2$, see Fig. 7.2, are given by

$$\phi_N(x^1, x^2, x^3) = \Big(\frac{2x^1}{1 - x^3}, \frac{2x^2}{1 - x^3}\Big),$$

$$\phi_S(x^1, x^2, x^3) = \Big(\frac{2x^1}{1 + x^3}, \frac{2x^2}{1 + x^3}\Big).$$

It can be shown as an exercise that the atlases $\mathcal{A}$ and $\mathcal{A}'$ are compatible, so they can be extended to the same complete atlas, i.e., the differential manifold structures induced by $\mathcal{A}$ and $\mathcal{A}'$ are the same.

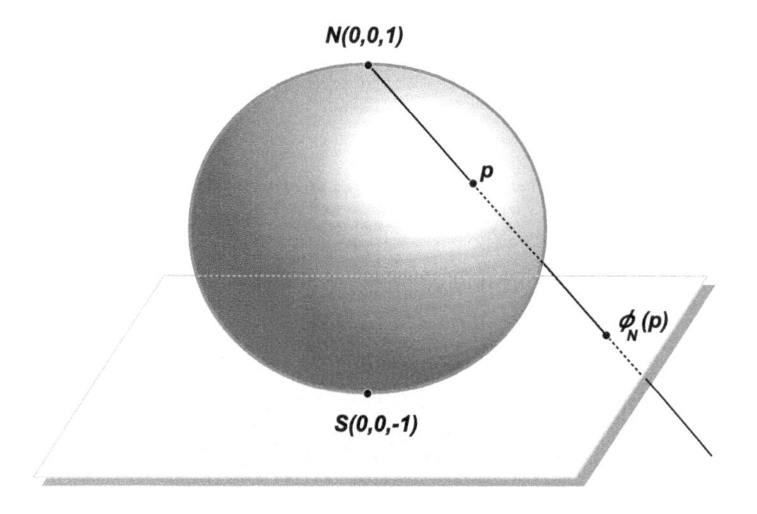


Figure 7.2: The stereographic projection from the north pole

5) Let $M = GL(n, \mathbb{R})$ be the set al all nonsingular $n \times n$ matrices. $M$ is a metric space with the metric

$$d(A, B) = \Big[\sum_{i,j}^{n}(a_{ij} - b_{ij})^2\Big]^{1/2}, \qquad \forall A, B \in M,$$

where $A = (a_{ij})$ and $B = (b_{ij})$. Then $M$ becomes a differential manifold with an atlas consisting of one chart, namely $\phi : M \to \mathbb{R}^{n^2}$,

$$\phi(A) = (a_{11}, a_{12}, \ldots, a_{1n}, a_{21}, \ldots, a_{2n}, \ldots, a_{n1}, \ldots, a_{nn}).$$

We note that $\phi(M)$ is open in $\mathbb{R}^{n^2}$. This follows from considering the continuous mapping $\rho : \mathbb{R}^{n^2} \to \mathbb{R}$ given by $\rho(a_{11}, \ldots, a_{nn}) = \det A$. Write $\phi(M) = \rho^{-1}(\mathbb{R}\backslash\{0\})$ for the pre-image of $\rho$ for all nonzero real numbers. Using Proposition 7.2.5, part $(a)$, implies $\phi(M)$ open in $\mathbb{R}^{n^2}$.

6) If $M$, $N$ are differentiable manifolds of dimensions $m$ and $n$, respectively, then $M \times N$ can be endowed with a structure of differentiable manifold, called the *product manifold.* If $\mathcal{A}_M$ and $\mathcal{A}_N$ are atlases on $M$ and $N$, respectively, then an atlas $\mathcal{A}_{M\times N}$ on $M \times N$ can be constructed by considering the charts $\big(U \times V, \Psi\big)$, with $\Psi : U \times V \to \mathbb{R}^{n+m}$, $\Psi(x, y) = \big(\phi(x), \psi(y)\big)$, where $(U, \phi) \in \mathcal{A}_M$ and $(V, \psi) \in \mathcal{A}_N$.

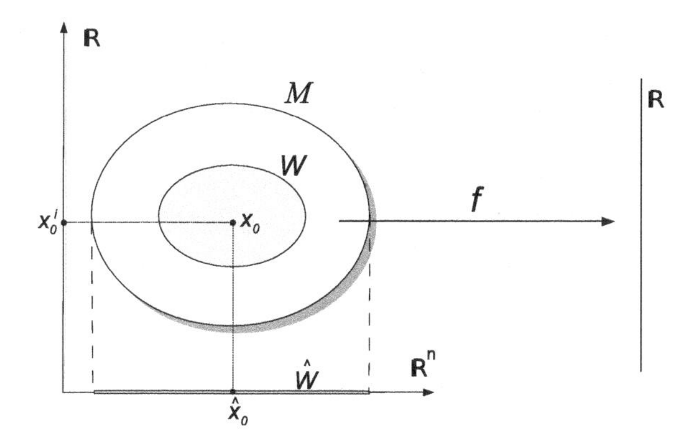


Figure 7.3: By the Implicit Functions Theorem $x^i = g(\hat{x})$ for any $x \in W$

The torus $\mathbb{T}^2 = \mathbb{S}^1 \times \mathbb{S}^1$ and the cylinder $\mathbb{S}^1 \times (0,1)$ are two usual examples of product manifolds.

7) Consider the set $M = f^{-1}(0) = \{x \in \mathbb{R}^{n+1}; f(x) = 0\}$, where $f : \mathbb{R}^{n+1} \to \mathbb{R}$ is a $C^\infty$-differentiable function (i.e., a function for which the partial derivatives exist for any order), such that

$$(grad\ f)(x) = \Big(\frac{\partial f}{\partial x^1}(x), \dots, \frac{\partial f}{\partial x^{n+1}}(x)\Big) \neq 0, \quad \forall x \in M.$$

Then $M$ is a differentiable manifold of dimension $n$, called the *hypersurface defined by* $f$.

The charts in this manifold are constructed as in the following. Consider a point $x^0 \in M$. Since $(grad\ f)(x^0) \neq 0$, there is an $i \in \{1, \dots, n+1\}$ such that $\dfrac{\partial f}{\partial x^i}(x^0) \neq 0$. By the Implicit Function Theorem, there is an open set $V$ around $x^0$ such that the equation $f(x^1, \dots, x^{n+1}) = 0$ can be solved uniquely for $x^i$ as $x^i = g(\hat{x})$, where $\hat{x} = (x^1, \dots, x^{i-1}, x^{i+1}, \dots, x^{n+1})$ and $g : \hat{V} \to \mathbb{R}$ is a differentiable function, see Fig. 7.3. Let $U = V \cap M$, and consider $\phi : I \to \mathbb{R}^n$ given by $\phi(x) = \hat{x}$. Then $(U, \phi)$ is a chart about the point $x^0$. The set of all charts of this type produces an atlas on $M$. The compatibility between these charts is left as an exercise to the reader.

This is an effective and practical way of constructing differentiable manifolds. For instance, if consider $f(x^1, \dots, x^{n+1}) = \sum_{k=1}^n (x^k)^2 - 1$, then $S^n = f^{-1}(0)$ is the $n$-dimensional sphere of radius 1.

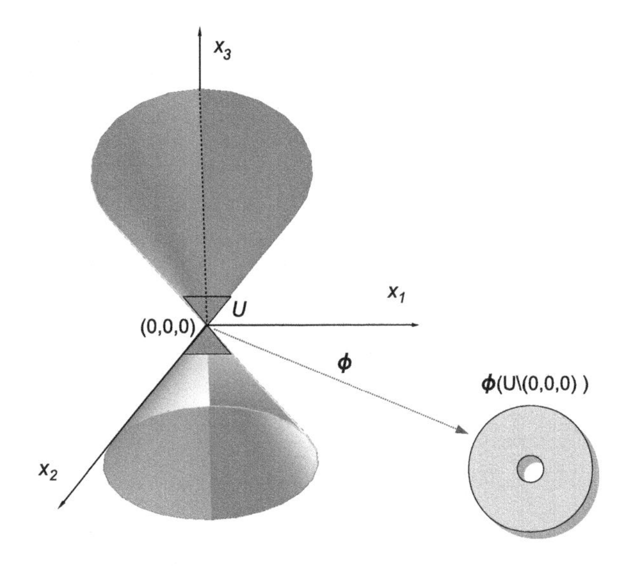


Figure 7.4: The cone is not a differentiable manifold

It is worth noting that the regularity condition $(grad\ f)(x) \neq 0$, for all $x \in M$, in general cannot be waived. For instance, if $f(x^1, x^2, x^3) = (x^1)^2 + (x^2)^2 - (x^3)^2$, then $\mathcal{C} = f^{-1}(0)$ is a cone in $\mathbb{R}^3$. We have that $(grad\ f)(x) = (2x^1, 2x^2, -2x^3)$ vanishes for $x = 0$. As a consequence, the cone $\mathcal{C} = \{(x^1)^2 + (x^2)^2 = (x^3)^2\}$ is not necessarily differentiable manifold. We investigate this by considering a chart $(U, \phi)$ around the origin $(0,0,0)$. Then $V = U\backslash(0,0,0)$ has two connected components, while $\phi(V) = \phi(U)\backslash\phi(0,0,0)$ has only one component, fact that leads to a contradiction, see Fig. 7.4. Hence the cone $\mathcal{C}$ is not a differentiable manifold.

## 7.4 Tangent Space

Before defining the concept of tangent vector, we need to introduce the notion of differentiable function on a manifold. We assume well-known from Calculus the concept of a differentiable function on $\mathbb{R}^n$. Since the differentiability has a local character, in the case of differentiable manifolds the function is required to be differentiable in a local chart.

**Definition 7.4.1** *A function $f : M \to \mathbb{R}$ is said to be differentiable if for any chart $(U, \phi)$ on $M$ the function $f \circ \phi^{-1} : \phi(U) \to \mathbb{R}$ is* *differentiable. The set of all differentiable functions on the manifold $M$ will be denoted by $\mathcal{F}(M)$.*

The notion of "differentiable" is not made too precise on the degree of smoothness. It can mean $C^\infty$ or just $C^k$-differentiable, for some $k \geq 1$, which depends on the nature of the problem.

Since a vector on $\mathbb{R}^n$ at a point can serve as a directional derivative of functions in $\mathcal{F}(\mathbb{R}^n)$, a similar idea can be used when defining the tangent vector on a manifold.

**Definition 7.4.2** *A tangent vector of $M$ at a point $p \in M$ is a function $X_p : \mathcal{F}(M) \to \mathbb{R}$ such that*

*i) $X_p$ is $\mathbb{R}$-linear*

$$X_p(af + bg) = aX_p(f) + bX_p(g), \forall a, b \in \mathbb{R}, \forall f, g \in \mathcal{F}(M);$$

*ii) the Leibniz rule is satisfied*

$$X_p(fg) = X_p(f)g(p) + f(p)X_p(g), \quad \forall f, g \in \mathcal{F}(M). \qquad (7.4.1)$$

**Definition 7.4.3** *Consider a differentiable curve $\gamma : (-\epsilon, \epsilon) \to M$ on the manifold $M$, with $\gamma(0) = p$. The tangent vector*

$$X_p(f) = \frac{d(f \circ \gamma)}{dt}(0), \qquad \forall f \in \mathcal{F}(M) \qquad (7.4.2)$$

*is called the tangent vector to $\gamma(-\epsilon, \epsilon)$ at $p = \gamma(0)$ and is denoted by $\dot{\gamma}(0)$.*

We note that the derivative in formula (7.4.2) is the usual derivative of the real-valued function $f \circ \gamma : (-\epsilon, \epsilon) \to \mathbb{R}$. Also, $X_p$ satisfies the conditions from the definition of the tangent vector. Condition *i)* follows from the linearity of the derivative $d/dt$, while condition *ii)* is an application of the product rule. Sometimes, the vector $\dot{\gamma}(0)$ is called the *velocity vector* of $\gamma$ at $p$.

Now consider the particular case of the $i$th *coordinate curve* $\gamma$. This means there is a chart $(U, \phi)$ around $p = \gamma(0)$ in which $\phi\big(\gamma(t)\big) = (x_0^1, \dots, x^i, \dots, x_0^n)$, where $\phi(p) = (x_0^1, \dots, x_0^i, \dots, x_0^n)$. Then the tangent vector to $\gamma$

$$\dot{\gamma}(0) = \frac{\partial}{\partial x^i}\bigg|_p$$

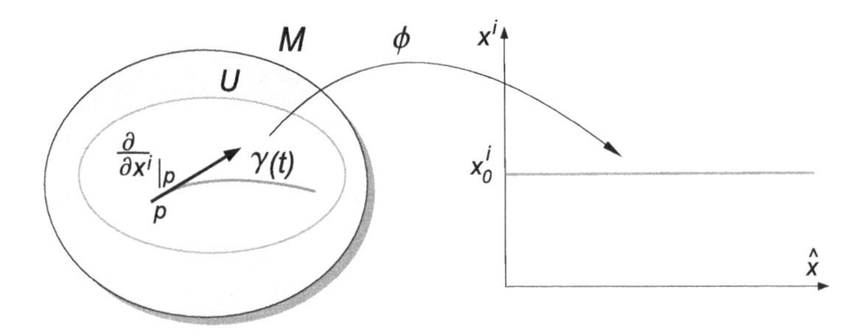


Figure 7.5: The geometric interpretation of the coordinate vector field $\dfrac{\partial}{\partial x^i}_{|p}$

is called a *coordinate tangent vector* at $p$, see Fig. 7.5. This can be defined equivalently as a derivation

$$\frac{\partial}{\partial x^i}\bigg|_p(f) = \frac{\partial(f\circ\phi^{-1})}{\partial u^i}(\phi(p)), \qquad \forall f \in \mathcal{F}(M), \tag{7.4.3}$$

where $\phi = (x^1,\dots,x^n)$ is a system of coordinates around $p$ and $u^1,\dots,u^n$ are the coordinate functions on $\mathbb{R}^n$.

**Definition 7.4.4** *The set of all tangent vectors at $p$ to $M$ is called the tangent space of $M$ at $p$, and is denoted by $T_pM$.*

$T_pM$ is a vectorial space of dimension $n$ with a basis given by the coordinate tangent vectors $\left\{\dfrac{\partial}{\partial x^1}\bigg|_p,\dots,\dfrac{\partial}{\partial x^n}\bigg|_p\right\}$. For a detailed proof of this fact the reader can consult, for instance, Millman and Parker [58]. The tangent space $T_pM$ can be also visualized geometrically as the set of velocities at $p$ along all curves passing through this point.

Using the aforementioned basis any vector $V \in T_pM$ can be written locally as $V = \sum_i V^i \frac{\partial}{\partial x^i}\big|_p$, where $V^i = V(x^i) \in \mathbb{R}$ are called the components of $V$ with respect to the system of coordinates $(x^1,\dots,x^n)$.

It is worth noting that if the vector $V$ is written with respect to a new system of coordinates $(\bar{x}^1,\dots,\bar{x}^n)$ as $V = \sum_i \bar{V}^i \frac{\partial}{\partial \bar{x}^i}\big|_p$, then the components in the two coordinates systems are related by

$$\bar{V}^k = \sum_{i=1}^n \frac{\partial \bar{x}^k}{\partial x^i} V^i. \tag{7.4.4}$$

It is also worthy to note that the change of coordinates matrix $\left(\frac{\partial \bar{x}^k}{\partial x^i}\right)_{i,k}$ is nonsingular, fact implied by the nonvanishing Jacobian[2] of a diffeomorphism, as stated by the Inverse Function Theorem.

The tangent vector $X_p$ acts on differentiable functions $f$ on $M$ as

$$X_p f = \sum_{i=1}^{n} X^i(p) \frac{\partial f}{\partial x^i}_{|p}.$$

**Definition 7.4.5** *A vector field $X$ on $M$ is a smooth map $X$ that assigns to each point $p \in M$ a vector $X_p$ in $T_pM$. For any function $f \in \mathcal{F}(M)$ we define the real-valued function $(Xf)_p = X_p f$. By "smooth" we mean the following: for each $f \in \mathcal{F}(M)$ then $Xf \in \mathcal{F}(M)$.*

Vector fields can be visualized as fields of forces on velocities for ocean currents, air currents, or convection currents, or river flows. They are important geometric objects used to model the dynamics on a manifold.
The set of all vector fields on $M$ will be denoted by $\mathcal{X}(M)$. In a local system of coordinates a vector field is given by $X = \sum X^i \frac{\partial}{\partial x^i}$, where the components $X^i \in \mathcal{F}(M)$ because they are given by $X^i = X(x^i)$, $1 \leq i \leq n$, where $x^i$ is the $i$th coordinate function of the chart.

We show next that to each vector field we can associate a family of non-intersecting curves. Given a vector field $X$, consider the ordinary differential equations system

$$\frac{dc^k}{dt}(t) = X^k_{c(t)}, \qquad 1 \leq k \leq n. \tag{7.4.5}$$

Standard theorems of existence and uniqueness of ODEs imply that the system (7.4.5) can be solved locally around any point $x_0 = c(0)$.

**Theorem 7.4.6** *Given $x_0 \in M$ and a nonzero vector field $X$ on an open set $U \subset M$, then there is an $\epsilon > 0$ such that the system (7.4.5) has a unique solution $c : [0, \epsilon) \to U$ satisfying $c(0) = x_0$.*

The solution $t \to c(t)$ is called the *integral curve* associated with the vector field $X$ through the point $x_0$. The integral curves play

[2] If $\phi(x) = (\phi^1(x), \ldots, \phi^n(x))$ is a function of $n$ variables $x^1, \ldots, x^n$, the Jacobian is the determinant of the matrix $\left(\frac{\partial \phi^j}{\partial x^k}\right)_{jk}$.

an important role in describing the evolution of a dynamical system modeled on the manifold. An effective description of the evolution of a dynamical system is usually done using conservation laws, i.e., relations whose value remains invariant along the integral curves of a vector field.

**Definition 7.4.7** *A function $f \in \mathcal{F}(M)$ is called a first integral of motion for the vector field $X$ if it remains constant along the integral curves of $X$, i.e,*

$$f\big(c(t)\big) = constant, \quad 0 \leq t \leq \epsilon,$$

*where $c(t)$ verifies (7.4.5).*

**Proposition 7.4.8** *Let $f \in \mathcal{F}(M)$, with $M$ differentiable manifold. Then $f$ is a first integral of motion for the vector field $X$ if and only if $X_{c(t)}(f) = 0$.*

*Proof:* Consider a local system of coordinates $(x^1, \dots, x^n)$ in which the vector field writes as $X = \sum_k X^k \frac{\partial}{\partial x^k}$. Then

$$\begin{aligned} X_{c(t)}(f) &= \sum_k X^k_{c(t)} \frac{\partial f}{\partial x^k} = \sum_k \frac{dc^k}{dt}(t) \frac{\partial f}{\partial x^k} \\ &= \frac{d}{dt} f\big(c(t)\big). \end{aligned}$$

Then $X_{c(t)}(f) = 0$ if and only if $\frac{d}{dt} f\big(c(t)\big) = 0$, which is equivalent to $f\big(c(t)\big) = \text{constant}$, $0 \leq t \leq \epsilon$, with $\epsilon$ small enough such that $c((0, \epsilon))$ is included in the initially considered chart. ■

## 7.5 Lie Bracket

This section deals with an important operation on vector fields, called the *Lie bracket*, which is given by $[\ ,\ ] : \mathcal{X}(M) \times \mathcal{X}(M) \to \mathcal{X}(M)$,

$$[X, Y]_p f = X_p(Yf) - Y_p(Xf), \quad \forall f \in \mathcal{F}(M),\ p \in M. \tag{7.5.6}$$

The Lie bracket will be used in later sections of the chapter to define the concepts of torsion and curvature of a linear connection, as well as the differential of a 1-form.

The vector fields $X$ and $Y$ *commute* if $[X, Y] = 0$. The Lie bracket $[X, Y]$, which at first sight looks to be a differential operator of second degree, turns out to be a vector field (a first order differential operator), which measures the noncommutativity between vector fields. In local coordinates, the Lie bracket takes the form (see Problem 7.3.)

$$[X,Y] = \sum_{i,j=1}^{n} \Big(\frac{\partial Y^i}{\partial x^j} X^j - \frac{\partial X^i}{\partial x^j} Y^j\Big) \frac{\partial}{\partial x^i}. \tag{7.5.7}$$

The bracket satisfies the following properties

1) $\mathbb{R}$-bilinearity:

$$\begin{aligned} [aX+bY,Z] &= a[X,Z]+b[Y,Z], \\ [Z,aX+bY] &= a[Z,X]+b[Z,Y], \qquad \forall a,b\in\mathbb{R}; \end{aligned}$$

2) Skew-symmetry:

$$[X,Y] = -[Y,X];$$

3) The cyclic sum is zero (Jacobi identity):

$$[X,[Y,Z]]+[Y,[Z,X]]+[Z,[X,Y]] = 0;$$

4) The Lie bracket is not $\mathcal{F}(M)$-linear, because $[fX,gY] \neq fg[X,Y]$. We have instead

$$[fX,gY] = fg[X,Y]+f(Xg)Y-g(Yf)X\ , \quad \forall f,g\in\mathcal{F}(M).$$

**Example 7.5.1** *Consider on $\mathbb{R}^2$ the vector fields $X = \partial_{x^1}$, $Y = x^1\partial_{x^2}$, called the Grushin vector fields. Then $[X,Y] = \partial_{x^2} \neq 0$, and hence $X$ and $Y$ do not commute.*

## 7.6 Differentiable Maps

The concept of differentiability on a manifold is defined locally with respect to charts.

**Definition 7.6.1** *A map $F: M \to N$ between two manifolds $M$ and $N$ is differentiable about $p \in M$ if for any charts $(U,\phi)$ on $M$ about $p$ and $(V,\psi) \in N$ about $F(p)$, the map $\psi\circ F\circ\phi^{-1}$ is differentiable from $\phi(U) \subset \mathbb{R}^m$ to $\psi(V) \subset \mathbb{R}^n$, see Fig. 7.6.*

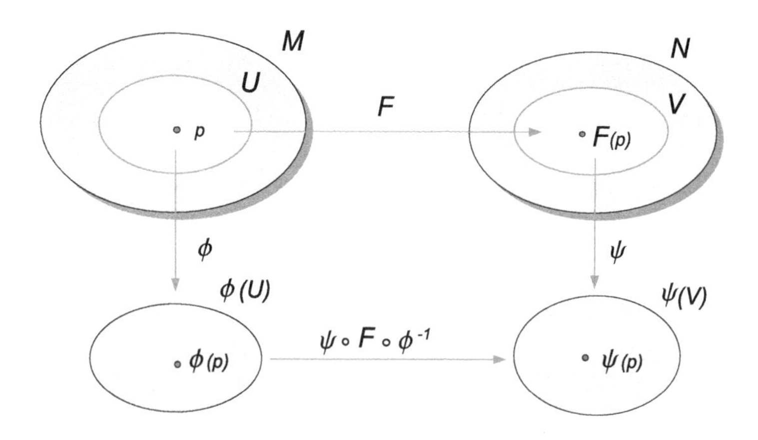


Figure 7.6: The diagram of a differentiable function

**Definition 7.6.2** *Let $F : M \to N$ be a differentiable map. For every $p \in M$, the differential map $dF$ at $p$ is defined by*

$$dF_p : T_pM \to T_{F(p)}N$$

$$(dF_p)(v)(f) = v(f \circ F)\,, \quad \forall v \in T_pM\,,\ \forall f \in \mathcal{F}(N). \tag{7.6.8}$$

The picture can be seen in Fig. 7.7. A few important properties of the differential of a map at a point, $dF_p$, are given in the following:

1) $dF_p$ is an $\mathbb{R}$-linear application between the tangent spaces $T_pM$ and $T_{F(p)}N$:

$$\begin{aligned} dF_p(v+w) &= dF_p(v) + dF_p(w), \quad \forall v, w \in T_pM; \\ dF_p(\lambda v) &= \lambda dF_p(v), \quad \forall v \in T_pM,\ \forall \lambda \in \mathbb{R}. \end{aligned}$$

2) Let $\{\frac{\partial}{\partial x^j}\big|_p\}$ and $\{\frac{\partial}{\partial y^j}\big|_{F(p)}\}$ be bases associated with the tangent spaces $T_pM$ and $T_{F(p)}N$. Consider the function $F{=}(F^1, \dots, F^n)$ and denote by $J_{kj} = \frac{\partial F^k}{\partial x^j}$ the Jacobian matrix of $F$ with respect to the charts $(x^1, \dots x^m)$ and $(y^1, \dots, y^n)$ on $M$ and $N$, respectively. Then $dF_p$ can be represented locally by

$$dF_p\Big(\frac{\partial}{\partial x^j}_{|p}\Big) = \sum_{k=1}^{n} J_{kj}(p)\frac{\partial}{\partial y^k}_{|F(p)}. \tag{7.6.9}$$

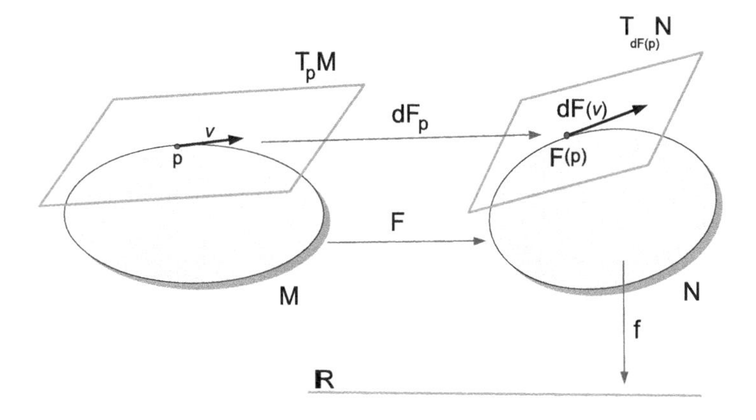


Figure 7.7: The differential of a map

3) Assume $\dim M = \dim N = n$. Then the following conditions are equivalent:

   (*i*) $dF_p : T_pM \to T_{F(p)}N$ is an isomorphism of vectorial spaces;

   (*ii*) $F$ is a local diffeomorphism in a neighborhood of $p$;

   (*iii*) There are two charts $(x^1, \dots, x^n)$ and $(y^1, \dots, y^n)$ on $M$ around $p$ and on $N$ around $F(p)$, respectively, such that the associated Jacobian is non-degenerate, i.e. $\det J_{kj}(p) \neq 0$.

   The foregoing assertion is usually called the Inverse Function Theorem on manifolds. For a proof the reader can consult the comprehensive book of Spivak [77].

4) Let $F : M \to N$ be a differentiable map. Then the differential $dF$ commutes with the Lie bracket

$$dF_p[v, w] = [dF_p(v), dF_p(w)], \qquad \forall v, w \in T_pM.$$

## 7.7 1-Forms

The differential of a function $f \in \mathcal{F}(M)$ is defined at any point $p$ by $(df)_p : T_pM \to \mathbb{R}$,

$$(df)_p(v) = v(f) \qquad \forall v \in T_pM. \qquad (7.7.10)$$

In local coordinates $(x^1, \dots, x^n)$ this takes the form $df = \sum_i \frac{\partial f}{\partial x^i} dx^i$, where $\{dx^i\}$ is the dual basis of $\{\frac{\partial}{\partial x^i}\}$ of $T_pM$, i.e.

$$dx^i\Big(\frac{\partial}{\partial x^j}\Big) = \delta^i_j,$$

where $\delta^i_j$ denotes the Kronecker symbol. The space spanned by $\{dx^1, \dots, dx^n\}$ is called the *cotangent space* of $M$ at $p$, and is denoted by $T^*_pM$. The elements of $T^*_pM$ are called *covectors*. The differential $df$ is an example of 1-form.

In general, a one form $\omega$ on the manifold $M$ is a mapping which assigns to each point $p \in M$ an element $\omega_p \in T^*_pM$. A 1-form can be written in local coordinates as

$$\omega = \sum_{i=1}^{n} \omega_i \, dx^i, \tag{7.7.11}$$

where $\omega_i = \omega(\frac{\partial}{\partial x^i})$ is the $i$th coordinate of the form with respect to the basis $\{dx^i\}$. The set of all 1-forms on the manifold $M$ will be denoted by $\mathcal{X}^*(M)$.

The interested reader can find more details about differential forms in DoCarmo [36].

## 7.8 Tensors

Let $T_pM$ and $T^*_pM$ be the tangent and the cotangent spaces of $M$ at $p$. We adopt the following useful notations

$$(T^*_pM)^r = \underbrace{T^*_pM \times \cdots \times T^*_pM}_{r \text{ times}}, \quad (T_pM)^s = \underbrace{T_pM \times \cdots \times T_pM}_{s \text{ times}}.$$

**Definition 7.8.1** *A tensor of type $(r, s)$ at $p \in M$ is an $\mathcal{F}(M)$-multilinear function $T : (T^*_pM)^r \times (T_pM)^s \to \mathbb{R}$.*
*A tensor field $\mathcal{T}$ of type $(r, s)$ is a differential map, which assigns to each point $p \in M$ an $(r, s)$−tensor $\mathcal{T}_p$ on $M$ at the point $p$.*

Since $\{dx^{j_1} \otimes \cdots \otimes dx^{j_r}\}_{j_1 < \cdots < j_r}$ and $\{\frac{\partial}{\partial x_{i_1}} \otimes \cdots \otimes \frac{\partial}{\partial x^{i_s}}\}_{i_1 < \cdots < i_s}$ are bases in the vectorial spaces $(T^*_pM)^r$ and $(T_pM)^s$, respectively, the tensor field $\mathcal{T}$ can be written using local coordinates as (with summation over repeated indices)

$$\mathcal{T} = \mathcal{T}^{i_1 i_2 \dots i_r}_{j_1 j_2 \dots j_s} \, dx^{j_1} \otimes \cdots \otimes dx^{j_s} \otimes \frac{\partial}{\partial x^{i_1}} \otimes \cdots \otimes \frac{\partial}{\partial x^{i_r}}, \tag{7.8.12}$$

where "$\otimes$" stands for the usual tensorial product. This means that $\mathcal{T}$ acts on $r$ 1-forms and $s$ vector fields as

$$\begin{aligned}
&\mathcal{T}(\omega^1,\dots,\omega^r,X_1,\dots,X_s)\\
&= \mathcal{T}^{i_1i_2\dots i_r}_{j_1j_2\dots j_s}\,dx^{j_1}(X_1)\otimes\cdots\otimes dx^{j_s}(X_s)\otimes\frac{\partial}{\partial x^{i_1}}(\omega^1)\otimes\dots\otimes\frac{\partial}{\partial x^{i_r}}(\omega^r)\\
&= \mathcal{T}^{i_1\dots i_r}_{j_1\dots j_s}X_1^{j_1}\dots X_s^{j_s}\omega^1_{i_1}\dots\omega^r_{i_r}.
\end{aligned}$$

We say the tensor $\mathcal{T}$ is $s$ covariant and $r$ contravariant. It is worth noting the following particular examples of tensors:

1. Any 1-form $\omega$ is a tensor of type $(0,1)$. For any vector field $X$

$$\omega(X)=\omega_i dx^i(X)=\omega_i dx^i(X^j\frac{\partial}{\partial x^j})=\omega_i X^i,$$

   with summation in the repeated index. In particular, the differential of a function, $df$, is a $(0,1)$-tensor.

2. Any vector field $X$ is a $(1,0)$-tensor on $M$, with

$$X(\omega)=\omega(X)=\omega_i X^i,\qquad \forall\omega.$$

3. An $s$-differentiable form is a skew-symmetric tensor of type $(0,s)$. In particular, a 2-form is a 2-covariant tensor $\Omega$ whose coordinates satisfy $\Omega_{ij}=-\Omega_{ji}$

4. A volume form on an $n$-dimensional manifold is an $n$-form, i.e., a skew-symmetric tensor of type $(0,n)$.

In order to show that $\mathcal{T}$ is a tensor, in practice we check the $\mathcal{F}(M)$-linearity in each argument. For instance, if $\mathcal{T}$ is 2-covariant, then we need to show that for any $f_1,f_2\in\mathcal{F}(M)$ and vector fields $X_1,X_2,Y_1,Y_2$ we have

$$\mathcal{T}(f_1X_1,f_2X_2)=f_1f_2\mathcal{T}(X_1,X_2)$$

$$\begin{aligned}
\mathcal{T}(X_1+Y_1,X_2)&=\mathcal{T}(X_1,X_2)+\mathcal{T}(Y_1,X_2)\\
\mathcal{T}(X_1,X_2+Y_2)&=\mathcal{T}(X_1,X_2)+\mathcal{T}(X_1,Y_2).
\end{aligned}$$

In the case of a symmetric tensor, $\mathcal{T}(X,Y)=\mathcal{T}(Y,X)$, it suffices to show the previous relations only in the first argument.

If we like to show that a tensor, or a tensorial expression vanishes, then in the virtue of the previous properties it suffices to show that it vanishes in just one system of coordinates.

## 7.9 Riemannian Manifolds

A Riemannian manifold is a manifold on which one is able to measure distances between points, angles between vectors, length of curves and volumes. Roughly speaking, it is a manifold endowed with a *metric structure.* The precise definitions are stated in the following.

**Definition 7.9.1** *A Riemannian metric $g$ on a differentiable manifold $M$ is a symmetric, positive definite 2-covariant tensor field. A Riemannian manifold is a differentiable manifold $M$ endowed with a Riemannian metric $g$.*

A Riemannian manifold will be denoted from now on by the pair $(M, g)$. The Riemannian metric $g$ can be considered as a positive definite scalar product $g_p : T_pM \times T_pM \to \mathbb{R}$ that depends differentially on the point $p \in M$. In local coordinates we write

$$g = g_{ij}\, dx^i dx^j, \tag{7.9.13}$$

with $g_{ij} = g_{ji} = g(\partial_i, \partial_j)$. The Riemannian metric $g$ acts on a pair of vector fields as $g(X,Y) = g_{ij}X^iY^j$, where we assume the summation convention over the repeated indices.

The most obvious example of Riemannian manifold is the $n$-dimensional Euclidean space $\mathbb{E}^n = (\mathbb{R}^n, \delta_{ij})$, which induces the scalar product $\langle X, Y\rangle = \sum_i X^iY^i$.

It can be proved that any differentiable manifold has a Riemannian metric structure. The idea of this construction is that a Riemannian manifold can be seen as a collection of local charts that resemble the Euclidean space $\mathbb{E}^n$. Using methods of global analysis, one can unify this local metrics into a global defined metric tensor, see, for instance, Auslander and MacKenzie [9].

A metric $g$ induces a natural bijective correspondence between 1-forms and vector fields on a Riemannian manifold $M$. If $X$ is a vector field, then one may associate with it the 1-form $\omega$ such that

$$\omega(Y) = g(Y, X), \qquad \forall Y \in \mathcal{X}(M). \tag{7.9.14}$$

In local coordinates this becomes $\omega_k = g_{jk}X^j$, where $\omega = \omega_i dx^i$ and $X = X^j \frac{\partial}{\partial x^j}$.

## 7.10 Linear Connections

A linear connection allows differentiation of a function, a vector field, or, in general, a tensor with respect to a given vector field. It can be seen as an extension of the directional derivative from the Euclidean case. The precise definition follows. Recall that $\mathcal{X}(M)$ denotes the set of vector fields on $M$.

**Definition 7.10.1** *A linear connection $\nabla$ on a differentiable manifold $M$ is a map $\nabla : \mathcal{X}(M) \times \mathcal{X}(M) \to \mathcal{X}(M)$ with the following properties:*

1) *$\nabla_X Y$ is $\mathcal{F}(M)$-linear in $X$;*

2) *$\nabla_X Y$ is $\mathbb{R}$-linear in $Y$;*

3) *it satisfies the Leibniz rule:*

$$\nabla_X(fY) = (Xf)Y + f\,\nabla_X Y, \quad \forall f \in \mathcal{F}(M).$$

For fixed vector fields $X$ and $Y$, the object $\nabla_X Y$ is also a vector field on $M$, which measures the vector rate change of $Y$ in the direction of $X$. In a local coordinates system $(x^1, \dots, x^n)$ we can write

$$\nabla_{\partial_i}\partial_j = \Gamma_{ij}^k \partial_k,$$

where $\Gamma_{ij}^k$ are the coordinates of the connection with respect to the local base $\{\partial_i\}$, where $\partial_i = \dfrac{\partial}{\partial x^i}$. If $X = X^i\partial_i$ and $Y = Y^j\partial_j$, then a straightforward computation provides the formula

$$\nabla_X Y = (\nabla_X Y)^k \partial_k,$$

where $(\nabla_X Y)^k = X^i\Big(\partial_i Y^k + Y^j\Gamma_{ij}^k\Big)$, with summation over $i$ and $j$.

An example of a linear connection on the Euclidean space $\mathbb{R}^n$ is given by $\overline{\nabla}_X Y = X(Y^j)e_j$, where $e_j = (0, \dots, 0, 1, 0, \dots, 0)$ is the $j$th basis vector on $\mathbb{R}^n$ and $Y = (Y^1, \dots, Y^n) = Y^j e_j$. The coordinates of this connection are zero, $\overline{\Gamma}_{ij}^k = 0$.

A connection can be also used to differentiate tensors. If $T$ is an $r$-covariant tensor field, we may differentiate it along a vector field $X$ with respect to the linear connection $\nabla$ as

$$(\nabla_X T)(Y_1, \dots, Y_r) = X\,T(Y_1, \dots, Y_r) - \sum_{i=1}^n T(Y_1, \dots, \nabla_X Y_i, \dots, Y_r). \tag{7.10.15}$$

In particular, we have the following concept:

**Definition 7.10.2** *Let $g$ be the Riemannian metric tensor. A linear connection $\nabla$ is called metric connection if $g$ is parallel with respect to $\nabla$, i.e.,*

$$\nabla_Z g = 0, \quad \forall\, Z \in \mathcal{X}(M). \tag{7.10.16}$$

*This can be stated equivalently as*

$$Z\, g(X,Y) = g(\nabla_Z X, Y) + g(X, \nabla_Z Y), \quad \forall\, X, Y, Z \in \mathcal{X}(M). \tag{7.10.17}$$

Let $X = \partial_i$, $Y = Y^j \partial_j$ and $Z = Z^k \partial_k$. Choosing $X = \dfrac{\partial}{\partial x^i}$, $Y = \dfrac{\partial}{\partial x^j}$, and $Z = \dfrac{\partial}{\partial x^k}$, a straightforward computation transforms (7.10.17) into

$$\partial_k g_{ij} = \Gamma^p_{ki} g_{pj} + \Gamma^r_{kj} g_{ir}. \tag{7.10.18}$$

It is worth noting that given the metric coefficients $g_{ij}$, there are $\frac{n^2(n+1)}{2}$ linear equations in $\Gamma^p_{ki}$ of type (7.10.18). The total number of unknowns $\Gamma^p_{ki}$ is $n^3$, where $n$ is the dimension of the manifold. The excess $\epsilon(n) = n^3 - \frac{n^2(n+1)}{2} = \frac{n^2(n-1)}{2}$ represents the number of arbitrary functions the family of linear connections depends on. For instance, on a curve there is only one linear connection, because $\epsilon(1) = 0$, but on a surface, the family of linear connections depends on $\epsilon(2) = 2$ arbitrary functions.

A linear connection is described by two other tensors, the *torsion* and *curvature*, which are defined shortly.

**Definition 7.10.3** *Let $\nabla$ be a linear connection. The torsion is defined as*

$$T : \mathcal{X}(M) \times \mathcal{X}(M) \to \mathcal{X}(M)$$

$$T(X,Y) = \nabla_X Y - \nabla_Y X - [X,Y]. \tag{7.10.19}$$

The torsion measures the noncommutativity of the derivation with respect to two vector fields. The last term, $[X,Y]$, is necessary because it confers tensorial properties to $T(\cdot,\cdot)$:

$$\begin{aligned} T(fX, hY) &= fhT(X,Y), \quad \forall X, Y, Z \in \mathcal{X}(M), \forall f, h \in \mathcal{F}(M) \\ T(X, Y+Z) &= T(X,Y) + T(X,Z). \end{aligned}$$

Since $T(X,Y) = -T(Y,X)$, then $T$ is a 2-covariant skew-symmetric tensor. Since in local coordinates we have

$$\begin{aligned} T_{ij} &= T(\partial_i, \partial_j) = \nabla_{\partial_i}\partial_j - \nabla_{\partial_j}\partial_i - \underbrace{[\partial_i, \partial_j]}_{=0} \\ &= \left(\Gamma_{ij}^k - \Gamma_{ji}^k\right)\partial_k, \end{aligned}$$

it follows that the torsion coordinates are given by $T_{ij}^k = \Gamma_{ij}^k - \Gamma_{ji}^k$. A connection $\nabla$ is called *torsion-free* if $T = 0$. This can be described equivalently as $\Gamma_{ij}^k = \Gamma_{ji}^k$, which is a symmetry relation for the connection coefficients. This is the reason why these type of connections are also called *symmetric*. There are exactly $\frac{n^2(n-1)}{2}$ equations of type $T_{ij}^k = T_{ji}^k$, which is exactly the excess $\epsilon(n)$. If these are considered as constraints applied to the linear system of equations (7.10.18), it follows that there is only one solution to this system. This leads to a unique linear connection, which is both symmetric and metric. We shall get in more detail regarding this issue later, when discussing the Levi–Civita connection.

**Definition 7.10.4** *The curvature of the linear connection* $\nabla$ *is given by*

$$R : \mathcal{X}(M) \times \mathcal{X}(M) \times \mathcal{X}(M) \to \mathcal{X}(M)$$

$$R(X,Y,Z) = \nabla_X\nabla_Y Z - \nabla_Y\nabla_X Z - \nabla_{[X,Y]}Z. \tag{7.10.20}$$

If we write the curvature as

$$R(X,Y,Z) = \Big([\nabla_X, \nabla_Y] - \nabla_{[X,Y]}\Big)Z,$$

it follows that $R$ is a measure of the noncommutativity of the connections with respect to $X$ and $Y$. It can be shown that $R$ satisfies the following properties

$$\begin{aligned} R(f_1X, f_2Y, f_3Z) &= f_1f_2f_3R(X,Y,Z) \\ R(X_1 + X_2, Y, Z) &= R(X_1,Y,Z) + R(X_2,Y,Z) \\ R(X, Y_1 + Y_2, Z) &= R(X,Y_1,Z) + R(X,Y_2,Z) \\ R(X,Y,Z_1 + Z_2) &= R(X,Y,Z_1) + R(X,Y,Z_2), \end{aligned}$$

for all $f_i \in \mathcal{F}(M)$ and $X_i, Y_j, Z_k \in \mathcal{X}(M)$, so that $R$ becomes a 3-covariant tensor field. The tensor $R$ is skew-symmetric in the first pair of arguments, i.e., $R(X,Y,Z) = -R(Y,X,Z)$. Since the first pair is more special, the curvature tensor is sometimes denoted by $R(X,Y)Z$. In a local system of coordinates we write

$$R\Big(\frac{\partial}{\partial x^i}, \frac{\partial}{\partial x^j}\Big)\frac{\partial}{\partial x^k} = R^p_{ijk}\frac{\partial}{\partial x^p}.$$

**Definition 7.10.5** *The Ricci curvature associated with the linear connection $\nabla$ is given by*

$$Ric : \mathcal{X}(M) \times \mathcal{X}(M) \to \mathcal{F}(M),$$

$$Ric(Y,Z) = Trace\Big(X \to R(X,Y)Z\Big).$$

This means that if $\{E_1, \dots E_n\}$ is an orthonormal set of tangent vectors at $p$, then $Ric(X,Y)_p = \sum_{j=1}^n g\big(R(E_j,X,Y),E_j\big)$. In local coordinates we write $R_{ij} = Ric(\partial_i, \partial_j)$. We can show that $R_{ij} = R^k_{ikj}$, with summation over $k$, see Problem 7.14. It is worth noting that $Ric$ is a 2-covariant tensor. It will play an important role in the study of equiaffine connections in Chap. 9.

## 7.11 Levi–Civita Connection

One of the most remarkable facts of Riemannian geometry is the existence and uniqueness of a metric connection that has zero torsion. This is called the *Levi–Civita connection* of the Riemannian manifold $(M,g)$, see, for instance, O'Neill [66]. Sometimes this is also called the *Riemannian connection* and will be denoted throughout the book by $\nabla^{(0)}$. For the purpose of this section we shall keep the notation $\nabla$.

The next theorem, also known as the fundamental lemma of Riemannian geometry, provides the Levi–Civita connection as an explicit expression in terms of the Riemannian metric $g$. This is an useful result that allows to eliminate the connection from a formula and write it in terms of the Riemannian metric only.

**Theorem 7.11.1** *On a Riemannian manifold there is a unique torsion-free, metric connection $\nabla$. Furthermore, $\nabla$ is given by the following Koszul formula*

$$\begin{aligned} 2g(\nabla_X Y, Z) &= X\,g(Y,Z) + Y\,g(X,Z) - Z\,g(X,Y) \\ &\quad + g([X,Y],Z) - g([X,Z],Y) - g([Y,Z],X). \end{aligned} \tag{7.11.21}$$

*Proof:* The proof has two parts, the existence and uniqueness.

*Existence:* We shall show that connection $\nabla$ defined by formula (7.11.21) is a metric and torsion-free connection.

First we need to show that $\nabla$ is a linear connection. Using the properties of vector fields and Lie brackets we can show by a direct computation that

$$2g(\nabla_{fX}Y, Z) = 2fg(\nabla_X Y, Z), \quad \forall Z \in \mathcal{X}(M),$$

so $\nabla_{fX}Y = f\nabla_X Y$, $\forall X, Y \in \mathcal{X}(M)$, i.e., $\nabla$ is $\mathcal{F}(M)$-linear in the first argument. Next we check the second property of connections:

$$\begin{aligned}
2g(\nabla_X(fY), Z) &= X\,g(fY, Z) + fY\,g(X, Z) - Z\,g(X, fY) \\
&\quad + g([X, fY], Z) - g([X, Z], fY) - g([fY, Z], X) \\
&= X(f)g(Y, Z) + fXg(Y, Z) + fYg(X, Z) \\
&\quad - Z(f)g(X, Y) - fZg(X, Y) \\
&\quad + fg([X, Y], Z) + X(f)g(Y, Z) - fg([X, Z], Y) \\
&\quad - fg([Y, Z], X) + Z(f)g(Y, X) \\
&= 2f\,g(\nabla_X Y, Z) + 2X(f)g(Y, Z) \\
&= 2g(f\nabla_X Y + X(f)Y, Z).
\end{aligned}$$

Dropping the $Z$-argument yields Leibniz formula. Therefore, $\nabla$ is a linear connection.

The next computation verifies that the connection is torsion-free. Using (7.11.21) yields

$$\begin{aligned}
2g(T(X, Y), Z)) &= 2g(\nabla_X Y, Z) - 2g(\nabla_Y X, Z) - 2g([X, Y], Z) \\
&= Xg(Y, Z) + Yg(X, Z) - Zg(X, Y) \\
&\quad + g([X, Y], Z) - g([X, Z], Y) - g([Y, Z], X) \\
&\quad - Yg(X, Z) - Xg(Y, Z) + Zg(Y, X) \\
&\quad - g([Y, X], Z) + g([Y, Z], X) + g([X, Z], Y) \\
&\quad - 2g([X, Y], Z) \\
&= g(2[X, Y] - 2[X, Y], Z) = 0, \qquad \forall Z \in \mathcal{X}(M).
\end{aligned}$$

Dropping the vector field $Z$ and using that $g(\cdot, \cdot)$ is non-degenerate yields $T(X, Y) = 0$, for all $X, Y \in \mathcal{X}(M)$.

Applying formula (7.11.21) twice and then cancelling in pairs, we have

$$\begin{aligned} & 2g(\nabla_Z X, Y) + 2g(X, \nabla_Z Y) \\ = \;& Z\,g(X,Y) + X\,g(Z,Y) - Y\,g(Z,X) + g([Z,X],Y) \\ & -g([Z,Y],X) - g([X,Y],Z) \\ & +Z\,g(Y,X) + Y\,g(Z,X) - X\,g(Z,Y) + g([Z,Y],X) \\ & -g([Z,X],Y) - g([Y,X],Z) \\ = \;& 2Zg(X,Y). \end{aligned}$$

Therefore $g(\nabla_Z X, Y) + g(X, \nabla_Z Y) = Zg(X,Y)$, i.e., $\nabla$ is a metric connection.

*Uniqueness:* We need to prove that any metric and symmetric connection $\nabla$ is given by formula (7.11.21). It suffices to do the verification in a local system of coordinates $(x^1, \dots, x^n)$. Let $X = \partial_i$, $Y = \partial_j$, $Z = \partial_k$. Using $\Gamma_{ij}^k = g(\nabla_{\partial_i}\partial_j, \partial_k)$ and $g_{ij} = g(\partial_i, \partial_j)$, then formula (7.11.21) becomes

$$2\Gamma_{ij}^p g_{pk} = \partial_i g_{jk} + \partial_j g_{ik} - \partial_k g_{ij}. \tag{7.11.22}$$

Writing that $\nabla$ is a metric connection in three different ways, using cyclic permutation of indices, see formula (7.10.18), we have

$$\begin{aligned} \partial_i g_{jk} &= \Gamma_{ij}^p g_{pk} + \Gamma_{ik}^r g_{jr} \\ \partial_j g_{ki} &= \Gamma_{jk}^p g_{pi} + \Gamma_{ji}^r g_{kr} \\ \partial_k g_{ij} &= \Gamma_{ki}^p g_{pj} + \Gamma_{kj}^r g_{ir}. \end{aligned}$$

Adding the first two equations and subtracting the last, using the symmetry $\Gamma_{ij}^k = \Gamma_{ji}^k$, yields exactly the Eq. (7.11.22). This ends the proof of uniqueness. ■

Solving for the connection coefficient in (7.11.22) we obtain

$$\Gamma_{ij}^p = \frac{1}{2} g^{pk}\big(\partial_i g_{jk} + \partial_j g_{ik} - \partial_k g_{ij}\big), \tag{7.11.23}$$

where $(g^{pk})$ denotes the inverse matrix of $(g_{ij})$. The coordinates $\Gamma_{ij}^p$ of the Levi–Civita connection, see (7.11.23), are called the *Christoffel symbols of second kind.* The *Christoffel symbols of first kind* are obtained lowering the indices

$$\Gamma_{ij,k} = \Gamma_{ij}^p g_{pk}.$$

Conversely, if the coordinates of a linear connection on a Riemannian manifold $(M, g)$ are given by formula (7.11.23), then the connection has to be the Levi–Civita connection.

The curvature tensor of type $(1, 3)$ associated with the Levi–Civita connection by formula (7.10.20) is called the *Riemann curvature tensor* of type $(1, 3)$. If in local coordinates we have $R(\partial_i, \partial_j)\partial_k = R^p_{ijk}\partial_p$, then the coordinate $R^P_{ijk}$ can be expressed in terms of Christoffel symbols as

$$R^r_{ijk} = \partial_i\Gamma^r_{jk} - \partial_j\Gamma^r_{ik} + \Gamma^r_{ih}\Gamma^h_{jk} - \Gamma^r_{jh}\Gamma^h_{ik}.$$

In Riemannian geometry the following $(0, 4)$-type curvature tensor is also useful

$$R : \mathcal{X}(M) \times \mathcal{X}(M) \times \mathcal{X}(M) \times \mathcal{X}(M) \to \mathcal{F}(M),$$

$$R(X, Y, Z, W) = g(R(X, Y, Z), W).$$

If in local coordinates we write $R(\partial_i, \partial_j, \partial_k, \partial_l) = R_{ijkl}$, then we have $R_{ijkl} = R^p_{ijk}g_{pl}$. The coordinates $R_{ijkl}$ satisfy several relations, the most useful being provided in the following:

1. Skew symmetry in the first and second pair:

$$R_{ijkl} = -R_{jikl} = -R_{ijlk}.$$

2. Interchange symmetry between pairs: $R_{ijkl} = R_{klij}$.

3. First Bianchi identity: $R_{ijkl} + R_{iklj} + R_{iljk} = 0$.

Another important 2-covariant tensor is the *Ricci tensor*, which is defined by the contraction

$$Ric(X, Y) = Trace\Big(V \to R(X, V, Y)\Big) = Trace\Big(V \to R(V, X, Y)\Big).$$

It can be shown that the Ricci tensor associated with the Levi–Civita connection is symmetric, $R(X, Y) = R(Y, X)$.

For more details about Calculus and Differential Geometry on differentiable manifolds the reader may consult Spivak [77, 78], and doCarmo [34, 35].

## 7.12 Problems

**7.1.** Let $p \in M$ be a point on the differentiable manifold $M$, and let $\mathcal{V}_p$ be a neighborhood of $p$. Show that there is a differentiable function $f \in \mathcal{F}(M)$ such that $f(p) = 1$ and $f(x) = 0$ if $x \notin \mathcal{V}_p$.

**7.2.** Let $p$ be a point on the differentiable manifold $M$. If $f \in \mathcal{F}(M)$ has a local extremum at $p$, then $X_p(f) = 0$, for any tangent vector $X_p$ at $p$.

**7.3.** (*a*) Let $X, Y \in \mathcal{X}(M)$ be two vector fields on the differentiable manifold $M$. Prove that the Lie bracket $[X, Y]$ is a vector field on $M$, which in local coordinates can be written as

$$[X, Y] = \sum_{i,j=1}^{n} \Big( \frac{\partial Y^i}{\partial x^j} X^j - \frac{\partial X^i}{\partial x^j} Y^j \Big) \frac{\partial}{\partial x^i},$$

where $X = \sum_i X^i \frac{\partial}{\partial x^i}$ and $Y = \sum_i Y^i \frac{\partial}{\partial x^i}$.

(*b*) Let $M = \mathbb{R}^2$ and consider the vector fields $X = x^1 x^2 \frac{\partial}{\partial x^1}$ and $Y = x^2 \frac{\partial}{\partial x^2}$. Show that $[X, Y] = -x^1 x^2 \frac{\partial}{\partial x^1}$.

**7.4.** Let $(M, g)$ be a Riemannian manifold. If $\omega = \omega_i dx^i$ is a 1-form, define the vector filed $\omega^{\#} = \omega^k \partial_{x^k}$, where $\omega^k g_{kr} = \omega_r$. Show that $g(\omega^{\#}, X) = \omega(X)$, $\forall X \in \mathcal{X}(M)$.

**7.5.** Show that the following properties of tangent vectors, $X_p \in T_pM$, hold:

(*i*) $X_p(c) = 0$, for any constant $c$;

(*ii*) $X_p(f^2) = 2fX_p(f)$, $\forall f \in \mathcal{F}(M)$;

(*iii*) If $f, g \in \mathcal{F}(M)$ such that $f(p) = g(p) = 0$, then $X_p(fg)$=0;

**7.6.** Let $M \simeq \mathbb{R}^{n^2}$ be the manifold of square $n \times n$-matrices, and $X_a(x) = a \cdot x$, $Y_a(x) = a \cdot x - x \cdot a$ be two vector fields on it, where $a, x \in M$.

(*a*) Compute the flow of the field $V_a$. Find first integrals for this flow.

(*b*) Compute the commutator $[X_a, X_b]$ for two vector fields $X_a$ and $X_b$, defined by two matrices $a, b \in M$.

(c) The same questions relative to the vector field $V_a(x)$.

**7.7.** A vector field $X$ on a manifold $M$ is called complete, if any of its trajectory can be infinitely continued forward and backward.

(a) Prove that on a compact manifold any vector field is complete.

(b) Show that on any manifold $M$ and any vector field $X$ on it, there exists a positive function $f \in C^1(M)$ such that the vector field $fX$ is complete.

**7.8.** Let $(M, g)$ be a Riemannian manifold and the corresponding volume form $\omega \in \Lambda^n(M)$. Prove that for any $2n$ vector fields $X_1, \cdots, X_n, Y_1, \cdots, Y_n$, we have

$$\omega(X_1, \cdots, X_n) \cdot \omega(Y_1, \cdots, Y_n) = \det[g(X_i, Y_j)].$$

**7.9.** (Hessian of Rosenbrok's banana function) Let us consider the Riemannian manifold $(\mathbb{R}^2, g)$, with the metric

$$g(x^1, x^2) = \begin{pmatrix} 1 + 4(x^1)^2 & -2x^1 \\ -2x^1 & 1 \end{pmatrix}.$$

Show that the Hessian of the Rosenbrok's banana function

$$f : \mathbb{R}^2 \to R,\ f(x^1, x^2) = 100(x^2 - (x^1)^2) + (1 - x^1)^2$$

is a Riemannian metric.

**7.10.** Let $f : \mathbb{R}^n \to \mathbb{R}$ be a $C^3$-function such that its Hessian $Hess(f)$ is positive definite. From the Euclidean space $(\mathbb{R}^n, \delta_{ij})$ we pass to the Riemannian manifold $(\mathbb{R}^2, Hess(f))$. Show that the equations of geodesics in this new manifold are

$$2\frac{\partial^2 f}{\partial x^i \partial x^k}(x(t))\, \ddot{x}^i(t) + \frac{\partial^3 f}{\partial x^i \partial x^j \partial x^k}(x(t))\, \dot{x}^i(t)\, \dot{x}^j(t) = 0.$$

**7.11.** (a) Find the Christoffel coefficients on the Riemannian manifold $(\mathbb{R}^2_+, g)$, where $g = \operatorname{diag}\left(\frac{1}{x^2}, \frac{1}{y^2}\right)$.

(*b*) Compute the Hessian of the function

$$f : \mathbb{R}^2_+ \to \mathbb{R}, \ f(x, y) = \frac{1}{x} + \sqrt{x} + \frac{1}{y} + \sqrt{y},$$

with respect to $g$.

(*c*) Find the geodesics of the Riemannian manifold $(\mathbb{R}^2_+, Hess_g(f))$.

**7.12.** Find the geodesics of the Riemannian manifold

$$(\mathbb{R}^2_+, g(x, y)),$$

when $g$ is a posinomial metric

$$g(x, y) = \begin{pmatrix} a_{11}x^{\alpha_{11}}y^{\beta_{11}} & a_{12}x^{\alpha_{12}}y^{\beta_{12}} \\ a_{12}x^{\alpha_{12}}y^{\beta_{12}} & a_{22}x^{\alpha_{22}}y^{\beta_{22}} \end{pmatrix}.$$

**7.13.** Let $(M, g)$ be a Riemannian manifold.

(*a*) Show that $R^r_{ijk} = -R^r_{jik}$.

(*b*) Assume $\dim M = 1$. Show that $M$ is flat, i.e., $R = 0$.

**7.14.** Let $R_{ij} = Ric(\partial_i, \partial_j)$ be the components of the Ricci tensor in local coordinates. Show that $R_{ij} = R^k_{ikj}$, with summation over $k$.

## 7.13 Historical Remarks

Differential Geometry started with the study of curves since around 1700s. Among the first mathematicians who had investigated the theory of curves were Euler, Monge, Venant, Serret, and Darboux. In 1827 Gauss published his celebrated work *Disquisitiones generales circa superficies curvas*, where he introduced the first and the second fundamental forms on surfaces in $\mathbb{R}^3$ and had shown that they characterize the surface up to a rigid motion. Gauss proved that the curvature is an intrinsic invariant of the surface, result that is called *Theorema Egregium*. The name emphasizes its profound philosophical implications, since the curvature is usually perceived as an extrinsic object.

Gauss' ideas of intrinsic geometry of a surface influenced his pupil, Riemann, who at only 28, presents his Ph.D. dissertation *Ueber die*

*Hypothesen welche der Geometrie zu Grunde liegen* at Göttingen in 1954. Riemann associated a metric with each hypersurface, fact that led to the concept of Riemannian manifold later. These results flourished into an elegant theory, which generalized Gauss' results on manifolds.

However, this theory requires laborious computations, fact that needed the construction of the tensorial formalism. Ricci developed the tensorial calculus on manifolds and Levi–Civita introduced the linear connection with the same name in 1900s.

Differential geometry has important consequences and applications. First, it closed the celebrated problem of the 5th postulate of Euclid. This was accomplished by finding examples of non-Euclidean spaces among Riemannian manifolds.

Another application is the use of differential geometry to General Theory of Relativity. Einstein's theory published in 1917 used tensorial calculus to write the equations of space-time invariantly. This way, the concept of inertial system from Newtonian mechanics is generalized and the new theory was able to explain the Mercury's perihelion advance and the light deflection about sun.

If Lorentz geometry, which is the geometry of a manifold endowed with a space-time type metric, is a good environment for relativity theory, then Riemannian geometry was proved to be suited for the Classical Mechanics, see Abraham and Marsden [1] or Calin and Chang [22]. The conservation laws of Newtonian Physics can be written in an elegant way in terms of the Riemannian Geometry language.

Another direction where Differential Geometry has recently been applied is the geometric theory of differential equations. Each differential operator is associated with a principal symbol, which can be considered as a Hamiltonian. This defines a metric on an associated manifold. The study of heat kernels and fundamental solutions can be geometrically based on the study of geodesics on the associated Riemannian manifold. The interested reader can consult this topic in Calin et al. [23] and [24]. For convex functions and optimization methods on Riemannian manifolds the reader is referred to Udriste [81].

Another related branch of Riemannian Geometry has been developed over the last several decades. It is known under the names of SubRiemannian Geometry, Non-holonomic geometry, or Carnot-Carathéodory geometry. It is related with Quantum Mechanics behavior of particles and Thermodynamics, see Calin and Chang [23].

The goal of the present book is to deal with one of the branches of Differential Geometry which applies to Information Theory, Probability and Statistics. This is known under the name of Information Geometry. Its main object of study is the statistical manifold, which is a Riemannian manifold that holds a dualistic structure and studies the relationship between dual geometric objects. All the next chapters deal with notions which culminate with the study of statistical manifolds.

# Chapter 8

# Dualistic Structure

*Statistical manifolds* are abstract generalizations of statistical models. Even if a statistical manifold is treated as a purely geometric object, however, the motivation for the definitions is inspired from statistical models. In this new framework, the manifold of density functions is replaced by an arbitrary Riemannian manifold $M$, and the Fisher information matrix is replaced by the Riemannian metric $g$ of the manifold $M$. The dual connections $\nabla^{(-1)}$ and $\nabla^{(1)}$ are replaced by a pair of dual connections $\nabla$ and $\nabla^*$. The skewness tensor, which measures the cummulants of the third order on a statistical model, is replaced by a 3-covariant skewness tensor.

There are at least three equivalent ways of defining a statistical manifold. One of them is to define a pair of dual connections $\nabla$ and $\nabla^*$ on a Riemannian manifold $(M, g)$, so the statistical manifold is represented by the quadruplet $(M, g, \nabla, \nabla^*)$. Another way is to first define a totally symmetric skewness tensor $C$ on a Riemannian manifold $(M, g)$, and then define the statistical manifold by the triplet $(M, g, C)$. In this case the dual connections are deduced from the skewness tensor $C$. A third way of introducing a statistical structure is to deduct the Riemannian metric and the conjugate connections from a given a contrast function, which is a non-symmetrical "proximity" measure between points. We shall deal with all these different approaches in the next few chapters of the book.

O. Calin and C. Udrişte, *Geometric Modeling in Probability and Statistics*,
DOI 10.1007/978-3-319-07779-6_8,

The main idea of this chapter is to study the relationship between geometric objects induced by dual connections, such as dual curvature tensors, dual Hessians, dual Laplacians, dual volume elements, dual divergences, etc. There are interesting properties emerging from the comparison of the aforementioned dual geometric objects, and this constitutes our main method of approach.

## 8.1 Dual Connections

In this section we shall define and investigate the main properties of *dual connections.* These connections were first introduced by A. P. Norden in affine differential geometry literature under the name of "conjugate connections," see Simon [76]. They had also been independently introduced by Nagaoka and Amari [61] and used by Lauritzen [54] in the definition of statistical structure. Recall that linear connections are introduced in Chap. 7, Definition 7.10.1.

**Definition 8.1.1** *Let $(M, g)$ be a Riemannian manifold. Two linear connections $\nabla$ and $\nabla^*$ on $M$ are called dual, with respect to the metric $g$, if*

$$Zg(X, Y) = g(\nabla_Z X, Y) + g(X, \nabla^*_Z Y), \qquad \forall X, Y, Z \in \mathcal{X}(M). \tag{8.1.1}$$

In local coordinates $(x^1, \ldots, x^n)$ the duality condition can be expressed as

$$\partial_{x^k} g_{ij} = \Gamma_{ki,j} + \Gamma^*_{kj,i}, \tag{8.1.2}$$

where $\Gamma_{ki,j} = g_{jm}\Gamma^m_{ki}$ and $\Gamma^*_{kj,i} = g_{im}\Gamma^{*m}_{kj}$ are the coordinate components of connections $\nabla$ and $\nabla^*$, respectively. We note that the PDE system (8.1.2) must be completely integrable.

We note that dual connections are a weaker version of metrical connections, see Definition 7.10.2, Chap. 7. In particular, a metrical linear connection is self-dual.

It is worth noting that since $g$ is parallel neither with respect to $\nabla$ nor to $\nabla^*$, then raising and lowering indices with $g$ does not commute with the covariant derivative produced by $\nabla$, respectively $\nabla^*$.

The triple $(g, \nabla, \nabla^*)$ is called a *dualistic structure* on $M$. A *statistical manifold* is a Riemannian manifold endowed with a dualistic structure, i.e., it is a quadruple $(M, g, \nabla, \nabla^*)$. For instance, any statistical model $\mathcal{S}$ with the Fisher metric $g$ and a pair of $\alpha$-connections, $\big(\nabla^{(\alpha)}, \nabla^{(-\alpha)}\big)$, is a statistical manifold.

The next few results deal with basic properties of dual connections, such as existence, uniqueness, involutivity, torsion, and curvature tensors.

**Proposition 8.1.2** *Given a linear connection $\nabla$ on the Riemannian manifold $(M, g)$, there is a unique connection $\nabla^*$ dual to $\nabla$.*

*Proof:* It suffices to prove the property locally, in a coordinates chart. *Existence:* Since the connection $\nabla$ is given, its components, $\Gamma_{ij,l}$, are known. Define $\Gamma^*_{ij,l} = \partial_{x^i} g_{lj} - \Gamma_{il,j}$ and $\Gamma^{*k}_{ij} = \Gamma^*_{ij,l} g^{lk}$, and construct the dual connection by

$$\nabla^*_{\partial_{x^i}} \partial_{x^j} = \Gamma^{*k}_{ij} \partial_{x^k}.$$

In the virtue of relation (8.1.2), it follows that $\nabla^*$ is dual to $\nabla$. *Uniqueness:* From relation (8.1.2) the connection components $\Gamma^*_{kj,i}$ of $\nabla^*$ are uniquely determined given the metric $g_{ij}$ and the connection coefficients $\Gamma_{ki,j}$ of $\nabla$. ■

**Proposition 8.1.3** *(i) Duality is involutive, i.e., $(\nabla^*)^* = \nabla$.*

*(ii) $\nabla$ is a metrical connection on the Riemannian manifold $(M, g)$ if and only if $\nabla = \nabla^*$.*

*Proof:*

(*i*) It follows from the symmetry in the indices $i$ and $j$ of the equation (8.1.2).

(*ii*) The fact that $\nabla$ is a metrical connection is written as

$$Zg(X, Y) = g(\nabla_Z X, Y) + g(X, \nabla_Z Y), \qquad \forall X, Y, Z \in \mathcal{X}(M), \tag{8.1.3}$$

Then subtracting (8.1.1) and (8.1.3) yields

$$g(X, \nabla_Z Y) = g(X, \nabla^*_Z Y), \qquad \forall X, Y, Z \in \mathcal{X}(M),$$

which is equivalent to $\nabla = \nabla^*$. ■

Let $\nabla$ be a general connection. Recall the $(1,2)$-torsion field and the $(1,3)$–curvature tensor field, which are defined by

$$\begin{aligned} T(X,Y) &= \nabla_X Y - \nabla_Y X - [X,Y], \\ R(X,Y,Z) &= \nabla_X \nabla_Y Z - \nabla_Y \nabla_X Z - \nabla_{[X,Y]} Z. \end{aligned}$$

These can be written locally as

$$T_{ij}^k = \Gamma_{ij}^k - \Gamma_{ji}^k \tag{8.1.4}$$

$$R_{ijk}^r = \partial_i \Gamma_{jk}^r - \partial_j \Gamma_{ik}^r + \Gamma_{ih}^r \Gamma_{jk}^h - \Gamma_{jh}^r \Gamma_{ik}^h, \tag{8.1.5}$$

with the tensor components given by

$$T(\partial_i, \partial_j) = T_{ij}^k \partial_k, \qquad R(\partial_i, \partial_j, \partial_k) = R_{ijk}^r \partial_r,$$

where we used the notation $\partial_i = \partial_{x^i}$.

Recall that a connection $\nabla$ has the constant curvature $K$ if for any vector fields $X$, $Y$, $Z$ the following relation holds

$$R(X,Y)Z = K\{g(Y,Z)X - g(X,Z)Y\}. \tag{8.1.6}$$

The relationship between the curvatures of two dual connections is given by the next result.

**Proposition 8.1.4** *Let $R$ and $R^*$ be the curvature tensors of $\nabla$ and $\nabla^*$, respectively. Then*

*(i)* $g(R(X,Y)Z,W) + g(R^*(X,Y)W,Z) = 0$.

*(ii)* $(M,g,\nabla)$ *has constant curvature if and only if* $(M,g,\nabla^*)$ *has constant curvature, and in this case the curvature tensors are equal. In particular,* $R = 0$ *if and only if* $R^* = 0$.

*Proof:*

(*i*) Since the relation is linear in the arguments $X, Y, W$, and $Z$, it suffices to prove it only on a basis. Therefore we assume $X, Y, W, Z \in \{\frac{\partial}{\partial x^1}, \dots, \frac{\partial}{\partial x^n}\}$, and take computational advantage of the following vanishing Lie brackets

$$[X,Y] = [Y,W] = [W,Z] = \cdots = 0.$$

From the definition of the curvature tensor and Eq. (8.1.1), we find the following equivalences

$$\begin{aligned}
0 &= g(R(X,Y)Z,W) + g(R^*(X,Y)W,Z) \Longleftrightarrow \\
0 &= g\big(\nabla_X\nabla_Y Z - \nabla_Y\nabla_X Z, W\big) + g\big(\nabla^*_X\nabla^*_Y W - \nabla^*_Y\nabla^*_X W, Z\big) \Longleftrightarrow \\
0 &= g\big(\nabla_X\nabla_Y Z, W\big) - g\big(\nabla_Y\nabla_X Z, W\big) + g\big(\nabla^*_X\nabla^*_Y W, Z\big) \\
&\quad - g\big(\nabla^*_Y\nabla^*_X W, Z\big) \Longleftrightarrow \\
0 &= Xg\big(\nabla_Y Z, W\big) - g\big(\nabla_Y Z, \nabla^*_X W\big) - Yg\big(\nabla_X Z, W\big) \\
&\quad + g\big(\nabla_X Z, \nabla^*_Y W\big) \\
&\quad + Xg\big(\nabla^*_Y W, Z\big) - g\big(\nabla^*_Y W, \nabla_X Z\big) - Yg\big(\nabla^*_X W, Z\big) \\
&\quad + g\big(\nabla^*_X W, \nabla_Y Z\big) \Longleftrightarrow \\
0 &= Xg\big(\nabla_Y Z, W\big) - Yg\big(\nabla_X Z, W\big) + Xg\big(\nabla^*_Y W, Z\big) \\
&\quad - Yg\big(\nabla^*_X W, Z\big) \Longleftrightarrow \\
0 &= XYg\big(Z,W\big) - Xg\big(Z, \nabla^*_Y W\big) - YXg\big(Z,W\big) - Yg\big(Z, \nabla^*_X W\big) \\
&\quad + Xg\big(\nabla^*_Y W, Z\big) - Yg\big(\nabla^*_X W, Z\big) \Longleftrightarrow \\
0 &= [X,Y]g\big(Z,W\big),
\end{aligned}$$

which holds true, since $[X,Y] = 0$.

(*ii*) Assume $(M, g, \nabla)$ has the constant curvature equal to $K$. Then using (*i*), we have

$$\begin{aligned}
-g\big(R^*(X,Y)W, Z\big) &= g\big(R(X,Y)Z, W\big) \\
&= K\Big(g(Y,Z)g(X,W) - g(X,Z)g(Y,W)\Big) \\
&= -K\Big(g(Y,W)g(X,Z) - g(X,W)g(Y,Z)\Big).
\end{aligned}$$

Dropping the argument $Z$ yields

$$R^*(X,Y)W = K\Big(g(Y,W)X - g(X,W)Y\Big),$$

which means that the tensor $(M, g, \nabla^*)$ has constant curvature $K$. ■

**Remark 8.1.5** (*i*) The connection $\nabla$ has zero curvature if and only if the dual connection $\nabla^*$ has zero curvature.

(*ii*) In local coordinates we have $R_{ijkl} = -R^*_{ijlk}$. We note that the antisymmetry works for the exchange in the first pair of indices, $R_{ijkl} = -R_{jikl}$, $R^*_{ijkl} = -R^*_{jikl}$, but it doesn't work for the second pair of indices.

## 8.2 Dual Flatness

A connection $\nabla$ is called *flat* in a given system of coordinates if its components vanish, i.e., $\Gamma_{ij}^k = 0$. Therefore, if $X = X^i \partial_i$ and $Y = Y^j \partial_j$ are two vector fields, then the covariant derivative with respect to a flat connection is

$$\nabla_X Y = X^i(\partial_i Y^k + Y^j \Gamma_{ij}^k)\partial_k = X^i \partial_i Y^k \partial_k = X(Y^k)\partial_k.$$

Relations (8.1.4)–(8.1.5) imply that the torsion and the curvature of a flat connection are zero. It can be shown that the converse is partially true in the following sense: if the torsion and curvature are zero, $T = 0$, $R = 0$, then for any point $p$, there is an open neighborhood $U$ of $p$ such that $\nabla$ is flat on $U$.

A statistical manifold $(\mathcal{S}, g, \nabla, \nabla^*)$ is called *dually flat* if both dual connections $\nabla$ and $\nabla^*$ are flat. Consequently, on a dually flat manifold we have $T = T^* = 0$ and $R = R^* = 0$.

**Proposition 8.2.1** *Let $(\mathcal{S}, g, \nabla, \nabla^*)$ be a dually flat statistical manifold.*

*(i) Then, in any local coordinate system, the metric coefficients $g_{ij}$ are constant.*

*(ii) If $\gamma$ is either a $\nabla$- or $\nabla^*$-autoparallel curve, then $\gamma^k(s) = \alpha^k s + \beta^k$, with $\alpha^k$, $\beta^k$ constants.*

*Proof:*

(i) Substituting $\Gamma_{ij}^k = \Gamma_{ij}^{*k} = 0$ in relation (8.1.2) yields $\partial_k g_{ij} = 0$, and hence the metric coefficients $g_{ij}$ do not depend on $x$.

(ii) The $\nabla$-autoparallel curves are characterized by the equation $\ddot{\gamma}^k(s) + \Gamma_{ij}^k(\gamma(s))\dot{\gamma}^i(s)\dot{\gamma}^j(s) = 0$, which becomes $\ddot{\gamma}^k(s) = 0$; this implies the degree one in $s$ for each component $\gamma^k(s)$. ■

## 8.3 Dual Connections in Strong Sense

Asking for two dual connections to be both torsion-free sounds like a strong requirement. This section tries to relax this condition by assuming that dual connections satisfy a dual symmetry relation that replaces the torsion-free conditions for each connection. However, we shall show that all dual connections, which are also dual symmetric, coincide with the Levi–Civita connection, and hence their study does not bring any novelty to the theory. This is the reason why the dual symmetry concept will be dropped in the next sections and replaced by torsion-free dual connections.

The following definition can be seen as an extension of the definition of the Levi–Civita connection.

**Definition 8.3.1** *The connection $\nabla$ is dual to $\nabla^*$ in strong sense, with respect to the metric $g$, if*

(1) $Zg(X,Y) = g\big(\nabla_Z X, Y\big) + g\big(X, \nabla^*_Z Y\big)$

(2) $\nabla_X Y - \nabla^*_Y X = [X,Y]$, *for all* $X, Y, Z \in \mathcal{X}(\mathcal{M})$.

The second condition can be interpreted as a dual symmetry between the connections $\nabla$ and $\nabla^*$.

It follows that the relations between the connection components of a pair of dual connections in strong sense is given by

$$\partial_{x^k} g_{ij} = \Gamma_{ki,j} + \Gamma^*_{kj,i} \tag{8.3.7}$$
$$\Gamma_{ij,k} = \Gamma^*_{ji,k}. \tag{8.3.8}$$

Let $T$ and $T^*$ be the torsion tensors associated with the connections $\nabla$ and $\nabla^*$. From the torsion formula and the definition of dual connections in strong sense, we have

$$\begin{aligned} T(X,Y) + T^*(X,Y) &= \nabla_X Y - \nabla_Y X - [X,Y] \\ &\quad + \nabla^*_X Y - \nabla^*_Y X - [X,Y] \\ &= \Big(\nabla_X Y - \nabla^*_Y X - [X,Y]\Big) \\ &\quad - \Big(\nabla_Y X - \nabla^*_X Y - [Y,X]\Big) \\ &= 0. \end{aligned}$$

In the following we shall show that, in fact, each of the foregoing torsions vanishes. This will imply that dual symmetry is a stronger condition than torsion-freeness for both connections.

**Proposition 8.3.2** *If $\nabla$ has a dual connection $\nabla^*$ in strong sense, then its torsion is zero.*

*Proof:* Let $\nabla^*$ be the dual connection of $\nabla$ in strong sense. Then

$$\begin{aligned} Zg(X,Y) &= g(\nabla_Z X, Y) + g(X, \nabla^*_Z Y) \\ &= g(\nabla_Z X, Y) + g(X, \nabla_Y Z + [Z,Y]) \end{aligned} \quad (8.3.9)$$

$$\begin{aligned} Zg(Y,X) &= g(\nabla_Z Y, X) + g(Y, \nabla^*_Z X) \\ &= g(\nabla_Z Y, X) + g(Y, \nabla_X Z + [Z,X]). \end{aligned} \quad (8.3.10)$$

Subtracting relations (8.3.9) and (8.3.10) yields

$$\begin{aligned} 0 &= g(\nabla_Z X - \nabla_X Z - [Z,X], Y) + g(X, \nabla_Y Z - \nabla_Z Y - [Y,Z]) \\ &= g(T(Z,X),Y) + g(X, T(Y,Z)). \end{aligned}$$

By cyclic permutations, we have

$$\begin{aligned} g(X,T(Y,Z)) + g(Y,T(Z,X)) &= 0 \\ g(Y,T(Z,X)) + g(Z,T(X,Y)) &= 0 \\ g(Z,T(X,Y)) + g(X,T(Y,Z)) &= 0. \end{aligned}$$

Adding the first two relations and subtracting the third one yields

$$2g(Y,T(Z,X)) = 0, \qquad \forall Y \in \mathcal{X}(M),$$

which implies $T(Z,X) = 0$ for any vector fields $X$ and $Z$. This leads to the desired result, $T = 0$. ■

**Corollary 8.3.3** *If $\nabla$ and $\nabla^*$ are dual in strong sense, then $T = T^* = 0$.*

The Levi–Civita connection $\nabla$ is strong auto-dual, i.e., $\nabla^* = \nabla$ in the strong sense. The next result shows that the converse statement also holds true.

**Theorem 8.3.4** *If $\nabla$ and $\nabla^*$ are dual connections in strong sense, then $\nabla = \nabla^*$, and hence $\nabla$ is the Levi–Civita connection.*

*Proof:* From Proposition 8.3.2 we have $T = 0$, or

$$\nabla_X Y - \nabla_Y X = [X,Y].$$

From the definition of dual connections in strong sense we have

$$\nabla_X Y - \nabla^*_Y X = [X,Y].$$

Subtracting the last two relations yields $\nabla_Y X = \nabla^*_Y X$, for all vector fields $X$ and $Y$. Hence $\nabla = \nabla^*$. Then $\nabla$ is a metrical and torsion-free connection, i.e., it is the Levi–Civita connection. ■

The previous proof was based on Proposition 8.3.2. In the following we supply a direct proof using local coordinates. Substituting (8.3.8) into (8.3.7) yields

$$\partial_{x^k} g_{ij} = \Gamma_{ki,j} + \Gamma_{jk,i}.$$

By circular permutations we have two similar relations

$$\partial_{x^i} g_{jk} = \Gamma_{ij,k} + \Gamma_{ki,j},$$
$$\partial_{x^j} g_{ki} = \Gamma_{jk,i} + \Gamma_{ij,k}.$$

Adding the last two relations and subtracting the first, after cancelations we get

$$\Gamma_{ij,k} = \frac{1}{2}\Big(\partial_{x^i} g_{jk} + \partial_{x^j} g_{ki} - \partial_{x^k} g_{ij}\Big),$$

which are the Christoffel symbols of first type associated with the metric $g$. This means that $\Gamma_{ij,k} = \Gamma^{(0)}_{ij,k}$, i.e., $\nabla$ is the Levi–Civita connection.

Theorem 8.3.4 inferres that the concept of strong duality is too restrictive since it always implies the connection to be Levi–Civita. This is the reason why in the following we shall assume that the connections are just dual, but not dual in strong sense.

## 8.4 Relative Torsion Tensors

Consider two connections $\nabla$ and $\nabla^*$, which are not dual in strong sense. Then it makes sense to define the nonzero amount

$$U(X,Y) = \nabla_X Y - \nabla^*_Y X - [X,Y].$$

Since $U$ is $\mathbb{R}$-linear in both variables, then for any smooth function $f \in \mathcal{F}(M)$ we have

$$\begin{aligned} U(fX,Y) &= \nabla_{fX} Y - \nabla^*_Y(fX) - [fX,Y] \\ &= f\nabla_X Y - f\nabla^*_Y X - Y(f)X - \big(fXY - Y(f)X - fYX\big) \\ &= f\Big(\nabla_X Y - \nabla^*_Y X - [X,Y]\Big) \\ &= fU(X,Y), \end{aligned}$$

$$U(X, fY) = fU(X, Y),$$

then the mapping $U$ becomes a $(1,2)$-type tensor on $M$.
The conjugate tensor to $U$ is defined by

$$U^*(X,Y) = \nabla^*_X Y - \nabla_Y X - [X,Y].$$

The tensors $U$ and $U^*$ are called the *relative torsion tensors* of connections $\nabla$ and $\nabla^*$.

**Proposition 8.4.1** *The following properties hold:*

*(i)* $U^*(X,Y) = -U(Y,X)$.

*(ii)* $U^{**} = U$.

*(iii)* $U + U^* = T + T^*$, *where* $T$ *and* $T^*$ *stand for the torsions of* $\nabla$ *and* $\nabla^*$.

*Proof:*

*(i)* We have

$$\begin{aligned} U^*(X,Y) &= \nabla^*_X Y - \nabla_Y X - [X,Y] \\ &= -\Big(\nabla_Y X - \nabla^*_X Y + [X,Y]\Big) \\ &= -U(Y,X). \end{aligned}$$

*(ii)* Using (1) yields

$$\begin{aligned} U^{**}(X,Y) &= \Big(-U(Y,X)\Big)^* = -U^*(Y,X) \\ &= U(X,Y). \end{aligned}$$

*(iii)* Subtracting the relations

$$\begin{aligned} U(X,Y) &= \nabla_X Y - \nabla^*_Y X - [X,Y] \\ U^*(X,Y) &= \nabla^*_X Y - \nabla_Y X - [X,Y], \end{aligned}$$

we find

$$\begin{aligned} (U+U^*)(X,Y) &= \Big(\nabla_X Y - \nabla_Y X - [X,Y]\Big) \\ &\quad + \Big(\nabla^*_X Y - \nabla^*_Y X - [X,Y]\Big) \\ &= (T+T^*)(X,Y), \end{aligned}$$

for any vector fields $X$ and $Y$.

■

**Corollary 8.4.2** *We have $U + U^* = 0$ if and only if $T + T^* = 0$.*

We note that the connections $\nabla$, $\nabla^*$ are dually symmetric if $U = 0$. Then Theorem 8.3.4 can be reformulated equivalently as:

**Proposition 8.4.3** *If $\nabla$ and $\nabla^*$ are dual connections and $U = 0$, then $\nabla = \nabla^*$, so $\nabla$ is the Levi–Civita connection.*

In the rest of the chapter we shall assume that connections $\nabla$ and $\nabla^*$ are both torsion-less. We have seen that two dual connections, which are dually symmetric, have zero torsions. The next result deals with a partial converse of this result.

**Proposition 8.4.4** *Let $\nabla$ and $\nabla^*$ be two dual torsion-free connections, $T = T^* = 0$. Then the following symmetry relation holds*

$$\begin{aligned} g\big(U^*(X,Y),Z\big) &= g\big(U^*(X,Z),Y\big) = g\big(U^*(Z,Y),X\big) \\ &= g\big(U^*(Y,X),Z\big). \end{aligned}$$

The proof follows from the formula $g\big(U^*(X,Y),Z\big) = C(X,Y,Z)$ and the total symmetry of $C$, facts proved at the end of Sect. 8.8.

## 8.5 Dual $\alpha$-Connections

Let $\nabla$ and $\nabla^*$ be two dual torsion-free connections, with respect to the metric $g$. Consider the one-parameter family of connections given by the convex combination of the foregoing dual connections, i.e.,

$$\nabla^{(\alpha)} = \frac{1+\alpha}{2}\nabla^* + \frac{1-\alpha}{2}\nabla, \qquad \alpha \in \mathbb{R}. \tag{8.5.11}$$

$\nabla^{(\alpha)}$ will be called the *$\alpha$-connection*, and will play a central role in the study of statistical manifolds.

**Proposition 8.5.1** *$\nabla^{(\alpha)}$ and $\nabla^{(-\alpha)}$ are dual connections with respect to the metric $g$.*

*Proof:* Using the duality of connections $\nabla$ and $\nabla^*$ and the definition (8.5.11), we find

$$\begin{aligned} g(\nabla_Z^{(\alpha)} X, Y) &= \frac{1+\alpha}{2} g(\nabla_Z^* X, Y) + \frac{1-\alpha}{2} g(\nabla_Z X, Y) \\ &= \frac{1+\alpha}{2}\Big(Zg(X,Y) - g(X, \nabla_Z Y)\Big) \\ &\quad + \frac{1-\alpha}{2} g(\nabla_Z X, Y) \\ &= \frac{1+\alpha}{2} Zg(X,Y) - \frac{1+\alpha}{2} g(X, \nabla_Z Y) \\ &\quad + \frac{1-\alpha}{2} g(\nabla_Z X, Y). \end{aligned} \tag{8.5.12}$$

Similarly,

$$g(\nabla_Z^{(-\alpha)} Y, X) = \frac{1-\alpha}{2} Zg(Y,X) - \frac{1-\alpha}{2} g(Y, \nabla_Z X) + \frac{1+\alpha}{2} g(\nabla_Z Y, X). \tag{8.5.13}$$

Adding (8.5.12) and (8.5.13) yields

$$g(\nabla_Z^{(\alpha)} X, Y) + g(\nabla_Z^{(-\alpha)} Y, X) = Zg(X,Y),$$

which shows that $\nabla^{(\alpha)}$ and $\nabla^{(-\alpha)}$ are dual connections. ■

Using (8.5.11), we have the following relations among the connection coefficients:

$$\begin{aligned} \Gamma_{ij,k}^{(\alpha)} &= \frac{1+\alpha}{2} \Gamma_{ij,k}^* + \frac{1-\alpha}{2} \Gamma_{ij,k} \\ &= \frac{1+\alpha}{2}\Big(\partial_{x^i} g_{jk} - \Gamma_{ik,j}\Big) + \frac{1-\alpha}{2} \Gamma_{ij,k} \\ &= \frac{1+\alpha}{2} \partial_{x^i} g_{jk} - \frac{1+\alpha}{2} \Gamma_{ik,j} + \frac{1-\alpha}{2} \Gamma_{ij,k}. \end{aligned} \tag{8.5.14}$$

If $\nabla$ is a flat connection, i.e., $\Gamma_{ij,k} = 0$, then the coefficients of the $\alpha$-connection become

$$\Gamma_{ij,k}^{(\alpha)} = \frac{1+\alpha}{2} \partial_{x^i} g_{jk}. \tag{8.5.15}$$

If $\nabla^*$ is flat, then using (8.1.2) yields $\Gamma_{ki,j} = \partial_{x^k} g_{ij}$ and hence, by using (8.5.14), we have

$$\begin{aligned} \Gamma_{ij,k}^{(\alpha)} &= \frac{1+\alpha}{2} \partial_{x^i} g_{jk} - \frac{1+\alpha}{2} \partial_{x^i} g_{kj} + \frac{1-\alpha}{2} \partial_{x^i} g_{jk} \\ &= \frac{1-\alpha}{2} \partial_{x^i} g_{jk}. \end{aligned} \tag{8.5.16}$$

The following particular values of $\alpha$ are of distinguished importance. Making $\alpha = -1, 0, 1$ in (8.5.11) yields the following connections:

$$\nabla^{(1)} = \nabla^*, \qquad \nabla^{(-1)} = \nabla, \qquad \nabla^{(0)} = \frac{1}{2}(\nabla + \nabla^*). \tag{8.5.17}$$

These connections can be used to generate any $\alpha$-connection. This can be seen from the next result.

**Proposition 8.5.2** *The $\alpha$-connection can be written in one of the following equivalent forms*

$$\nabla^{(\alpha)} = (1-\alpha)\nabla^{(0)} + \alpha\nabla^{(1)} \tag{8.5.18}$$

$$= (1+\alpha)\nabla^{(0)} - \alpha\nabla^{(-1)} \tag{8.5.19}$$

$$= \nabla^{(0)} + \frac{1}{2}\alpha\big(\nabla^{(1)} - \nabla^{(-1)}\big). \tag{8.5.20}$$

*Proof:* A straightforward computation, using (8.5.11) and the aforementioned connections given by (8.5.17) produces the following relations

$$\begin{aligned} \nabla^{(\alpha)} &= \frac{1+\alpha}{2}\nabla^* + \frac{1-\alpha}{2}\nabla \\ &= \frac{1+\alpha}{2}\nabla^* + \Big(\frac{1-\alpha}{2}\nabla + \frac{1-\alpha}{2}\nabla^*\Big) - \frac{1-\alpha}{2}\nabla^* \\ &= \alpha\nabla^* + (1-\alpha)\nabla^{(0)}, \end{aligned}$$

and hence (8.5.18) holds true. Also

$$\begin{aligned} \nabla^{(\alpha)} &= \frac{1+\alpha}{2}\nabla^* + \frac{1-\alpha}{2}\nabla \\ &= \Big(\frac{1+\alpha}{2}\nabla^* + \frac{1+\alpha}{2}\nabla\Big) - \frac{1+\alpha}{2}\nabla + \frac{1-\alpha}{2}\nabla \\ &= (1+\alpha)\nabla^{(0)} - \alpha\nabla, \end{aligned}$$

which leads to (8.5.19). Then

$$\begin{aligned} \nabla^{(\alpha)} &= \frac{1+\alpha}{2}\nabla^* + \frac{1-\alpha}{2}\nabla = \frac{1}{2}(\nabla^* + \nabla) + \frac{\alpha}{2}(\nabla^* - \nabla) \\ &= \nabla^{(0)} + \frac{\alpha}{2}(\nabla^{(1)} - \nabla^{(-1)}), \end{aligned}$$

and hence we have shown relation (8.5.20). ■

**Proposition 8.5.3** *(i) If $\nabla$ and $\nabla^*$ are dual connections, with respect to the metric $g$, then $\nabla^{(0)}$ is a metrical connection.*

*(ii) If $\nabla$ and $\nabla^*$ are torsion-free dual connections, then $\nabla^{(0)}$ is the Levi–Civita connection.*

*Proof:*

(*i*) Adding the following two relations

$$\begin{aligned} g(\nabla_Z^{(0)}X, Y) &= \frac{1}{2}g(\nabla_Z X, Y) + \frac{1}{2}g(\nabla_Z^* X, Y) \\ g(X, \nabla_Z^{(0)}Y) &= \frac{1}{2}g(X, \nabla_Z Y) + \frac{1}{2}g(X, \nabla_Z^* Y) \end{aligned}$$

after using (8.1.1) yields

$$\begin{aligned} g(\nabla_Z^{(0)}X, Y) + g(X, \nabla_Z^{(0)}Y) &= \frac{1}{2}Zg(X, Y) + \frac{1}{2}Zg(X, Y) \\ &= Zg(X, Y), \end{aligned}$$

which means that $\nabla^{(0)}$ is a metrical connection.

(*ii*) The fact that $\nabla$ and $\nabla^*$ are torsion-free implies that $\nabla^{(0)}$ has the same property. Then using that the Levi–Civita connection is the only torsion-free and metrical connection, see Theorem 7.11.1, we obtain the desired result.

■

## 8.6 Difference Tensor

Let $\nabla$ and $\nabla^*$ be two torsion-free dual connections. We define the *difference* $(1, 2)$*-tensor* by

$$K(X, Y) = \nabla_X^* Y - \nabla_X Y. \tag{8.6.21}$$

For the definition and properties of tensors the reader is referred to Sect. 7.8, Chap. 1.

The tensor $K$ is symmetric since

$$\begin{aligned} K(X, Y) - K(Y, X) &= (\nabla_X^* Y - \nabla_Y^* X) + (\nabla_Y X - \nabla_X Y) \\ &= [X, Y] + [Y, X] = 0. \end{aligned}$$

The symmetry and the tensor properties of $K$ can be combined as

$$K(fX,Y) = K(Y,fX) = K(X,fY) = K(fY,X) = fK(X,Y),$$

for any smooth function $f$.

The difference tensor $K$ can be also expressed in terms of the Levi–Civita connection $\nabla^{(0)} = \frac{1}{2}(\nabla + \nabla^*)$ as in the following

$$K(X,Y) = 2(\nabla_X^{(0)} Y - \nabla_X Y) = 2(\nabla_X^* Y - \nabla_X^{(0)} Y).$$

Using

$$\begin{aligned} \nabla^{(-\alpha)} - \nabla^{(\alpha)} &= \Big(\frac{1-\alpha}{2}\nabla + \frac{1+\alpha}{2}\nabla^*\Big) - \Big(\frac{1+\alpha}{2}\nabla + \frac{1-\alpha}{2}\nabla^*\Big) \\ &= \alpha(\nabla^* - \nabla) = \alpha K, \end{aligned}$$

it follows that the difference tensor can be written as

$$K(X,Y) = \frac{1}{\alpha}\Big(\nabla_X^{(-\alpha)} Y - \nabla_X^{(\alpha)} Y\Big).$$

Taking $\alpha \to 0$ yields

$$K(X,Y) = -2\frac{d}{d\alpha}\nabla^{(\alpha)}|_{\alpha=0},$$

which expresses the difference tensor as a derivative with respect to parameter $\alpha$.

The components of the difference tensor can be expressed in local coordinates as $K(\partial_i, \partial_j) = K_{ij}^k \partial_k$, where

$$K_{ij}^k = \Gamma_{ij}^{*\,k} - \Gamma_{ij}^k = 2\Big(\Gamma_{ij}^{*\,k} - \Gamma_{ij}^{(0)\,k}\Big) = 2\Big(\Gamma_{ij}^{(0)\,k} - \Gamma_{ij}^k\Big).$$

Using (8.1.2) the previous relation can be further expressed as

$$K_{ij}^k = g^{lk}\partial_{x^i} g_{jl} - \Gamma_{ij}^k - g^{lk}\Gamma_{il,j}.$$

## 8.7 Curvature Vector Field

The *curvature vector field* $K$ associated with a pair of dual connections $(\nabla, \nabla^*)$ is defined as the trace of the difference tensor

$$K = Trace\Big((X,Y) \to K(X,Y)\Big). \qquad (8.7.22)$$

If $K = K^\ell \partial_{x^\ell}$ is a representation of the curvature vector field in local coordinates, then

$$K^\ell = g^{ij} K_{ij}^\ell = g^{ij}\Big(\Gamma_{ij}^{*\,\ell} - \Gamma_{ij}^\ell\Big). \qquad (8.7.23)$$

The vector field $K$ will be used later in the formula of the $\alpha$-Laplacian.

## 8.8 Skewness Tensor

The $(0, 3)$-*skewness tensor* on the Riemannian manifold $(M, g)$ with respect to the linear connection $\nabla$ is defined by $C = \nabla g$, i.e.,

$$C(X, Y, Z) = (\nabla g)(X, Y, Z) = Xg(Y, Z) - g(\nabla_X Y, Z) - g(Y, \nabla_X Z). \tag{8.8.24}$$

Componentwise, we have

$$C_{ijk} = \partial_{x^i} g_{jk} - \Gamma_{ij,k} - \Gamma_{ik,j}, \tag{8.8.25}$$

where $C_{ijk} = C(\partial_{x_i}, \partial_{x_j}, \partial_{x_k})$.

The following result deals with the tensorial relationship between the difference tensor $K$ and the skewness tensor $C$.

**Proposition 8.8.1** *The skewness tensor and the difference tensor are related by*

$$C(X, Y, Z) = g\big(K(X, Y), Z\big). \tag{8.8.26}$$

*Proof:* We have on components

$$\begin{aligned} K_{ij}^k &= \Gamma_{ij}^{*k} - \Gamma_{ij}^k \\ C_{ijk} &= \partial_{x^i} g_{jk} - \Gamma_{ij,k} - \Gamma_{ik,j}. \end{aligned}$$

Contracting with the metric tensor yields

$$g_{kl} K_{ij}^k = \Gamma_{ij,l}^* - \Gamma_{ij,l} = \partial_{x^i} g_{jl} - \Gamma_{il,j} - \Gamma_{ij,l} = C_{ijl},$$

which is relation (8.8.26) in local coordinates. ■

**Proposition 8.8.2** *The skewness tensor is totally symmetric, i.e.,*

$$C_{ijk} = C_{ikj} = C_{jik} = C_{jki} = C_{kij} = C_{kji}.$$

*Proof:* It suffices to prove only the symmetry in the first pair, $C_{ijk} = C_{jik}$, and the last pair, $C_{ijk} = C_{ikj}$. The symmetry in the first pair is a consequence of the torsion-freeness of the dual connections. Since the difference tensor $K$ is symmetric, we have

$$\begin{aligned} C_{ijk} - C_{jik} &= g\big(K(\partial_i, \partial_j), \partial_k\big) - g\big(K(\partial_j, \partial_i), \partial_k\big) \\ &= g\big(K(\partial_i, \partial_j) - K(\partial_j, \partial_i), \partial_k\big) = 0. \end{aligned}$$

In order to show the symmetry in the last pair, we write

$$\begin{aligned} C_{ijk} - C_{ikj} &= g\big(K(\partial_i, \partial_j), \partial_k\big) - g\big(K(\partial_i, \partial_k), \partial_j\big) \\ &= g\big(\nabla^*_{\partial_i}\partial_j, \partial_k\big) - g\big(\nabla_{\partial_i}\partial_j, \partial_k\big) - g\big(\nabla^*_{\partial_i}\partial_k, \partial_j\big) \\ &\quad + g\big(\nabla_{\partial_i}\partial_k, \partial_j\big) \\ &= \Big[g\big(\nabla^*_{\partial_i}\partial_j, \partial_k\big) + g\big(\partial_j, \nabla_{\partial_i}\partial_k\big)\Big] \\ &\quad - \Big[g\big(\nabla_{\partial_i}\partial_j, \partial_k\big) + g\big(\partial_j, \nabla^*_{\partial_i}\partial_k\big)\Big] \\ &= \partial_i g(\partial_j, \partial_k) - \partial_i g(\partial_j, \partial_k) = 0, \end{aligned}$$

where we used the definitions of dual connections and difference tensor. ■

The total symmetry of the skewness tensor $C$ can be written also as

$$C(X, Y, Z) = C(X, Z, Y) = C(Z, Y, X) = C(Y, X, Z),$$

for all tangent vector fields $X, Y, Z \in \mathcal{X}(M)$.

**Corollary 8.8.3** *We have*

$$K^p_{ik} g_{pj} = K^p_{ij} g_{pk}, \tag{8.8.27}$$

*with summation over* $p$.

*Proof:* The expression on the left is equal to $C_{ikj}$, while the one on the right to $C_{ijk}$. Using the total symmetry of tensor $C$ yields the desired result. ■

Now we go back to the proof of Proposition 8.4.4. It is worthy to make the point that the relative torsion tensor and the difference tensor are related in the following way

$$U^*(X, Y) = \nabla^*_X Y - \nabla_Y X - [X, Y] = \nabla^*_Y X - \nabla_Y X = K(Y, X).$$

Therefore

$$g\big(U^*(X, Y), Z\big) = g\big(K(Y, X), Z\big) = C(Y, X, Z).$$

Since $C$ is totally symmetric, we obtain

$$\begin{aligned} g\big(U^*(X, Y), Z\big) &= g\big(U^*(X, Z), Y\big) = g\big(U^*(Z, Y), X\big) \\ &= g\big(U^*(Y, X), Z\big), \end{aligned}$$

which is the relation claimed by Proposition 8.4.4.

**Proposition 8.8.4** *Let $\nabla$ be a torsion-free connection. Then $\nabla^*$ defined by*

$$g(Y, \nabla^*_X Z) = Xg(Y, Z) - g(\nabla_X Y, Z) \tag{8.8.28}$$

*is a torsion-free linear connection, dual to $\nabla$. Furthermore, $\nabla^*$ satisfies*

$$(\nabla^* g)(X, Y, Z) = -C(X, Y, Z).$$

*Proof:* First we show that $\nabla^*$ is a linear connection. For any smooth function $f$ we have

$$\begin{aligned} g(Y, \nabla^*_{fX} Z) &= fXg(Y, Z) - g(\nabla_{fX} Y, Z) \\ &= f\Big(Xg(Y, Z) - g(\nabla_X Y, Z)\Big) \\ &= f\, g(Y, \nabla^*_X Z) = g(Y, f\nabla^*_X Z), \end{aligned}$$

which yields $\nabla^*_{fX} Z = f\nabla^*_X Z$. Next we check the Leibniz property in the second argument

$$\begin{aligned} g(Y, \nabla^*_X fZ) &= Xg(Y, fZ) - g(\nabla_X Y, fZ) \\ &= X(f)g(Y, Z) + f\, Xg(Y, Z) - f\, g(\nabla_X Y, Z) \\ &= g\big(Y, X(f)Z\big) + f\, g(Y, \nabla^*_X Z) \\ &= g\big(Y, X(f)Z + f\nabla^*_X Z\big). \end{aligned}$$

Since $\nabla$ is torsion-free, $\Gamma_{ij,k} = \Gamma_{ji,k}$, then the symmetry of the tensor $C$ in the first two arguments yields

$$\begin{aligned} C_{ijk} &= C_{jik} \Longleftrightarrow \\ \partial_{x^i} g_{jk} - \Gamma_{ij,k} - \Gamma_{ik,j} &= \partial_{x^j} g_{ik} - \Gamma_{ji,k} - \Gamma_{jk,i} \Longleftrightarrow \\ \partial_{x^i} g_{jk} - \Gamma_{ik,j} &= \partial_{x^j} g_{ik} - \Gamma_{jk,i} \Longleftrightarrow \\ \Gamma^*_{ij,k} &= \Gamma^*_{ji,k}, \end{aligned}$$

which shows that $\nabla^*$ is torsion-free. The duality follows from relation (8.8.28).

The last relation will be also shown in local coordinates. Substituting

$$\begin{aligned} \Gamma_{ij,k} &= \partial_{x^i} g_{jk} - \Gamma^*_{ik,j} \\ \Gamma_{ik,j} &= \partial_{x^i} g_{kj} - \Gamma^*_{ij,k} \end{aligned}$$

into

$$C_{ijk} = \partial_{x^i} g_{jk} - \Gamma_{ij,k} - \Gamma_{ik,j},$$

after canceling the term $\partial_{x^i} g_{jk}$, yields

$$C_{ijk} = -(\partial_{x^i} g_{kj} - \Gamma^*_{ik,j} - \Gamma^*_{ij,k}) = -(\nabla^* g)_{ijk},$$

and hence $\nabla^* g = -C$. ■

**Corollary 8.8.5** $\frac{1}{2}(\nabla + \nabla^*)$ *is a metrical connection.*

*Proof:* We have $\frac{1}{2}(\nabla + \nabla^*)g = \frac{1}{2}\nabla g + \frac{1}{2}\nabla^* g = \frac{1}{2}C - \frac{1}{2}C = 0$. ■

The following two results presents the relation between the $\alpha$-connection and the skewness tensor.

**Proposition 8.8.6** *We have the following relation*

$$g(\nabla^{(\alpha)}_X Y, Z) = g(\nabla^{(0)}_X Y, Z) + \frac{\alpha}{2} C(X, Y, Z). \tag{8.8.29}$$

*Proof:* From Proposition 8.5.2

$$\nabla^{(\alpha)} = \nabla^{(0)} + \frac{\alpha}{2}(\nabla^* - \nabla)$$

and hence

$$\nabla^{(\alpha)}_X Y = \nabla^{(0)}_X Y + \frac{\alpha}{2} K(X, Y).$$

Then using (8.8.26) yields

$$g(\nabla^{(\alpha)}_X Y, Z) = g(\nabla^{(0)}_X Y, Z) + \frac{\alpha}{2} C(X, Y, Z).$$

■

**Proposition 8.8.7** *We have* $\nabla^{(\alpha)} g = -\alpha C$, *i.e.,*

$$-\alpha C(X, Y, Z) = X g(Y, Z) - g(\nabla^{(\alpha)}_X Y, Z) - g(Y, \nabla^{(\alpha)}_X Z).$$

*Proof:* Using the definition of the $\alpha$-connection (8.5.11) and Proposition 8.8.4 yields

$$\begin{aligned} \nabla^{(\alpha)} g &= \frac{1+\alpha}{2}\nabla^* g + \frac{1-\alpha}{2}\nabla g \\ &= \frac{1+\alpha}{2}(-C) + \frac{1-\alpha}{2} C \\ &= -\alpha C. \end{aligned}$$

■

## 8.9 Relative Curvature Tensor

In this section we define the relative curvature tensor of two connections $\nabla^{(\alpha)}$ and $\nabla^{(\beta)}$ and study its main properties. This can be seen as an extension to $\alpha$-connections of the curvature tensor introduced in Chap. 7, Definition 7.10.4.

For any $\alpha, \beta \in \mathbb{R}$ define

$$\begin{aligned} R^{(\alpha,\beta)}(X,Y,Z) &= [\nabla^{(\alpha)}_X, \nabla^{(\beta)}_Y]Z - \nabla^{(\alpha)}_{[X,Y]}Z \\ &= \nabla^{(\alpha)}_X \nabla^{(\beta)}_Y Z - \nabla^{(\beta)}_X \nabla^{(\alpha)}_Y Z - \nabla^{(\alpha)}_{[X,Y]}Z, \end{aligned} \tag{8.9.30}$$

which measures the non-commutativity of connections $\nabla^{(\alpha)}$ and $\nabla^{(\beta)}$. It is easy to check that $R^{(\alpha,\beta)}$ is $\mathbb{R}$-linear and $\mathcal{F}(M)$-linear in each of the arguments. For instance, we shall check the $\mathcal{F}(M)$-linearity in the first argument. Using $[fX, Y] = f[X,Y] - Y(f)X$, we have

$$\begin{aligned} R^{(\alpha,\beta)}(fX,Y,Z) &= f\nabla^{(\alpha)}_X \nabla^{(\beta)}_Y Z - \nabla^{(\beta)}_Y (f\nabla^{(\alpha)}_X Z) - \nabla^{(\alpha)}_{[fX,Y]} \\ &= f\nabla^{(\alpha)}_X \nabla^{(\beta)}_Y Z - Y(f)\nabla^{(\alpha)}_X Z - f\nabla^{(\beta)}_Y \nabla^{(\alpha)}_X Z \\ &\quad -\nabla^{(\alpha)}_{f[X,Y]} + \nabla^{(\alpha)}_{Y(f)X} Z \\ &= f\Big\{\nabla^{(\alpha)}_X \nabla^{(\beta)}_Y Z - \nabla^{(\beta)}_X \nabla^{(\alpha)}_Y Z - \nabla^{(\alpha)}_{[X,Y]}\Big\} \\ &= fR^{(\alpha,\beta)}(X,Y,Z). \end{aligned}$$

Similar computations lead to $R^{(\alpha,\beta)}(X,fY,Z) = fR^{(\alpha,\beta)}(X,Y,Z)$ and $R^{(\alpha,\beta)}(X,Y,fZ) = fR^{(\alpha,\beta)}(X,Y,Z)$. Hence $R^{(\alpha,\beta)}$ is a $(3,1)$-tensor, called the *relative curvature tensor* of $\nabla^{(\alpha)}$ with respect to $\nabla^{(\beta)}$.

A relation which shows the lack of discrimination between indices $\alpha$ and $\beta$ is

$$R^{(\alpha,\beta)}(X,Y,Z) + R^{(\beta,\alpha)}(X,Y,Z) = -\left(\nabla^{(\alpha)}_{[X,Y]} + \nabla^{(\beta)}_{[X,Y]}\right) Z.$$

A few familiar curvature tensors can be retrieved as particular cases:

$$R^{(\alpha,\alpha)} = R^{(\alpha)}, \quad R^{(1,1)} = R^{(1)} = R^*$$

$$R^{(0,0)} = R^{(0)}, \quad R^{(-1,-1)} = R^{(-1)} = R.$$

The following two relative curvature tensors will play a distinguished role:

$(i)$ The relative curvature tensor of $\nabla$, with respect to its dual connection $\nabla^*$, that is

$$R^{(-1,1)}(X,Y,Z) = \nabla_X \nabla^*_Y Z - \nabla^*_X \nabla_Y Z - \nabla_{[X,Y]}.$$

$(ii)$ The relative curvature tensor of $\nabla^*$, with respect to $\nabla$, that is

$$R^{(1,-1)}(X,Y,Z) = \nabla^*_X \nabla_Y Z - \nabla_X \nabla^*_Y Z - \nabla^*_{[X,Y]}.$$

The role of the aforementioned tensors is given by the following result.

**Proposition 8.9.1** *Any relative curvature tensor can be written as a combination of the aforementioned particular curvature tensors as*

$$\begin{aligned} 4R^{(\alpha,\beta)} &= (1+\alpha)(1+\beta)R^* + (1-\alpha)(1-\beta)R \\ &\quad +(1+\alpha)(1-\beta)R^{(1,-1)} + (1-\alpha)(1+\beta)R^{(-1,1)}. \end{aligned}$$

*Proof:* The following computation is based on the linear algebra of connections. We have

$$\begin{aligned} \nabla^{(\alpha)}_X \nabla^{(\beta)}_Y Z &= \Big(\frac{1+\alpha}{2}\nabla^*_X + \frac{1-\alpha}{2}\nabla_X\Big)\Big(\frac{1+\beta}{2}\nabla^*_Y + \frac{1-\beta}{2}\nabla_Y\Big)Z \\ &= \frac{(1+\alpha)(1+\beta)}{4}\nabla^*_X\nabla^*_Y Z + \frac{(1+\alpha)(1-\beta)}{4}\nabla^*_X\nabla_Y Z \\ &\quad + \frac{(1-\alpha)(1+\beta)}{4}\nabla_X\nabla^*_Y Z + \frac{(1-\alpha)(1-\beta)}{4}\nabla_X\nabla_Y Z. \end{aligned}$$

$$\begin{aligned} \nabla^{(\beta)}_Y \nabla^{(\alpha)}_X Z &= \Big(\frac{1+\beta}{2}\nabla^*_Y + \frac{1-\beta}{2}\nabla_Y\Big)\Big(\frac{1+\alpha}{2}\nabla^*_X + \frac{1-\alpha}{2}\nabla_X\Big)Z \\ &= \frac{(1+\alpha)(1+\beta)}{4}\nabla^*_Y\nabla^*_X Z + \frac{(1-\alpha)(1+\beta)}{4}\nabla^*_Y\nabla_X Z \\ &\quad + \frac{(1+\alpha)(1-\beta)}{4}\nabla_Y\nabla^*_X Z + \frac{(1-\alpha)(1-\beta)}{4}\nabla_Y\nabla_X Z. \end{aligned}$$

$$\nabla^{(\alpha)}_{[X,Y]} Z = \frac{1+\alpha}{2}\nabla^*_{[X,Y]} Z + \frac{1-\alpha}{2}\nabla_{[X,Y]} Z.$$

Then

$$\begin{aligned}
R^{(\alpha,\beta)}(X,Y,Z) &= \nabla_X^{(\alpha)}\nabla_Y^{(\beta)}Z - \nabla_X^{(\beta)}\nabla_Y^{(\alpha)}Z - \nabla_{[X,Y]}^{(\alpha)} \\
&= \frac{(1+\alpha)(1+\beta)}{4}(\nabla_X^*\nabla_Y^* - \nabla_Y^*\nabla_X^*)Z \\
&\quad + \frac{(1+\alpha)(1-\beta)}{4}(\nabla_X^*\nabla_Y - \nabla_Y\nabla_X^*)Z \\
&\quad + \frac{(1-\alpha)(1+\beta)}{4}(\nabla_X\nabla_Y^* - \nabla_Y^*\nabla_X)Z \\
&\quad + \frac{(1-\alpha)(1-\beta)}{4}(\nabla_X\nabla_Y - \nabla_Y\nabla_X)Z \\
&\quad - \frac{1+\alpha}{2}\nabla_{[X,Y]}^*Z - \frac{1-\alpha}{2}\nabla_{[X,Y]}Z.
\end{aligned}$$

Writing the expression in the parenthesis in terms of curvature tensors, we obtain

$$\begin{aligned}
R^{(\alpha,\beta)}(X,Y,Z) &= \frac{(1+\alpha)(1+\beta)}{4}\Big[R^*(X,Y,Z) + \nabla_{[X,Y]}^*Z\Big] \\
&\quad + \frac{(1+\alpha)(1-\beta)}{4}\Big[R^{(1,-1)}(X,Y,Z) + \nabla_{[X,Y]}^*Z\Big] \\
&\quad + \frac{(1-\alpha)(1+\beta)}{4}\Big[R^{(-1,1)}(X,Y,Z) + \nabla_{[X,Y]}Z\Big] \\
&\quad + \frac{(1-\alpha)(1-\beta)}{4}\Big[R^{(-1,-1)}(X,Y,Z) + \nabla_{[X,Y]}^*Z\Big] \\
&\quad - \frac{1+\alpha}{2}\nabla_{[X,Y]}^*Z - \frac{1-\alpha}{2}\nabla_{[X,Y]}Z.
\end{aligned}$$

Performing all cancelations, we get

$$\begin{aligned}
R^{(\alpha,\beta)}(X,Y,Z) &= \frac{(1+\alpha)(1+\beta)}{4}R^*(X,Y,Z) \\
&\quad + \frac{(1-\alpha)(1-\beta)}{4}R(X,Y,Z) \\
&\quad + \frac{(1+\alpha)(1-\beta)}{4}R^{(1,-1)}(X,Y,Z) \\
&\quad + \frac{(1-\alpha)(1+\beta)}{4}R^{(-1,1)}(X,Y,Z),
\end{aligned}$$

which is the desired result. ■

Making $\alpha = 1$ and then $\beta = 1$, we obtain the following two particular cases of relative curvatures:

$$\begin{aligned} R^{(\alpha,1)}(X,Y,Z) &= \frac{1+\alpha}{2}R^*(X,Y,Z) + \frac{1-\alpha}{2}R^{(-1,1)}(X,Y,Z) \\ R^{(1,\beta)}(X,Y,Z) &= \frac{1+\beta}{2}R^*(X,Y,Z) + \frac{1-\beta}{2}R^{(1,-1)}(X,Y,Z). \end{aligned}$$

**Corollary 8.9.2** *In the case of dually flat connections, i.e., $R = R^* = 0$, we have*

$$4R^{(\alpha,\beta)} = (1+\alpha)(1-\beta)R^{(1,-1)} + (1-\alpha)(1+\beta)R^{(-1,1)}.$$

The next goal is to write the relative curvature tensor $R^{(\alpha,\beta)}$ in terms of the difference tensor $K(X,Y)$. Recall the generalized difference tensor (1.12.37)

$$K^{(\alpha,\beta)}(X,Y) = \nabla^{(\beta)}_X Y - \nabla^{(\alpha)}_X Y. \qquad (8.9.31)$$

This can be written in terms of the regular difference tensor $K(X,Y)= \nabla^*_X Y - \nabla_X Y$ as in the following:

**Lemma 8.9.3** *(i) We have*

$$K^{(\alpha,\beta)}(X,Y) = \frac{\beta-\alpha}{2}K(X,Y).$$

*(ii) The iterated difference tensor is given by*

$$K^{(\alpha,\beta)}\Big(X, K^{(\alpha,\beta)}(Y,Z)\Big) = \frac{(\alpha-\beta)^2}{4}K\Big(X, K(Y,Z)\Big).$$

*Proof:*

(*i*) Using the linear combination (8.5.11), we obtain

$$\begin{aligned} K^{(\alpha,\beta)}(X,Y) &= \nabla^{(\beta)}_X Y - \nabla^{(\alpha)}_X Y \\ & \Big[\frac{1+\beta}{2}\nabla^*_X Y + \frac{1-\beta}{2}\nabla_X Y\Big] \\ & -\Big[\frac{1+\alpha}{2}\nabla^*_X Y + \frac{1-\alpha}{2}\nabla_X Y\Big] \\ &= \frac{\beta-\alpha}{2}\nabla^*_X Y - \frac{\beta-\alpha}{2}\nabla_X Y = \frac{\beta-\alpha}{2}K(X,Y). \end{aligned}$$

(*ii*) Using (*i*) twice leads to

$$\begin{aligned} K^{(\alpha,\beta)}\Big(X, K^{(\alpha,\beta)}(Y,Z)\Big) &= \frac{\beta-\alpha}{2} K\Big(X, K^{(\alpha,\beta)}(Y,Z)\Big) \\ &= \frac{(\beta-\alpha)^2}{4} K\Big(X, K(Y,Z)\Big). \end{aligned}$$

■

**Proposition 8.9.4** *The following relation holds:*

$$\begin{aligned} R^{(\alpha)}(X,Y,Z) + R^{(\beta)}(X,Y,Z) &= R^{(\alpha,\beta)}(X,Y,Z) + R^{(\beta,\alpha)}(X,Y,Z) \\ &+ \frac{(\alpha-\beta)^2}{4}\Big\{K\Big(X, K(Y,Z)\Big) \\ &- K\Big(Y, K(X,Z)\Big)\Big\}. \end{aligned}$$

*Proof:* From (8.9.31), we have

$$\begin{aligned} K^{(\beta,\alpha)}\Big(X, K^{(\beta,\alpha)}(Y,Z)\Big) &= K^{(\beta,\alpha)}\Big(X, \nabla_Y^{(\alpha)} Z - \nabla_Y^{(\beta)} Z\Big) \\ &= K^{(\beta,\alpha)}\Big(X, \nabla_Y^{(\alpha)} Z\Big) - K^{(\beta,\alpha)}\Big(X, \nabla_Y^{(\beta)} Z\Big) \\ &= \nabla_X^{(\alpha)}\nabla_Y^{(\alpha)} Z - \nabla_X^{(\beta)}\nabla_Y^{(\alpha)} Z - \nabla_X^{(\alpha)}\nabla_Y^{(\beta)} Z \\ &\quad + \nabla_X^{(\beta)}\nabla_Y^{(\beta)} Z. \end{aligned}$$

From Lemma 8.9.3, part (*ii*), the previous relation becomes

$$\begin{aligned} \frac{(\alpha-\beta)^2}{4} K\Big(X, K(Y,Z)\Big) &= \nabla_X^{(\alpha)}\nabla_Y^{(\alpha)} Z - \nabla_X^{(\beta)}\nabla_Y^{(\alpha)} Z \\ &\quad - \nabla_X^{(\alpha)}\nabla_Y^{(\beta)} Z + \nabla_X^{(\beta)}\nabla_Y^{(\beta)} Z. \end{aligned} \tag{8.9.32}$$

Swapping $X$ and $Y$ yields

$$\begin{aligned} \frac{(\alpha-\beta)^2}{4} K\Big(Y, K(X,Z)\Big) &= \nabla_Y^{(\alpha)}\nabla_X^{(\alpha)} Z - \nabla_Y^{(\beta)}\nabla_X^{(\alpha)} Z \\ &\quad - \nabla_Y^{(\alpha)}\nabla_X^{(\beta)} Z + \nabla_Y^{(\beta)}\nabla_X^{(\beta)} Z. \end{aligned} \tag{8.9.33}$$

Subtract (8.9.32) and (8.9.33) and then use the curvature definition to get

$$\begin{aligned}
&\frac{(\alpha-\beta)^2}{4}\Big\{K\Big(X, K(Y,Z)\Big) - K\Big(Y, K(X,Z)\Big)\Big\} \\
&= R^{(\alpha)}(X,Y,Z) + R^{(\beta)}(X,Y,Z) - \{[\nabla^{(\alpha)}_X, \nabla^{(\beta)}_Y]Z - \nabla^{(\alpha)}_{[X,Y]}Z\} \\
&\quad -\{[\nabla^{(\beta)}_X, \nabla^{(\alpha)}_Y]Z - \nabla^{(\beta)}_{[X,Y]}Z\} \\
&= R^{(\alpha)}(X,Y,Z) + R^{(\beta)}(X,Y,Z) - R^{(\alpha,\beta)}(X,Y,Z) \\
&\quad -R^{(\beta,\alpha)}(X,Y,Z).
\end{aligned}$$

■

Making $\alpha = 1$ and $\beta = -1$ in Proposition 8.9.4, we obtain the following consequence that will be useful in later applications.

**Corollary 8.9.5** *The following relation holds*

$$\begin{aligned}
R^{(1,-1)}(X,Y,Z) + R^{(-1,1)}(X,Y,Z) &= R^*(X,Y,Z) + R(X,Y,Z) \\
&\quad -\Big\{K\Big(X, K(Y,Z)\Big) \\
&\quad -K\Big(Y, K(X,Z)\Big)\Big\}.
\end{aligned}$$

## 8.10 Curvature of $\alpha$-Connections

This section deals with the relationship among curvatures $R^{(\alpha)}$, $R$ and $R^*$ associated, respectively, with connections $\nabla^{(\alpha)}$, $\nabla$ and $\nabla^*$. The following formula for $R^{(\alpha)}$ was computed in Zhang [87].[1] We shall present it here as a consequence of the relative curvature developed in the previous section.

**Proposition 8.10.1** *The curvature $R^{(\alpha)}$ satisfies the equation*

$$\begin{aligned}
R^{(\alpha)}(X,Y,Z) &= \frac{1+\alpha}{2}R^*(X,Y,Z) + \frac{1-\alpha}{2}R(X,Y,Z) \\
&\quad +\frac{1-\alpha^2}{4}\Big(K\big(Y, K(X,Z)\big) - K\big(X, K(Y,Z)\big)\Big).
\end{aligned}$$

*Proof:* Making $\alpha = \beta$ in Proposition 8.9.1 yields

$$\begin{aligned}
4R^{(\alpha)} &= (1+\alpha)^2 R^* + (1-\alpha)^2 R + (1-\alpha^2)R^{(1,-1)} \\
&\quad +(1-\alpha^2)R^{(-1,1)} \\
&= (1+\alpha)^2 R^* + (1-\alpha)^2 R + (1-\alpha^2)\Big(R^{(1,-1)} + R^{(-1,1)}\Big).
\end{aligned}$$

[1]In our case the roles of $R$ and $R^*$ are reversed.

Substituting the last term from Corollary 8.9.5, we obtain

$$\begin{aligned} 4R^{(\alpha)}(X,Y,Z) &= (1+\alpha)^2 R^*(X,Y,Z) + (1-\alpha)^2 R(X,Y,Z) \\ &\quad +(1-\alpha^2)R^*(X,Y,Z) + (1-\alpha^2)R(X,Y,Z) \\ &\quad -(1-\alpha^2)\Big\{K\Big(X,K(Y,Z)\Big) - K\Big(Y,K(X,Z)\Big)\Big\} \\ &= 2(1+\alpha)R^*(X,Y,Z) + 2(1-\alpha)R(X,Y,Z) \\ &\quad +(1-\alpha^2)\Big(K\big(Y,K(X,Z)\big) - K\big(X,K(Y,Z)\big)\Big), \end{aligned}$$

which after dividing by 4 leads to the desired relation. ∎

**Corollary 8.10.2** *The following relation holds*

$$R^{(\alpha)}(X,Y,X) - R^{(-\alpha)}(X,Y,X) = \alpha\Big(R^*(X,Y,X) - R(X,Y,X)\Big). \tag{8.10.34}$$

*Proof:* Writing the equations for $R^{(\alpha)}$ and $R^{(-\alpha)}$, after subtraction we have

$$\begin{aligned} R^{(\alpha)}(X,Y,X) - R^{(-\alpha)}(X,Y,X) &= \Big(\frac{1+\alpha}{2} - \frac{1-\alpha}{2}\Big)R^*(X,Y,X) \\ &\quad +\Big(\frac{1-\alpha}{2} - \frac{1+\alpha}{2}\Big)R(X,Y,Z) \\ &= \alpha\Big(R^*(X,Y,X) - R(X,Y,X)\Big). \end{aligned}$$

∎

It is worth noting that if connections $\nabla$ and $\nabla^*$ have zero curvature tensors, $R = R^* = 0$, then $R^{(\alpha)} = R^{(-\alpha)}$. However, the curvature tensor of the connection $\nabla^{(\alpha)}$ does not necessarily vanish, its expression being given by

$$R^{(\alpha)}(X,Y,Z) = \frac{1-\alpha^2}{4}\Big(K\big(Y,K(X,Z)\big) - K\big(X,K(Y,Z)\big)\Big).$$

It follows that a necessary condition for all $\alpha$-connections to have zero curvature tensors is that

$$K\big(Y,K(X,Z)\big) = K\big(X,K(Y,Z)\big). \tag{8.10.35}$$

This can be written in terms of the skewness tensor as

$$C(K(X,Z),Y,W) = C(X,K(Y,Z),W), \qquad \forall X,Y,Z,W \in \mathcal{X}(M). \tag{8.10.36}$$

The next result is an extension of Proposition 8.1.4 to $\alpha$-connections.

**Proposition 8.10.3** *We have*

$$
\begin{aligned}
& g\Big(R^{(\alpha)}(X,Y,Z),W\Big)+g\Big(R^{(-\alpha)}(X,Y,W),Z\Big)\\
= \; & \frac{1-\alpha^2}{4}\Big\{C\Big(K(X,Z),Y,W\Big)+C\Big(K(X,W),Y,Z\Big)\\
& -C\Big(X,K(Y,Z),W\Big)-C\Big(X,K(Y,W),Z\Big)\Big\}.
\end{aligned}
$$

*Proof:* Applying Proposition 8.10.1 for $\alpha$ and $-\alpha$, we have, respectively,

$$
\begin{aligned}
& g\Big(R^{(\alpha)}(X,Y,Z),W\Big)\\
& =\frac{1+\alpha}{2}g\Big(R^*(X,Y,Z),W\Big)+\frac{1-\alpha}{2}g\Big(R(X,Y,Z),W\Big)\\
& +\frac{1-\alpha^2}{4}\Big\{C\Big(Y,K(X,Z),W\Big)-C\Big(X,K(Y,Z),W\Big)\Big\};
\end{aligned}
$$

$$
\begin{aligned}
& g\Big(R^{(-\alpha)}(X,Y,W),Z\Big)\\
& =\frac{1-\alpha}{2}g\Big(R^*(X,Y,W),Z\Big)+\frac{1+\alpha}{2}g\Big(R(X,Y,W),Z\Big)\\
& +\frac{1-\alpha^2}{4}\Big\{C\Big(Y,K(X,W),Z\Big)-C\Big(X,K(Y,W),Z\Big)\Big\}.
\end{aligned}
$$

Using

$$
\begin{aligned}
g\Big(R(X,Y,Z),W\Big)+g\Big(R^*(X,Y,W),Z\Big) &= 0\\
g\Big(R^*(X,Y,Z),W\Big)+g\Big(R(X,Y,W),Z\Big) &= 0,
\end{aligned}
$$

see Proposition 8.1.4, part $(i)$, then adding the previous two expressions leads to the formula claimed by the proposition. ■

## 8.11 Statistical Manifolds

Statistical manifolds have been introduced by several authors (see Lauritzen [54] and Simon [76]) in a few equivalent ways at different moments in time. In this section we show how starting from the skewness tensor $C$ one can construct the geometry of dual connections and the family of $\alpha$-connections of a statistical manifold.

A *statistical structure* is a triple $(M, g, C)$, where $C$ is a 3-covariant, totally symmetric tensor on the Riemannian manifold $(M, g)$, called the skewness tensor. We shall show how to construct a statistical manifold $(M, g, \nabla, \nabla^*)$ starting from the triple $(M, g, C)$.

First, we consider the Levi–Civita connection $\nabla^{(0)}$ associated with the Riemannian metric $g$, which is the unique metrical and torsion-free linear connection on $(M, g)$. This is given by the Koszul formula, see Theorem 7.11.1,

$$\begin{aligned} 2g(\nabla^{(0)}_X Y, Z) &= X(g(Y,Z)) + Y(g(Z,X)) - Z(g(X,Y)) \\ &\quad + g([X,Y],Z) - g([Y,Z],X) - g([X,Z],Y). \end{aligned}$$

The previous formula is equivalent locally with the fact that the connection components can be expressed in terms of the metric coefficients by

$$\Gamma^l_{ij} = \frac{1}{2} g^{kl}\Big(\partial_{x^i} g_{kj} + \partial_{x^j} g_{ik} - \partial_{x^k} g_{ij}\Big),$$

which is nothing else but the formula for the Christoffel symbols.

The next result introduces a pair of dual connections.

**Proposition 8.11.1** *The geometric objects $\nabla$, $\nabla^*$ defined by*

$$\begin{aligned} g(\nabla_X Y, Z) &= g(\nabla^{(0)}_X Y, Z) - \frac{1}{2} C(X,Y,Z) \\ g(\nabla^*_X Y, Z) &= g(\nabla^{(0)}_X Y, Z) + \frac{1}{2} C(X,Y,Z), \ \forall X, Y, Z \in \mathcal{X}(M) \end{aligned}$$

*are torsion-less dual connections.*

*Proof:* We need to show first that $\nabla$ and $\nabla^*$ are torsion-less linear connections. It suffices to show this for the first connection, since a similar argument applies to the second one. Obviously $\nabla_X Y$ is $\mathbb{R}$–linear in both arguments. For any smooth function $f$ on $M$, we have

$$\begin{aligned} g(\nabla_{fX} Y, Z) &= g(\nabla^{(0)}_{fX} Y, Z) - \frac{1}{2} C(fX, Y, Z) \\ &= g(f\nabla^{(0)}_X Y, Z) - \frac{1}{2} f C(X,Y,Z) \\ &= f\{g(\nabla^{(0)}_X Y, Z) - \frac{1}{2} C(X,Y,Z)\} \\ &= f g(\nabla_X Y, Z), \qquad \forall Z \in \mathcal{X}(M), \end{aligned}$$

so $\nabla_{fX}Y = f\nabla_X Y$. Next we check the Leibniz property in the second argument,

$$\begin{aligned} g(\nabla_X(fY), Z) &= g(\nabla^{(0)}_X(fY), Z) - \frac{1}{2}C(X, fY, Z) \\ &= g(f\nabla^{(0)}_X Y + X(f)Y, Z) - \frac{1}{2}fC(X, Y, Z) \\ &= f\{g(\nabla^{(0)}_X Y, Z) - \frac{1}{2}C(X, Y, Z)\} + g(X(f)Y, Z) \\ &= fg(\nabla_X Y, Z) + g(X(f)Y, Z) \\ &= g(f\nabla_X Y + X(f)Y, Z), \qquad \forall Z \in \mathcal{X}(M), \end{aligned}$$

which implies $\nabla_X(fY) = f\nabla_X Y + X(f)Y$.

The torsion-less property of $\nabla$ follows from the next computation:

$$\begin{aligned} & g(\nabla_X Y - \nabla_Y X - [X, Y], Z) \\ = \; & g(\nabla_X Y, Z) - g(\nabla_Y X, Z) - g([X, Y], Z) \\ = \; & g(\nabla^{(0)}_X Y, Z) - \frac{1}{2}C(X, Y, Z) - \{g(\nabla^{(0)}_Y X, Z) - \frac{1}{2}C(Y, X, Z)\} \\ & -g([X, Y], Z) \\ = \; & g(\nabla^{(0)}_X Y - \nabla^{(0)}_Y X - [X, Y], Z) = 0, \end{aligned}$$

where the last identity uses that $\nabla^{(0)}$ is torsion-less.

In the following we show that $\nabla$ and $\nabla^*$ are dual connections. Since $\nabla^{(0)}$ is a metrical connection and $C$ is a totally symmetric tensor, we have

$$\begin{aligned} g(\nabla_X Y, Z) + g(Y, \nabla^*_X Z) &= g(\nabla^{(0)}_X Y, Z) - \frac{1}{2}C(X, Y, Z) \\ & \quad +g(Y, \nabla^{(0)}_X Z) + \frac{1}{2}C(X, Y, Z) \\ &= Xg(Y, Z) + \frac{1}{2}\{C(X, Z, Y) - C(X, Y, Z)\} \\ &= Xg(Y, Z). \end{aligned}$$

■

**Corollary 8.11.2** *The dual connections $\nabla$, $\nabla^*$ and the skewness tensor $C$ are related by*

$$g(\nabla^*_X Y, Z) = g(\nabla_X Y, Z) + C(X, Y, Z).$$

The skewness tensor $C$ can be used to introduce the $\alpha$-connection as in the following

$$\begin{aligned} g(\nabla^{(\alpha)}_X Y, Z) &= g\Big(\frac{1-\alpha}{2}\nabla_X Y + \frac{1+\alpha}{2}\nabla^*_X Y, Z\Big) \\ &= \frac{1-\alpha}{2} g(\nabla_X Y, Z) + \frac{1+\alpha}{2} g(\nabla^*_X Y, Z) \\ &= \frac{1-\alpha}{2}\{g(\nabla^{(0)}_X Y, Z) - \frac{1}{2}C(X,Y,Z)\} \\ &\quad + \frac{1+\alpha}{2}\{g(\nabla^{(0)}_X Y, Z) + \frac{1}{2}C(X,Y,Z)\} \\ &= g(\nabla^{(0)}_X Y, Z) + \frac{\alpha}{2}C(X,Y,Z). \end{aligned}$$

Hence the $\alpha$-connection can be introduced by the formula

$$g(\nabla^{(\alpha)}_X Y, Z) = g(\nabla^{(0)}_X Y, Z) + \frac{\alpha}{2}C(X,Y,Z). \tag{8.11.37}$$

**Proposition 8.11.3** *The covariant derivatives of the metric tensor $g$, with respect to connections $\nabla$, $\nabla^*$, and $\nabla^{(\alpha)}$, respectively, are related to the skewness tensor $C$ by*

(*i*) $\nabla g = C$;

(*ii*) $\nabla^* g = -C$;

(*iii*) $\nabla^{(\alpha)} g = -\alpha C$.

*Proof:* Using the definition of the covariant derivative of a tensor and Corollary 8.11.2, we have

$$\begin{aligned} (\nabla g)(X,Y,Z) &= Xg(Y,Z) - g(\nabla_X Y, Z) - g(Y, \nabla_X Z) \\ &= g(Y, \nabla^*_X Z) - g(Y, \nabla_X Z) \\ &= g(Y, \nabla^*_X Z) - g(Y, \nabla_X Z) = C(X,Y,Z). \end{aligned}$$

The other two relations can be proved using a similar argument. ■

## 8.12 Problems

**8.1.** Let $\nabla$ be a metrical connection on the Riemannian manifold $(M,g)$. If $\gamma(s)$ is a curve on $M$ and $X, Y \in \mathcal{X}(M)$ such that $X$ and $Y$ are parallely transported along the curve $\gamma$ with respect to $\nabla$, show that

$$g(X_{|\gamma(t)}, Y_{|\gamma(t)}) = g(X_{|\gamma(0)}, Y_{|\gamma(0)}).$$

**8.2.** Let $(M, g, \nabla, \nabla^*)$ be a statistical manifold and $\gamma : [0, T] \to M$ be a differentiable curve. Consider the vector fields $X, Y \in \mathcal{X}(M)$ such that

$$\nabla_{\dot{\gamma}} X = 0, \qquad \nabla^*_{\dot{\gamma}} Y = 0. \tag{8.12.38}$$

Show that

$$g(X_{|\gamma(t)}, Y_{|\gamma(t)}) = g(X_{|\gamma(0)}, Y_{|\gamma(0)}). \tag{8.12.39}$$

**8.3.** Let $\gamma : [0, T] \to M$ be a smooth curve on a manifold, and consider the vector field $X \in \mathcal{X}(M)$ such that $\nabla_{\dot{\gamma}} X = 0$. Show that

$$X^j_{|\gamma(t)} = X^j_{|\gamma(0)} - t\Big(\Gamma^j_{\ell p} \dot{\gamma}^\ell X^p_{|\gamma(t)}\Big)_{|t=0} + O(t^2).$$

**8.4.** Show that if relation (8.12.39) holds for any two vector fields $X, Y \in \mathcal{X}(M)$ that satisfy (8.12.38), then $\nabla^*$ is the conjugate connection of $\nabla$.

**8.5.** Let $R$ and $R^*$ be the Riemannian curvature tensors associated with dual connections $\nabla$ and $\nabla^*$. Show that

(a) $R_{ijkl} = R^*_{ikjl} + R^*_{ilkj}$;

(b) $R^*_{ijlk} = R_{iklj} + R^*_{iljk}$.

**8.6.** Two coordinate systems $(x^i)$ and $(\zeta_j)$ on a Riemannian manifold $(M, g)$ are called dual if $g(\partial_{x^i}, \partial_{\zeta_j}) = \delta^i_j$, where $\partial_{x^i} = \frac{\partial}{\partial x^i}$ and $\partial_{\zeta_j} = \frac{\partial}{\partial \zeta_j}$ are the coordinate vector fields associated with the systems $(x^i)$ and $(\zeta_j)$, respectively. Let $(x^i)$ and $(\zeta_j)$ be dual coordinate systems on $(M, g)$.

(a) Show that we have

$$\partial_{x^i} = \frac{\partial \zeta_j}{\partial x^i} \partial_{\zeta_j}, \qquad \partial_{\zeta_j} = \frac{\partial x^k}{\partial \zeta_j} \partial_{x^k}.$$

(b) Denote by $g_{ij}(x) = g(\partial_{x^i}, \partial_{x^j})$ and $g^{ij}(x) = g(\partial_{\zeta_i}, \partial_{\zeta_j})$. Show that

$$g_{ij}(x) = \frac{\partial \zeta_j}{\partial x^i}, \qquad g_{ij}(\zeta) = \frac{\partial x^j}{\partial \zeta_i}.$$

(c) Show that $g_{ij}(x)$ and $g_{ij}(\zeta)$ are each other matrix inverse, i.e. $\sum_j g_{ij}(x)g_{jk}(\zeta) = \delta_{ik}$.

(d) Show that there are two functions $\psi(x)$ and $\varphi(\zeta)$ such that

$$x^i = \partial_{\zeta_i}\varphi(\zeta), \qquad \zeta_j = \partial_{x^j}\psi(x).$$

The functions $\psi(x)$ and $\varphi(\zeta)$ are called potentials.

(e) Show that $\psi$ and $\phi$ are related by the following Legendre transform

$$\varphi(\zeta) = x^i\zeta_i - \psi(x).$$

(f) Verify that the Riemannian metric can be written as a Hessian:

$$g_{ij}(x) = \partial_{x^i}\partial_{x^j}\psi(x), \qquad g_{ij}(\zeta) = \partial_{\zeta_i}\partial_{\zeta_j}\varphi(\zeta).$$

(g) Give a reason why the potential functions $\psi(x)$ and $\varphi(\zeta)$ are convex functions.

(e) Prove the following maximization property

$$\psi(x) = \max_{\zeta}\big(x^i\zeta_i - \varphi(\zeta)\big), \qquad \varphi(\zeta) = \max_{x}\big(x^i\zeta_i - \psi(x)\big).$$

**8.7.** Let $(M, g)$ be a Riemannian manifold. Assume there are locally defined two convex functions $\psi(x)$ and $\varphi(\zeta)$ such that

$$g_{ij}(x) = \partial_{x^i}\partial_{x^j}\psi(x), \qquad g_{ij}(\zeta) = \partial_{\zeta_i}\partial_{\zeta_j}\varphi(\zeta).$$

Show that $x^i = \partial_{\zeta_i}\varphi(\zeta)$ and $\zeta_j = \partial_{x^j}\psi(x)$ are dual coordinate systems.

**8.8.** Let $(M, g, \nabla, \nabla^*)$ be a dually flat statistical manifold. Prove that there is a pair of dual coordinate systems $(x^i)$ and $(\zeta_\alpha)$ such that $(x^i)$ is $\nabla$-affine and $(\zeta_\alpha)$ is $\nabla^*$-affine.

**8.9.** Two coordinate systems $(x^i)$ and $(\zeta_\alpha)$ are called affine if $\zeta_\alpha = a_{j\alpha}x^j + b_\alpha$, where $(a_{j\alpha})$ is an $n \times n$ real matrix and $b_\alpha$ are constants.

Let $\Gamma_{ij}^k(x)$ and $\Gamma_{\alpha\beta}^\gamma(\zeta)$ be the coefficients of the connection $\nabla$ in the aforementioned affine systems of coordinates, i.e.,

$$\nabla_{\partial_{x^i}}\partial_{x^j} = \Gamma_{ij}^k(x)\partial_{x^k}, \qquad \nabla_{\partial_{\zeta_\alpha}}\partial_{\zeta_\beta} = \Gamma_{\alpha\beta}^\gamma(\zeta)\partial_{\zeta_\gamma}.$$

(*a*) Verify the following change of coefficients formula

$$\Gamma^{\gamma}_{\alpha\beta}\partial_{\zeta_\gamma}x^k = (\partial_{\zeta_\alpha}x^i)(\partial_{\zeta_\beta}x^j)\Gamma^k_{ij} + \partial_{\zeta_\alpha}\partial_{\zeta_\beta}x^k.$$

(*b*) Show that if the connection $\nabla$ is flat with respect to both coordinates systems, then $(x^i)$ and $(\zeta_\alpha)$ are affine coordinates systems.

**8.10.** Let $(x^i)$ and $(\zeta_\alpha)$ be dual coordinate systems.

(*a*) Show that if a connection $\nabla$ is flat with respect to the coordinate system $(x^i)$, then the coefficients of the dual connection $\nabla^*$ are given by

$$\Gamma^*_{ij,k} = \partial_{x^i}\partial_{x^j}\partial_{x^k}\psi(x),$$

where $\psi(x)$ is the potential associated with $\zeta$.

(*b*) Find the coefficients of connection $\nabla$ in terms of $(x^i)$, $(\zeta_\alpha)$ and $\psi(x)$.

**8.11.** Let $(x^i)$ and $(\zeta_\alpha)$ be dual coordinate systems, such that the connection $\nabla$ is flat in $(x^i)$. Show that the dual connection $\nabla^*$ is flat with respect to $(\zeta_\alpha)$.

**8.12.** Show that

$$\Gamma^{(\alpha)}_{ij,k} + \Gamma^{(-\alpha)}_{ij,k} = \partial_{x^i}g_{jk} - \Gamma_{ik,j} + \Gamma_{ij,k}.$$

**8.13.** If $\omega$ is a 1-form on the Riemannian manifold $(M, g)$ and $\nabla$ a linear connection, define $\nabla\omega$ by

$$(\nabla_Y\omega)(X) = Y\omega(X) - \omega(\nabla_Y X), \qquad \forall X, Y \in \mathcal{X}(M).$$

Let $\nabla^*$ be the dual connection of $\nabla$. Show the following relations:

(*a*) $(\nabla_Y\omega)(X) = g(\nabla^*_Y\omega^\#, X)$;

(*b*) $(\nabla_Y\omega)^\# = \nabla^*_Y\omega^\#$;

(*c*) $\omega$ is $\nabla$-parallel if and only if $\omega^\#$ is $\nabla^*$-parallel.

# Chapter 9

# Dual Volume Elements

This chapter defines the volume elements associated with two dual connections and investigates their relationship. First, we define the Riemannian volume element and show that it is parallel with respect to the Levi–Civita connection. Since the converse is also true, this provides an alternate definition for the volume element used in defining volume elements associated with other connections. In particular, we define and study the volume element associated with an $\alpha$-connection. The volume elements for the exponential model and mixture model are computed, as examples of distinguished importance in the theory.

The necessary and sufficient condition for a torsion-free connection to admit a parallel volume form was found by Nomizu and Sasaki [62]. They described this condition in terms of the symmetry of the associated Ricci tensor. These type of connections are called equiaffine connections. A sufficient condition was found by Takeuki and Amari [79], who also developed an expression for the $\alpha$-parallel volume form for the exponential family. Further sufficient conditions have been investigated by Matsuzoe et al. [55].

It is also worth noting the relation with Bayesian statistics. From the differential geometry point of view, Jeffrey's prior in Bayesian statistics is the parallel volume form with respect to the Levi–Civita connection of the Fisher metric. This explains why the Riemannian volume form has traditionally been taken as Jeffrey's prior. Following this idea, Takeuki and Amari [79], and later Matsuzoe et al. [55] advanced the idea of an $\alpha$-parallel prior, which is a volume form that is parallel with respect to the $\nabla^{(\alpha)}$ connection, for general $\alpha$. The value $\alpha = 0$ recovers Jeffrey's prior from Bayesian inference.

O. Calin and C. Udrişte, *Geometric Modeling in Probability and Statistics*,
DOI 10.1007/978-3-319-07779-6_9,

## 9.1 Riemannian Volume Element

Let $(M, g)$ be an oriented Riemannian manifold. For any point $p \in M$, let $\{e_i, i = 1, \dots, n\}_p$ be an orthonormal basis in $T_pM$, with respect to the metric $g$. Denote by $\{e_i^*\}$ the associated dual basis, i.e., $e_i^*(e_j)=\delta^i_j$. The $n$-form $dv$ defined by

$$dv_{|p} = e_1^*(p) \wedge \cdots \wedge e_n^*(p)$$

is called the *Riemannian volume form* on $M$.

It is useful to write the form $dv$ in local coordinates. Let $(U, x)$ be a local chart. Then $\{\partial_{x^1}(p), \dots, \partial_{x^n}(p)\}$ is a local basis of $T_pM$ and the coefficients of the Riemannian metric are given by

$$g_{ij}(p) = g(\partial_{x^i}(p), \partial_{x^j}(p)).$$

Let $\partial_{x^i}(p) = a_i^k e_k^*$ with the matrix $A = (a_i^k)$. Using the definition of the wedge product we have

$$\begin{aligned} dv_p\big(\partial_{x^1}(p), \dots, \partial_{x^n}(p)\big) &= e_1^*(p) \wedge \cdots \wedge e_n^*(p)\Big(\partial_{x^1}(p), \dots, \partial_{x^n}(p)\Big) \\ &= \det\Big(e_i^*\big(\partial_{x^j}(p)\big)\Big)_{(p)} = \det g_p\Big(e_i^*, \partial_{x^j}(p)\Big) \\ &= \det g_p\Big(e_i^*, a_j^k e_k^*\Big) = \det\Big(a_j^k g_p(e_i^*, e_k^*)\Big) \\ &= \det(a_j^k \delta_k^i) = \det(a_j^i) = \det A. \end{aligned} \qquad (9.1.1)$$

On the other side

$$\begin{aligned} \det(g_{ij}) &= \det g\big(\partial_{x_i}(p), \partial_{x_j}(p)\big) = \det\Big(a_i^k a_j^l g(e_k^*, e_l^*)\Big) = \det(a_i^k a_j^l \delta_{kl}) \\ &= \det(AA^T) = (\det A)^2. \end{aligned} \qquad (9.1.2)$$

From (9.1.1) and (9.1.2) we obtain

$$dv\big(\partial_{x^1}(p), \dots, \partial_{x^n}(p)\big) = \sqrt{\det(g_{ij})}\,,$$

and hence the Riemannian volume element in local coordinates is

$$dv = \sqrt{\det g}\; dx^1 \wedge \cdots \wedge dx^n\,, \qquad (9.1.3)$$

where $g = (g_{ij})$.

The next result shows that the volume element form $dv$ is parallel with respect to the covariant derivation realized by the Levi–Civita connection. This will provide in the sequel an alternate way of defining volume elements associated with linear connections.

**Lemma 9.1.1** *The Riemannian volume element is parallel with respect to the Levi–Civita connection, i.e.,* $\nabla^{(0)}_Y dv = 0$ *for all* $Y \in \mathcal{X}(M)$.

*Proof:* Since $\nabla^{(0)}_Y = Y^i \nabla^{(0)}_{\partial_{x^i}}$ it suffices to show that

$$\nabla^{(0)}_{\partial_{x^i}} dv = 0, \qquad \forall i = 1, \dots, n.$$

The following computation is based on the definition of the covariant derivative of an $n$-form and a manipulation of the matrix $(g_{ij})$ and the relation between its inverse and its determinant. We have

$$\begin{aligned}
&(\nabla^{(0)}_{\partial_{x^i}} dv)(\partial_{x^1}, \dots, \partial_{x^n}) \\
&= \partial_{x^i} dv(\partial_{x^1}, \dots, \partial_{x^n}) - dv(\nabla^{(0)}_{\partial x^i} \partial_{x^1}, \partial_{x^2}, \dots, \partial_{x^n}) \\
&\quad - \dots - dv(\partial_{x^1}, \dots, \partial_{x^{n-1}}, \nabla^{(0)}_{\partial x^i} \partial_{x_n}) \\
&= \partial_{x^i} \sqrt{\det g} - dv\big(\Gamma^{(0)k_1}_{i1} \partial_{x^{k_1}}, \partial_{x^2}, \dots, \partial_{x^n}\big) \\
&\quad - \dots - dv\big(\partial_{x^1}, \dots, \partial_{x^{n-1}}, \Gamma^{(0)k_n}_{in} \partial_{x_{k^n}}\big) \\
&= \partial_{x^i} \sqrt{\det g} - \Gamma^{(0)k_1}_{i1} \delta_{1,k_1} \sqrt{\det g} - \dots - \Gamma^{(0)k_n}_{in} \delta_{n,k_n} \sqrt{\det g} \\
&= \partial_{x^i} \sqrt{\det g} - \Big(\Gamma^{(0)1}_{i1} + \dots + \Gamma^{(0)n}_{in}\Big) \sqrt{\det g} \\
&= \partial_{x^i} \sqrt{\det g} - \Gamma^{(0)j}_{ij} \sqrt{\det g} \\
&= \partial_{x^i} \sqrt{\det g} - \frac{1}{2} g^{jp} \Big( \frac{\partial g_{ip}}{\partial x^j} + \frac{\partial g_{jp}}{\partial x^i} - \frac{\partial g_{ij}}{\partial x^p} \Big) \sqrt{\det g} \\
&= \partial_{x^i} \sqrt{\det g} - \frac{1}{2} g^{jp} \frac{\partial g_{jp}}{\partial x^i} \sqrt{\det g} \\
&= \frac{1}{2} \frac{1}{\sqrt{\det g}} \frac{\partial (\det g)}{\partial x^i} - \frac{1}{2} g^{jp} \frac{\partial g_{jp}}{\partial x^i} \sqrt{\det g} \\
&= \frac{1}{2} \frac{1}{\sqrt{\det g}} \frac{\partial (\det g)}{\partial x^i} - \frac{1}{2(\det g)} \frac{\partial (\det g)}{\partial g_{jp}} \frac{\partial g_{jp}}{\partial x^i} \sqrt{\det g} = 0.
\end{aligned}$$

■

The next result states that the only $n$-form that is parallel to the Levi–Civita connection is proportional to the Riemannian volume element. Let $\Lambda^n(M)$ denote the space of $n$-forms on the manifold $M$.

**Lemma 9.1.2** *If* $\omega \in \Lambda^n(M)$ *and* $\nabla^{(0)} \omega = 0$, *then there is a constant* $C$ *such that* $\omega = C\, dv$.

*Proof:* Since $\omega$ and $dv$ are both $n$-forms, they are proportional, i.e., there is a function $f \in \mathcal{F}(M)$ such that

$$\omega = f\,dv.$$

For any vector field $Y \in \mathcal{X}(M)$, we have

$$0 = \nabla^{(0)}_Y \omega = \nabla^{(0)}_Y (f dv) = f \underbrace{\nabla^{(0)}_Y dv}_{=0} + Y(f)dv,$$

where the first term vanishes in the virtue of Lemma 9.1.1. Hence $Y(f) = 0$, for all $Y \in \mathcal{X}(M)$, i.e., $f$ is a constant function. ∎

**Theorem 9.1.3** *Let $\omega \in \Lambda^n(M)$. Then*

$$\nabla^{(0)}_Y \omega = 0, \qquad \forall Y \in \mathcal{X}(M)$$

*if and only if $\omega = C dv$, with $C \in \mathbb{R}$.*

*Proof:* The proof is a straightforward consequence of Lemmas 9.1.1 and 9.1.2. However, for the sake of completeness we shall give next a direct proof.

Since $\omega \in \Lambda^n(M)$, there is a function $f \in \mathcal{F}(M)$ such that locally we have

$$\omega = f dx^1 \wedge \cdots \wedge dx^n.$$

A computation similar with the one done in the proof of Lemma 9.1.1 yields

$$\begin{aligned} 0 &= (\nabla^{(0)}_{\partial_{x^i}} \omega)(\partial_{x^1}, \dots, \partial_{x^n}) = \partial_{x^i} f(x) - f(x)\Gamma^{(0)j}_{ij} \\ &= \partial_{x^i} f - \frac{1}{2} f \frac{1}{g} \frac{\partial(\det g)}{\partial g_{jp}} \frac{\partial g_{jp}}{\partial x^i} = \partial_{x^i} f - \frac{1}{2} f \partial_{x^i}(\ln(\det g)). \end{aligned}$$

Hence, the function $f$ satisfies the following PDE

$$\partial_{x^i} f - \frac{1}{2} f \partial_{x^i}(\ln(\det g)) = 0.$$

Multiplying by the integrating factor $g^{-1/2}$, we find

$$\partial_{x^i}\Big(\frac{f}{\sqrt{\det g}}\Big) = 0, \qquad \forall i = 1, \dots, n,$$

so $f = C\sqrt{g}$, with $C$ constant. Hence

$$\omega = f dx = C\sqrt{\det g}\, dx^1 \wedge \cdots \wedge dx^n = C dv, \quad C \in \mathbb{R}.$$

■

Inspired by the previous theorem, we shall introduce the following volume concept.

**Definition 9.1.4** *The $\alpha$-volume element is an $n$-form $\omega \in \Lambda^n(M)$ such that*

$$\nabla^{(\alpha)}\omega = 0.$$

We notice that $\omega$ is defined up to a scaling factor.

Taking $\alpha = 1$ and $\alpha = -1$ we obtain the following two distinguished cases

$$\nabla^*\omega^* = 0, \qquad \nabla\omega = 0.$$

The next two sections provide explicit calculations for the $\alpha$-volume elements in the case of the following distributions

(*i*) the exponential model

$$p(x,\xi) = e^{C(x)+\xi^i F_i(x)-\psi(\xi)};$$

(*ii*) the mixture model

$$p(x;\xi) = C(x) + \xi^i F_i(x).$$

## 9.2 $\alpha$-Volume Element for Exponential Model

Since any two $n$-forms are proportional, it suffices to determine a function $f \in \mathcal{F}(M)$ such that $\nabla^{(\alpha)}\omega = 0$ with $\omega = f\, dv$. The equation $\nabla^{(\alpha)}_{\partial_{x_i}}\omega = 0$ becomes the following PDE

$$(\partial_{x^i} f)dv + f\nabla^{(\alpha)}_{\partial_{x^i}} dv = 0. \tag{9.2.4}$$

Since $\nabla^{(\alpha)}_{\partial_{x^i}} : \Lambda^n(M) \to \Lambda^n(M)$, there are $n$ functions $h_i^\alpha$ such that

$$\nabla^{(\alpha)}_{\partial_{x^i}} dv = h_i^\alpha\, dv, \tag{9.2.5}$$

and hence (9.2.4) becomes $\partial_{x^i} f + f h_i^\alpha = 0$ for any $1 \le i \le n$. This can be written as

$$\partial_{x^i}(\ln f) = -h_i^\alpha, \qquad i = 1, \ldots, n. \tag{9.2.6}$$

Before integrating the Eq. (9.2.6), we shall compute the functions $h_i^\alpha$ in terms of functions $h_i^1$. Using the decomposition

$$\nabla^{(\alpha)} = (1-\alpha)\nabla^{(0)} + \alpha\nabla^{(1)}, \tag{9.2.7}$$

and applying Lemma 9.1.1, we have

$$\nabla^{(\alpha)}_{\partial_{x^i}} dv = \alpha\nabla^{(1)}_{\partial_{x^i}} dv. \tag{9.2.8}$$

Substituting in (9.2.5), we find

$$\alpha\nabla^{(1)}_{\partial_{x^i}} dv = h_i^\alpha \, dv. \tag{9.2.9}$$

Making $\alpha = 1$ in (9.2.5), we have

$$\nabla^{(1)}_{\partial_{x_i}} dv = h_i^1 \, dv \tag{9.2.10}$$

and by comparison with (9.2.9), we get

$$h_i^\alpha = \alpha h_i^1. \tag{9.2.11}$$

Hence, it suffices to compute only $h_i^1$. We shall do this by applying both terms of Eq. (9.2.10) on the $n$-uple $(\partial_{x^1}, \dots, \partial_{x^n})$, i.e.,

$$(\nabla^{(1)}_{\partial_{x^i}} dv)(\partial_{x^1}, \dots, \partial_{x^n}) = h_i^1 \, dv(\partial_{x^1}, \dots, \partial_{x^n}).$$

The right side writes

$$(h_i^1 dv)(\partial_{x^1}, \dots, \partial_{x^n}) = h_i^1 \sqrt{\det g}, \tag{9.2.12}$$

while the left side can be computed as

$$\begin{aligned}
(\nabla^{(1)}_{\partial_{x^i}} dv)(\partial_{x^1}, \dots, \partial_{x^n}) &= \partial_{x^i} dv(\partial_{x^1}, \dots, \partial_{x^n}) \\
&\quad - dv\big(\nabla^{(1)}_{\partial_{x^i}} \partial_{x^1}, \partial_{x^2}, \dots, \partial_{x^n}\big) \\
&\quad - \dots - dv\big(\partial_{x^1}, \dots, \partial_{x^{n-1}}, \nabla^{(1)}_{\partial_{x^i}} \partial_{x^n}\big) \\
&= \partial_{x^i} \sqrt{\det g} - \Gamma^{*j}_{ij} \sqrt{\det g} \\
&= \Big(\frac{1}{2}\frac{1}{\det g} \partial_{x^i}(\det g) - \Gamma^{*j}_{ij}\Big)\sqrt{\det g} \\
&= \Big(\partial_{x^i}\big(\ln \sqrt{\det g}\big) - \Gamma^{*j}_{ij}\Big)\sqrt{\det g}.
\end{aligned} \tag{9.2.13}$$

Equating (9.2.12) and (9.2.13) yields

$$h_i^1 = \partial_{x^i}(\ln \sqrt{\det g}) - \Gamma_{ij}^{*j}. \tag{9.2.14}$$

Using (9.2.11) relation (9.2.6) implies

$$\partial_{x^i}(\ln f) = -\alpha \partial_{x_i}(\ln \sqrt{\det g}) + \alpha \Gamma_{ij}^{*j},$$

which can be also written as

$$\partial_{x^i} \ln(f\, (\det g)^{\alpha/2}) = \alpha \Gamma_{ij}^{*j}. \tag{9.2.15}$$

In the case of exponential model the right side of the above expression is zero. This follows from Example 1.12.1, part $(i)$, which states that exponential models are $\nabla^*$-flat, so $\Gamma_{ij}^{*j} = 0$. Therefore (9.2.15) becomes the exact equation

$$\partial_{x^i} \ln(f(\det g)^{\alpha/2}) = 0, \qquad \forall i = 1, \ldots, n.$$

with solution $f = C(\det g)^{-\alpha/2}$, where $C$ is a nonzero constant. We obtain the $\alpha$-volume element for the exponential model

$$\begin{aligned} \omega &= f\,dv = C(\det g)^{-\alpha/2}(\det g)^{1/2} dx_1 \wedge \cdots \wedge dx_n \\ &= C(\det g)^{\frac{1-\alpha}{2}} dx_1 \wedge \cdots \wedge dx_n. \end{aligned}$$

When $C = 1$ and $\alpha = 0$, we obtain the Riemannian volume element $dv = (\det g)^{1/2} dx^1 \wedge \cdots \wedge dx^n$. Replacing $x^i$ by the local coordinates $\xi^i$, we conclude with the following result:

**Proposition 9.2.1** *The $\alpha$-volume element on an exponential model is locally given by*

$$\omega = (\det g)^{\frac{1-\alpha}{2}} d\xi^1 \wedge \cdots \wedge d\xi^n.$$

## 9.3 α-Volume Element for Mixture Model

Formulas (9.2.4)–(9.2.6) developed in the previous section are still valid in the case of the mixture model. The decomposition formula (9.2.7) is replaced in this case by

$$\nabla^{(\alpha)} = (1+\alpha)\nabla^{(0)} - \alpha \nabla^{(-1)} \tag{9.3.16}$$

and (9.2.8) becomes

$$\nabla^{(\alpha)}_{\partial_{x^i}} dv = -\alpha \nabla^{(-1)}_{\partial_{x_i}} dv. \tag{9.3.17}$$

Substituting in (9.2.5), we get

$$-\alpha \nabla^{(-1)}_{\partial_{x_i}} dv = h_i^{\alpha} dv. \tag{9.3.18}$$

Let $\alpha = -1$ in (9.2.5) to obtain

$$\nabla^{(-1)}_{\partial_{x^i}} dv = h_i^{-1}\, dv.$$

Comparing with (9.3.18) yields

$$h_i^{\alpha} = -\alpha h_i^{-1}. \tag{9.3.19}$$

In the following we shall compute the function $h_i^{-1}$. Relation (9.2.12) becomes

$$(h_i^{-1})(\partial_{x^1}, \dots, \partial_{x^n}) = h_i^{-1}\sqrt{\det g}, \tag{9.3.20}$$

while (9.2.13) is reduced to

$$\big(\nabla^{(-1)}_{\partial_{x^i}} dv\big)(\partial_{x^1}, \dots, \partial_{x^n}) = \big(\partial_{x^i}(\ln \sqrt{\det g}) - \Gamma^j_{ij}\big)\sqrt{\det g}. \tag{9.3.21}$$

From (9.3.20) and (9.3.21), we have

$$h_i^{-1} = \partial_{x^i}(\ln \sqrt{\det g}) - \Gamma^j_{ij}. \tag{9.3.22}$$

Substituting in (9.3.19) yields

$$\begin{aligned} h_i^{\alpha} &= -\alpha \partial_{x^i}(\ln \sqrt{\det g}) + \alpha \Gamma^j_{ij} \\ &= \partial_{x^i}\big(\ln(\det g)^{-\alpha/2}\big) + \alpha \Gamma^j_{ij} \end{aligned}$$

and hence (9.2.5) becomes

$$\begin{aligned} \partial_{x^i}(\ln f) &= -\partial_{x^i}(\ln(\det g)^{-\alpha/2}) - \alpha \Gamma^j_{ij} \iff \\ \partial_{x^i} \ln(f(\det g)^{-\alpha/2}) &= -\alpha \Gamma^j_{ij}. \end{aligned} \tag{9.3.23}$$

Since a mixture model is $\nabla^{-1}$-flat, see Example 1.12.2, part $(ii)$, then we have $\Gamma^j_{ij} = g^{jk}\Gamma^{(-1)}_{ij,k} = 0$. Hence, (9.3.23) can be written as

$$\partial_{x^i} \ln(f(\det g)^{-\alpha/2}) = 0, \qquad \forall 1 \le i \le n,$$

with the solution $f = C(\det g)^{\alpha/2}$, $C$ nonzero constant. Replacing $x^i$ by $\xi^i$ we arrive at the following result:

**Proposition 9.3.1** *The $\alpha$-volume element for a mixture model is given by*

$$\omega = (\det g)^{\frac{1+\alpha}{2}} d\xi^1 \wedge \cdots \wedge d\xi^n.$$

## 9.4 Dual Volume Elements

We have found that the volume elements associated with the exponential and mixture models are proportional to the form $dx^1 \wedge \cdots \wedge dx^n$, with the proportionality functions $C_1(\det g)^{\frac{1+\alpha}{2}}$ and $C_2(\det g)^{\frac{1-\alpha}{2}}$, respectively. We note that the product of these functions is independent of $\alpha$, i.e.,

$$C_1(\det g)^{\frac{1+\alpha}{2}} \cdot C_2(\det g)^{\frac{1-\alpha}{2}} = C\sqrt{\det g}.$$

The goal of this section is to prove a similar relation for the general case of two dual connections on a statistical manifold.

**Theorem 9.4.1** *Let $\nabla$ and $\nabla^*$ be two dual connections on the Riemannian manifold $M$. Let $\omega$ and $\omega^*$ be volume elements that are parallel to the above connections, i.e.,*

$$\nabla\omega = 0, \qquad \nabla^*\omega^* = 0.$$

*Then there is a nonzero function $f$ and a constant $C > 0$ such that*

$$\omega = f\,dv, \qquad \omega^* = \frac{C}{f}dv.$$

*Proof:* Since $\omega, \omega^*$ are $n$-forms, there are two functions $f, f^* \in \mathcal{F}(M)$ such that $\omega = fdv$ and $\omega^* = f^*dv$. Applying $\nabla_{\partial_{x^i}}$ to $\omega$ and $\nabla^*_{\partial_{x^i}}$ to $\omega^*$ yields

$$\begin{aligned}(\partial_{x^i}f)dv + f\nabla_{\partial_{x^i}}dv &= 0\\ (\partial_{x^i}f^*)dv + f^*\nabla^*_{\partial_{x^i}}dv &= 0.\end{aligned}$$

Let $h_i, h_i^*$ be such that

$$\nabla_{\partial_{x^i}}dv = h_i\,dv, \qquad \nabla^*_{\partial_{x^i}}dv = h_i^*\,dv,$$

so the aforementioned equations become

$$\partial_{x^i}f + f\,h_i = 0, \qquad \partial_{x^i}f^* + f^*\,h_i = 0,$$

or, equivalently

$$\partial_{x^i}(\ln f) = -h_i, \qquad \partial_{x^i}(\ln f^*) = -h_i^*.$$

Adding yields

$$\partial_{x^i}\big(\ln(f\,f^*)\big) = -(h_i + h_i^*). \tag{9.4.24}$$

Since $\nabla^{(0)} = \dfrac{1}{2}(\nabla + \nabla^*)$ is the Levi–Civita connection, using Lemma 9.1.1 we have

$$\begin{aligned}0 &= \nabla^0_{\partial_{x^i}} dv = \frac{1}{2}\Big(\nabla_{\partial_{x^i}} dv + \nabla^*_{\partial_{x_i}} dv\Big)\\ &= \frac{1}{2}(h_i + h_i^*)dv,\end{aligned}$$

so $h_i + h_i^* = 0$. Substituting in (9.4.24) yields $\ln(ff^*) = k$, i.e., $ff^* = e^k > 0$, constant. Hence $f^* = C/f$, with $C = e^k$. ■

## 9.5 Existence and Uniqueness

Let $\nabla$ be a given linear connection on the manifold $M$ and $\omega$ be an $n$-form. We shall investigate the conditions under which the equation

$$\nabla\omega = 0$$

has solutions. Using that the $n$-forms $\omega$ and $dv = \sqrt{\det g}\, dx^1\wedge\cdots\wedge dx^n$ are proportional, we write $\omega = f dv$, where $f$ is a function subject to be found out. We have

$$0 = \nabla_{\partial_{x^i}}\omega = \nabla_{\partial_{x^i}}(f dv) = (\partial_{x^i} f)dv + f(\nabla_{\partial_{x^i}} dv).$$

Applying the above equation to the $n$-uple $(\partial_{x^1}, \ldots, \partial_{x^n})$ yields

$$0 = (\partial_{x^i} f)\sqrt{\det g} + f(\nabla_{\partial_{x^i}})(\partial_{x^1}, \ldots, \partial_{x^n}). \tag{9.5.25}$$

By the first part of the proof of Lemma 9.1.1, we find

$$(\nabla_{\partial_{x^i}} dv)(\partial_{x^1}, \ldots, \partial_{x^n}) = \partial_{x^i}\sqrt{\det g} - \sqrt{\det g}\,\Gamma_{ij}^j,$$

so (9.5.25) becomes

$$0 = (\partial_{x_i} f)\sqrt{\det g} + f\Big(\partial_{x_i}\sqrt{\det g} - \sqrt{\det g}\,\Gamma_{ij}^j\Big).$$

Dividing by $f\sqrt{\det g}$ yields

$$\begin{aligned}\frac{\partial_{x^i} f}{f} + \frac{\partial_{x^i}\sqrt{\det g}}{\sqrt{\det g}} &= \Gamma_{ij}^j \Longleftrightarrow\\ \partial_{x^i}(\ln f) + \partial_{x^i}(\ln\sqrt{\det g}) &= \Gamma_{ij}^j \Longleftrightarrow\\ \partial_{x^i}\Big(\ln(f\sqrt{\det g})\Big) &= \Gamma_{ij}^j.\end{aligned}$$

Let $\varphi = \ln(f\sqrt{\det g})$ and $\rho_i = \sum_j \Gamma_{ij}^j$. Then the last equation becomes

$$\partial_{x_i}\varphi = \rho_i, \qquad i = 1, \ldots, n. \tag{9.5.26}$$

Equation (9.5.26) has solutions if and only if the following exactness conditions hold locally

$$\partial_{x_k}\rho_i = \partial_{x_i}\rho_k,$$

which can be written equivalently as

$$\partial_k\Big(\sum_j \Gamma_{ij}^j\Big) = \partial_i\Big(\sum_j \Gamma_{kj}^j\Big). \tag{9.5.27}$$

**Lemma 9.5.1** *We have*

$$R_{ij} = R_{ji} \Leftrightarrow \partial_j\big(\sum_r \Gamma_{ir}^r\big) = \partial_i\big(\sum_r \Gamma_{jr}^r\big).$$

*Proof:* See Problem 9.2. ■

From Lemma 9.5.1 and formula (9.5.27) we obtain that $R_{ij} = R_{ji}$, i.e., the Ricci tensor is symmetric.

Solving for the function $f$ we obtain $f = \dfrac{e^{\varphi}}{\sqrt{\det g}}$ and hence the volume element is

$$\omega = f dv = e^{\varphi} dx_1 \wedge \cdots \wedge dx_n.$$

We conclude the previous computation with the following existence result:

**Theorem 9.5.2** *The equation $\nabla\omega = 0$ has solutions in the space of $n$-forms if and only if the Ricci tensor associated with the connection $\nabla$ is symmetric.*

**Corollary 9.5.3** *The Ricci tensor of $\nabla$ is symmetric if and only if the Ricci tensor of $\nabla^*$ is symmetric, i.e.,*

$$R_{ij} = R_{ji} \Leftrightarrow R_{ij}^* = R_{ji}^*.$$

*Proof:* "$\Rightarrow$" Assume $R_{ij} = R_{ji}$ and let $\omega = f dv$ be a solution of $\nabla\omega = 0$. Then $\omega^* = \frac{C}{f} dv$ is a solution of $\nabla^*\omega^* = 0$ and by Theorem 9.5.2 we get $R_{ij}^* = R_{ji}^*$. The converse has a similar proof. ■

The uniqueness is stated by the following proposition.

**Proposition 9.5.4** *Let $M$ be a connected manifold. Then the equation $\nabla\omega = 0$ has at most one solution, which is unique up to a scaling constant.*

*Proof:* Assume the Eq. (9.5.26) has two solutions $\varphi_1$ and $\varphi_2$. Then $\tilde{\varphi} = \varphi_2 - \varphi_1$ satisfies the equation $\partial_{x_i}\tilde{\varphi} = 0$ and hence $\tilde{\varphi} = C$ constant. Then $\omega_1 = e^{\varphi_1}dx$ and $\omega_2 = e^{\varphi_2}dx = e^C e^{\varphi_1}dx = e^C\omega_1$. It follows that the solution is unique up to a multiplicative positive constant. ■

## 9.6 Equiaffine Connections

If there is an $n$-form $\omega$ that is parallel with respect to the connection $\nabla$, i.e., $\nabla\omega = 0$, then the connection $\nabla$ is called *equiaffine* and the pair $(\nabla^{(\alpha)}, \omega)$ is called an *equiaffine structure.* The study of the equiaffine structures is done in Nomizu and Sasaki [62].

Therefore, Theorem 9.5.2 can be equivalently stated by saying that a necessary and sufficient condition for a connection to be equiaffine is that its Ricci curvature tensor is symmetric. Therefore, in order to check the equiaffinity of $\nabla^{(\alpha)}$-connections it suffices to verify the symmetry of the associated Ricci tensor $Ric^{\alpha}$.

We start by recalling the relation provided by Proposition 8.10.1

$$\begin{aligned} R^{(\alpha)}(X,Y,Z) &= \frac{1+\alpha}{2}R^*(X,Y,Z) + \frac{1-\alpha}{2}R(X,Y,Z) \\ &\quad + \frac{1-\alpha^2}{4}\Big(K\big(Y,K(X,Z)\big) - K\big(X,K(Y,Z)\big)\Big). \end{aligned}$$

Zhang [87] used this formula to show that connection $\nabla^{(\alpha)}$ is equiaffine on a dually flat statistically manifold. Applying the contractions

$$\begin{aligned} \mathrm{Tr}\,\{X \to R(X,Y)Z\} &= \mathrm{Tr}\,R(\cdot,Y)Z = Ric(Y,Z) \\ \mathrm{Tr}\,\{X \to R^*(X,Y)Z\} &= \mathrm{Tr}\,R^*(\cdot,Y)Z = Ric^*(Y,Z), \end{aligned}$$

we obtain an analogous relation in terms of Ricci curvature tensors

$$\begin{aligned} Ric^{(\alpha)}(Y,Z) &= \frac{1+\alpha}{2}Ric^*(Y,Z) + \frac{1-\alpha}{2}Ric(Y,Z) \\ &\quad + \frac{1-\alpha^2}{4}Q(Y,Z), \end{aligned} \qquad (9.6.28)$$

where

$$Q(Y,Z) = \mathrm{Tr}\,K\big(Y,K(\cdot,Z)\big) - \mathrm{Tr}\,K\big(\cdot,K(Y,Z)\big). \qquad (9.6.29)$$

**Lemma 9.6.1** *The 2-covariant tensor $Q$ is symmetric, i.e., $Q(Y,Z) = Q(Z,Y)$, for any two vector fields $Y$ and $Z$.*

*Proof:* Since the difference tensor $K$ is symmetric, then

$$\operatorname{Tr} K\big(\cdot, K(Y,Z)\big) = \operatorname{Tr} K\big(\cdot, K(Z,Y)\big).$$

Therefore, it remains to show only the symmetry of the first term, i.e.

$$\operatorname{Tr} K\big(Y, K(\cdot, Z)\big) = \operatorname{Tr} K\big(Z, K(\cdot, Y)\big).$$

By linearity, it suffices to show this relation only on a basis. In local coordinates

$$K\big(\partial_i, K(\partial_j, \partial_l)\big) = K\big(\partial_i, K_{jl}^k \partial_k\big) = K_{jl}^k K(\partial_i, \partial_k) = K_{jl}^k K_{ik}^p \partial_p,$$

with summation over $k$ and $p$. Taking the contraction

$$\begin{aligned} \operatorname{Tr} K\big(\partial_i, K(\cdot, \partial_l)\big) &= \sum_{j,p} g(K_{jl}^k K_{ik}^p \partial_p, \partial_j) = K_{jl}^k K_{ik}^p g_{pj} \\ &= K_{jl}^k K_{ij}^p g_{pk}, \end{aligned} \tag{9.6.30}$$

in the virtue of relation (8.8.27). Similarly, we arrive at

$$\operatorname{Tr} K\big(\partial_l, K(\cdot, \partial_i)\big) = K_{ji}^k K_{lj}^p g_{pk}, \tag{9.6.31}$$

with summation over $p, k, j$. The symmetry of tensors $K$ and $g$ show the identity between relations (9.6.30) and (9.6.31)

$$K_{jl}^k K_{ij}^p g_{pk} = K_{lj}^k K_{ji}^p g_{pk} = K_{lj}^p K_{ji}^k g_{kp} = K_{lj}^p K_{ji}^k g_{pk} = K_{ji}^k K_{lj}^p g_{pk}.$$

Therefore $\operatorname{Tr} K\big(\partial_i, K(\cdot, \partial_l)\big) = \operatorname{Tr} K\big(\partial_l, K(\cdot, \partial_i)\big)$, and hence the tensor $Q$ is symmetric. ■

**Proposition 9.6.2** *The following equivalencies hold:*

$$\begin{aligned} \exists \alpha_0 \neq 0 \text{ with } Ric^{(\alpha_0)}(Y,Z) &= Ric^{(\alpha_0)}(Z,Y) \Longleftrightarrow \\ \forall \alpha \quad Ric^{(\alpha)}(Y,Z) &= Ric^{(\alpha)}(Z,Y) \Longleftrightarrow \\ Ric(Y,Z) &= Ric(Z,Y) \Longleftrightarrow \\ Ric^*(Y,Z) &= Ric^*(Z,Y), \end{aligned}$$

*for any vector fields $Y$, $Z$.*

*Proof:* From relation (9.6.28) and symmetry of $Q$, see Lemma 9.6.1, we obtain by subtraction

$$\begin{aligned} Ric^{(\alpha)}(Y,Z) - Ric^{(\alpha)}(Z,Y) &= \frac{1+\alpha}{2}\Big(Ric^*(Y,Z) - Ric^*(Z,Y)\Big) \\ &+\frac{1-\alpha}{2}\Big(Ric(Y,Z) - Ric(Z,Y)\Big). \end{aligned}$$

Let $\alpha = 0$ to obtain

$$\begin{aligned} Ric^{(0)}(Y,Z) - Ric^{(0)}(Z,Y) &= \frac{1}{2}\Big(Ric(Y,Z) - Ric(Z,Y)\Big) \\ &+\frac{1}{2}\Big(Ric^*(Y,Z) - Ric^*(Z,Y)\Big). \end{aligned} \tag{9.6.32}$$

Since the Ricci tensor, $Ric^{(0)}$, associated with the Levi–Civita connection $\nabla^{(0)}$ is symmetric, the previous relation becomes

$$Ric(Y,Z) - Ric(Z,Y) = Ric^*(Z,Y) - Ric^*(Y,Z). \tag{9.6.33}$$

This implies that $Ric$ is symmetric if and only if $Ric^*$ is symmetric. Denote by $\beta = \beta(Y,Z) = -\beta(Z,Y)$ the value of expression (9.6.33). Substituting in (9.6.32) yields

$$\begin{aligned} Ric^{(\alpha)}(Y,Z) - Ric^{(\alpha)}(Z,Y) &= \frac{1+\alpha}{2}\beta + \frac{1-\alpha}{2}(-\beta) \\ &= \frac{\alpha\beta}{2}. \end{aligned} \tag{9.6.34}$$

If there is $\alpha_0 \neq 0$ such that $Ric^{(\alpha_0)}$ is symmetric, substituting in (9.6.34) yields $\dfrac{\alpha\beta}{2} = 0$, and hence $\beta = 0$. This means that both Ricci tensors $Ric$ and $Ric^*$ are symmetric. Substituting back $\beta = 0$ into relation (9.6.34) provides

$$Ric^{(\alpha)}(Y,Z) - Ric^{(\alpha)}(Z,Y) = 0, \qquad \alpha \neq 0,$$

i.e., $Ric^{(\alpha)}$ is symmetric. The symmetry of $Ric^{(\alpha)}$ for $\alpha = 0$ is obvious. ■

Even if, in general, neither $Ric$ nor $Ric^*$ are symmetric tensors, their sum is always symmetric:

**Corollary 9.6.3** *For any vector fields $Y$, $Z$ we have*

$$Ric(Y,Z) + Ric^*(Y,Z) = Ric(Z,Y) + Ric^*(Z,Y). \tag{9.6.35}$$

*Proof:* It follows from relation (9.6.33). ■

Since a necessary and sufficient condition for a connection to be equiaffine is the symmetry of the Ricci tensor, the previous result can be stated equivalently as in the following, see Takeuchi and Amari [79] and Matsuzoe et al. [55]:

**Theorem 9.6.4** *The following conditions are equivalent:*

(*i*) $\nabla^{(\alpha)}$ *is equiaffine for any* $\alpha$;

(*ii*) *there is an* $\alpha_0 \neq 0$ *such that* $\nabla^{(\alpha_0)}$ *is equiaffine;*

(*iii*) $\nabla$ *is equiaffine;*

(*iv*) $\nabla^*$ *is equiaffine.*

**Corollary 9.6.5** *If* $\nabla$, $\nabla^*$ *are dually flat, then* $\nabla^{(\alpha)}$ *is equiaffine for any* $\alpha$.

*Proof:* If $\nabla$, $\nabla^*$ are dually flat, then the curvature tensors vanish, and hence the Ricci tensors also vanish, $Ric = Ric^* = 0$. Since a zero tensor is symmetric by default, if follows that $\nabla$ and $\nabla^*$ are equiaffine. Then Theorem 9.6.4 implies that $\nabla^{(\alpha)}$ is equiaffine for any $\alpha$.

A variant of proof can be done using the formula (9.6.28) in local coordinates

$$R_{ij}^{(\alpha)} = \frac{1+\alpha}{2} R_{ij}^* + \frac{1-\alpha}{2} R_{ij} + \frac{1-\alpha^2}{4} Q_{ij}. \tag{9.6.36}$$

If the connections $\nabla, \nabla^*$ are dually flat, then $R_{ij}^* = R_{ij} = 0$, so $R_{ij}^{(\alpha)} = \frac{1-\alpha^2}{4} Q_{ij}$. Since the tensor $Q$ is symmetric, see Lemma 9.6.1, then $R^{(\alpha)}$ is also symmetric. Then applying Theorem 9.6.4 yields that $\nabla^{(\alpha)}$ is equiaffine for any $\alpha$. ■

The following necessary condition for equiaffinity can be found in Min et al. [59].

**Theorem 9.6.6** *If there is an* $\alpha_0 \neq 0$ *such that* $Ric^{(\alpha_0)} = Ric^{(-\alpha_0)}$, *then*

(*i*) $Ric^{(\alpha)} = Ric^{(-\alpha)}$ *for any* $\alpha$;

(*ii*) *the connection* $\nabla^{(\alpha)}$ *is equiaffine for any* $\alpha$.

*Proof:*

(*i*) Using $Q_{ij} = Q_{ji}$, from (9.6.36) we obtain by subtraction

$$R_{ij}^{(\alpha)} - R_{ij}^{(-\alpha)} = \alpha(R_{ij}^* - R_{ij}). \tag{9.6.37}$$

Therefore, if there is an $\alpha_0 \neq 0$ for which the left side vanishes, then $R_{ij}^* = R_{ij}$, and hence the left side has to be zero for any $\alpha \neq 0$, i.e.,

$$R_{ij}^{(\alpha)} - R_{ij}^{(-\alpha)} = 0.$$

The identity for $\alpha = 0$ follows from the symmetry of the Ricci tensor associated with the Levi–Civita connection $\nabla^{(0)}$.

(*ii*) Substituting $R_{ij} = R_{ij}^*$ back into formula (9.6.36) leads to

$$R_{ij}^{(\alpha)} = R_{ij} + \frac{1-\alpha^2}{4} Q_{ij}, \quad \forall \alpha. \tag{9.6.38}$$

Then making $\alpha = 0$ and solving for $R_{ij}$ we obtain

$$R_{ij} = R_{ij}^{(0)} - \frac{1}{4} Q_{ij}.$$

Since $R_{ij}^{(0)}$ and $Q_{ij}$ are symmetric, then so will be $R_{ij}$. Looking back to formula (9.6.38), it follows that $R_{ij}^{(\alpha)}$ is symmetric, for any $\alpha$. Hence the connection $\nabla^{(\alpha)}$ is equiaffine for all $\alpha$.

■

The next notion was introduced by Lauritzen [54]:

**Definition 9.6.7** *A statistical manifold $(M, g, \nabla, \nabla^*)$ is said to be conjugate symmetric if the curvatures of the pair of conjugate connections are equal, i.e.*

$$R(X, Y, Z) = R^*(X, Y, Z),$$

*for any vector fields $X, Y, Z$.*

**Proposition 9.6.8** *Let $M$ be a conjugate symmetric manifold. Then $R^{(\alpha)}$ is an even function of $\alpha$, i.e.*

$$R^{(\alpha)}(X, Y)Z = R^{(-\alpha)}(X, Y)Z,$$

*for any vector fields $X, Y, Z$ on $M$.*

*Proof:* Writing the relation provided by Proposition 8.10.1 for $\alpha$ and $-\alpha$, we have

$$\begin{aligned} R^{(\alpha)}(X,Y,Z) &= \frac{1+\alpha}{2}R^*(X,Y,Z) + \frac{1-\alpha}{2}R(X,Y,Z) \\ &+\frac{1-\alpha^2}{4}\Big(K\big(Y,K(X,Z)\big) - K\big(X,K(Y,Z)\big)\Big) \end{aligned}$$

$$\begin{aligned} R^{(-\alpha)}(X,Y,Z) &= \frac{1-\alpha}{2}R^*(X,Y,Z) + \frac{1+\alpha}{2}R(X,Y,Z) \\ &+\frac{1-\alpha^2}{4}\Big(K\big(Y,K(X,Z)\big) - K\big(X,K(Y,Z)\big)\Big), \end{aligned}$$

and then subtracting yields

$$R^{(\alpha)}(X,Y,Z) - R^{(-\alpha)}(X,Y,Z) = \alpha\Big(R^*(X,Y,Z) - R(X,Y,Z)\Big). \tag{9.6.39}$$

If the manifold is conjugate symmetric, the right side is equal to zero, and hence $R^{(\alpha)} = R^{(-\alpha)}$. ■

The following more restrictive concept was introduced in Min et al. [59].

**Definition 9.6.9** *A statistical manifold $(M, g, \nabla, \nabla^*)$ is called conjugate Ricci-symmetric if*

$$Ric(Y,Z) = Ric^*(Y,Z)$$

*for all $Y, Z$ vector fields on $M$.*

The above condition can be written in local coordinates as $R_{ij} = R^*_{ij}$. It is worth noting that if $M$ is conjugate symmetric, then it is conjugate Ricci-symmetric.

**Theorem 9.6.10** *Let $(M, g, \nabla, \nabla^*)$ be a conjugate Ricci-symmetric statistical manifold. Then the connection $\nabla^{(\alpha)}$ is equiaffine for any $\alpha$.*

*Proof:* If $M$ is conjugate Ricci-symmetric manifold, then formula (9.6.37) implies

$$R^{(\alpha)}_{ij} - R^{(-\alpha)}_{ij} = \alpha(R^*_{ij} - R_{ij}) = 0 \tag{9.6.40}$$

for any $\alpha \neq 0$, and hence $R^{(\alpha)}_{ij} = R^{(-\alpha)}_{ij}$. Applying Theorem 9.6.6 yields that $\nabla^{(\alpha)}$ is equiaffine for any $\alpha \neq 0$. The case $\alpha = 0$ is covered by the fact that the Levi–Civita connection $\nabla^{(0)}$ is equiaffine (the Riemannian volume element is parallel with respect to $\nabla^{(0)}$). ■

**Corollary 9.6.11** *Let $(M, g, \nabla, \nabla^*)$ be a conjugate symmetric statistical manifold. Then the connection $\nabla^{(\alpha)}$ is equiaffine for any $\alpha$.*

The next section deals with a distinguished particular type of conjugate symmetric manifolds.

## 9.7 Manifolds with Constant Curvature

Recall that a connection $\nabla$ on the manifold $(M, g)$ has the constant curvature $K$ if relation (8.1.6) holds, i.e.,

$$R(X,Y)Z = K\{g(Y,Z)X - g(X,Z)Y\}, \quad \forall X, Y, Z \in \mathcal{X}(M). \tag{9.7.41}$$

Let $(M, g, \nabla, \nabla^*)$ be a statistical manifold. As it had been shown in Proposition 8.1.4, part $(ii)$, if the connection $\nabla$ has constant curvature, then its dual connection, $\nabla^*$, also has constant curvature, and the curvatures of $\nabla$ and $\nabla^*$ are equal. On this basis we consider the following concept.

**Definition 9.7.1** *$(M, g, \nabla, \nabla^*)$ is a statistical manifold of constant curvature if the dual connections $\nabla$ and $\nabla^*$ have equal constant curvatures.*

The main properties of these type of manifolds are contained in the following result.

**Theorem 9.7.2** *Let $(M, g, \nabla, \nabla^*)$ be a statistical manifold of constant curvature. Then*

$(i)$ *$M$ is a conjugate symmetric manifold;*

$(ii)$ *$M$ is a conjugate Ricci-symmetric manifold;*

$(iii)$ *$Ric^{(\alpha)} = Ric^{(-\alpha)}$, for any $\alpha$;*

$(iv)$ *$Ric^{*(\alpha)} = Ric^{*(-\alpha)}$, for any $\alpha$;*

$(v)$ *$\nabla^{(\alpha)}$ is equiaffine for any $\alpha$.*

*Proof:*

$(i)$ Writing relation (9.7.41) for both tensors $R$ and $R^*$ and using that $K = K^*$, see Proposition 8.1.4, part $(ii)$, we have

$$\begin{aligned} R(X,Y)Z &= K\{g(Y,Z)X - g(X,Z)Y\} \\ &= K^*\{g(Y,Z)X - g(X,Z)Y\} \\ &= R^*(X,Y)Z, \quad \forall X, Y, Z \in \mathcal{X}(M). \end{aligned}$$

(*ii*) It follows by contracting the relation $R(X,Y)Z = R^*(X,Y)Z$ over $X$ to obtain $Ric(Y,Z) = Ric^*(Y,Z)$.

(*iii*) It is an obvious application of Proposition 9.6.8.

(*iv*) It is implied by (*iii*) and formula (9.6.39).

(*v*) Apply Theorem 9.6.6.

■

The following result was previously noticed by Takeuki and Amari [79]. We obtain it here as a consequence of the previous analysis.

**Corollary 9.7.3** *Let $(M, g, \nabla, \nabla^*)$ be a dually flat statistical manifold. Then $\nabla^{(\alpha)}$ is equiaffine for any $\alpha \in \mathbb{R}$.*

*Proof:* A dually flat statistical manifold has the constant curvature equal to zero. Then apply Theorem 9.7.2, part (*v*). ■

In the end of this topic we make a few concluding remarks. The subject of equiaffine connections is of outgrowing interest and several other authors have brought their contributions to this topic.

Uohashi [84] introduced in 2002 the notion of *$\alpha$-transitive flat connections*. A connection $\nabla^{(\alpha)}$ is called $\alpha$-transitive flat if $\nabla = \nabla^{(-1)}$ is curvature-free (we also note that $\nabla^* = \nabla^{(1)}$ is curvature-free). Then Uohashi result can be restated as "all $\alpha$-transitive flat connections are equiaffine."

Zhang [87] proves that if two torsion-free connections $\nabla$, $\tilde{\nabla}$ are equiaffine, with corresponding parallel volume forms $\omega$, $\tilde{\omega}$, then the connection given by their convex combination $a\nabla + b\tilde{\nabla}$, $a + b = 1$, is equiaffine for all $a \in \mathbb{R}$, with parallel volume form given by (up to a scaling constant) $\omega^a \tilde{\omega}^b$. It is worth noting that the result follows from the formula $\Gamma^l_{il} = \partial_i(\ln \omega)$ and the logarithm properties. As an application, Zhang shows that the $\alpha$-volume form of $\nabla^{(\alpha)}$ satisfies

$$\omega^{(\alpha)} = (\omega)^{\frac{1-\alpha}{2}} (\omega)^{*\frac{1+\alpha}{2}},$$

where $\nabla^* = \nabla^{(1)}$, $\nabla = \nabla^{(-1)}$. An alternative expression is provided by Matsuzoe et al. [55] in terms of the Levi–Civita connection $\nabla^{(0)}$ via formula

$$\omega^{(\alpha)} = e^{-\frac{\alpha}{2}\phi} \omega^{(0)},$$

where $\phi = \log \frac{\omega^*}{\omega}$. From here one can easily infer that

$$\omega^{(\alpha)}\omega^{(-\alpha)} = \omega\omega^* = (\omega^{(0)})^2,$$

which is a result first stated in Simon [76], p.913.

Another issue investigated in [87] is the relation between the $\alpha$-scalar curvature and the difference tensor $K_{ij}$. More precisely, if the scalar curvature is defined as the contraction of the Ricci tensor, $\sigma = g^{jl}R_{lj}$, then the scalar curvature of $\nabla^{(\alpha)}$ is related to $\nabla$ via formula

$$\sigma^{(\alpha)} = \sigma + \frac{1-\alpha^2}{4} g^{ij}(K_{ik}^m K_{jm}^k - K_{ij}^m K_{km}^k).$$

Last, but not least, dual connections support several generalizations, see Calin et al. [25]. For instance, connections $\nabla$ and $\overline{\nabla}^*$ are called *generalized conjugate* if there is a 1-form $\tau$ such that

$$Xg(Y, Z) = g(\nabla_X Y, Z) + g(Y, \overline{\nabla}^*_X Z) - \tau(X)g(Y, Z)$$

for all vector fields $X, Y, Z$. It is shown in [25] that if $\nabla$ and $\overline{\nabla}^*$ are torsion-free, then $\nabla$ is equiaffine (or equivalently $\overline{\nabla}^*$ is equiaffine) if and only if $\tau$ is an exact 1-form. If $\tau = d\varphi$, let $\omega$, $\overline{\omega}^*$ be the parallel volume elements with respect to the generalized conjugate connections $\nabla$ and $\overline{\nabla}^*$. Then there is a constant $C > 0$ such that

$$\omega\overline{\omega}^* = Ce^{n\varphi},$$

where $n$ is the dimension of the statistical manifold.

## 9.8 Divergence of a Vector Field

The concept of *divergence* of a vector field helps with modeling the evolution of the volume element along the integral curves of the vector field. In other words, the divergence of a vector field defines the speed of contraction–dilation of volumes by the corresponding local flow. This section is concerned with the divergence taken with respect to a pair of dual connections and the relations between them. The relation between the divergence and the volume element is also emphasized.

**Definition 9.8.1** *Let $\nabla$ be a linear connection on the Riemannian manifold $(M, g)$ and let $X \in \mathcal{X}(M)$ be a $C^1$-vector field. The divergence of $X$ is defined as the trace of the covariant derivative $\nabla X$, i.e.,*

$$divX = Trace\Big(Y \to g(\nabla_Y X, Y)\Big). \tag{9.8.42}$$

This can be expressed in local coordinates as

$$div X = \frac{\partial X^i}{\partial x^i} + \Gamma^i_{ij} X^j, \tag{9.8.43}$$

where $X(x) = X^k(x)\partial_{x_k}$.

Let $\Gamma^{(0)k}_{ij}$ denote the Christoffel symbols of second kind associated with the Levi–Civita connection $\nabla^{(0)}$. The divergence with respect to $\nabla^{(0)}$ is called the *Riemannian divergence* and is denoted by $div^{(0)}$.

The following equivalent formulas hold for Riemannian divergences, see Calin and Chang [22], p.19.

**Lemma 9.8.2** *The Riemannian divergence can be expressed in the following equivalent ways*

$$\begin{aligned} div^{(0)} X &= \frac{1}{\sqrt{\det g}} \frac{\partial}{\partial x^j}(\sqrt{\det g} X^j) \\ &= \frac{\partial X^i}{\partial x^i} + \frac{1}{\sqrt{\det g}} X(\sqrt{\det g}) \\ &= \frac{\partial X^i}{\partial x^i} + \Gamma^{(0)i}_{ij} X^j. \end{aligned}$$

The next result holds for any divergence.

**Lemma 9.8.3** *For any $C^1$-vector field $X$ and any $C^1$-function $f$, we have*

$$\begin{aligned} div(fX) &= f div(X) + g(X, grad\, f) \\ &= f div(X) + X(f). \end{aligned}$$

*Proof:* Since $\nabla_Y(fX) = f\nabla_Y X + Y(f)X$, we find

$$g\big(\nabla_Y(fX), Y\big) = fg(\nabla_Y X, Y) + Y(f)g(X, Y).$$

Taking the trace yields

$$\begin{aligned} div(fX) &= Trace\big(Y \to g(\nabla_Y(fX), Y)\big) \\ &= Trace\big(Y \to fg(\nabla_Y X, Y)\big) + Trace\big(Y \to Y(f)g(X, Y)\big) \\ &= f\, Trace\big(Y \to g(\nabla_Y X, Y)\big) + g\Big(X, Trace\big(Y \to Y(f)Y\big)\Big) \\ &= f div(X) + g(X, grad\, f). \end{aligned}$$

■

Denote by $div X$ and $div^* X$ the divergence of the vector field $X$ with respect to the dual connections $\nabla$ and $\nabla^*$. The following result will be useful shortly.

**Lemma 9.8.4** *We have*

$$\frac{1}{2}g^{ij}\partial_k g_{ij} = \Gamma^{(0)i}_{ki}.$$

*Proof:* Since

$$\Gamma^{(0)i}_{jk} = \frac{1}{2}g^{il}\Big(\frac{\partial g_{jl}}{\partial x^k} + \frac{\partial g_{kl}}{\partial x^j} - \frac{\partial g_{jk}}{\partial x^l}\Big),$$

using the symmetry in the lower indices, yields

$$\Gamma^{(0)i}_{ji} = \frac{1}{2}g^{is}\Big(\frac{\partial g_{js}}{\partial x^i} + \frac{\partial g_{is}}{\partial x^j} - \frac{\partial g_{ji}}{\partial x^s}\Big) = \frac{1}{2}g^{is}\frac{\partial g_{is}}{\partial x^j}.$$

■

**Proposition 9.8.5** *The Riemannian divergence is the average of dual divergences*

$$div^{(0)}X = \frac{1}{2}(divX) + \frac{1}{2}(div^* X).$$

*Proof:* Taking the trace in the formula $\nabla^{(0)} = \frac{1}{2}\nabla + \frac{1}{2}\nabla^*$ and using (9.8.42) yields the desired result.

For the sake of completeness, we shall perform in the following a computation in local coordinates using formula (9.8.43). Since $\nabla$ and $\nabla^*$ are dual connections, the relation between the connection components is

$$\partial_{x^k} g_{ij} = \Gamma_{ki,j} + \Gamma^*_{kj,i}.$$

Contracting by $\frac{1}{2}g^{ij}$ and applying Lemma 9.8.4 we obtain

$$\Gamma^{(0)i}_{ki} = \frac{1}{2}\Gamma^i_{ki} + \frac{1}{2}\Gamma^{*\ i}_{ki}$$

and hence for any vector field $X$ we have

$$\Gamma^{(0)i}_{ki}X^k = \frac{1}{2}\Gamma^i_{ki}X^k + \frac{1}{2}\Gamma^{*\ i}_{ki}X^k.$$

Adding $\frac{\partial X^i}{\partial x^i}$ on both sides yields

$$\begin{aligned}\frac{\partial X^i}{\partial x^i} + \Gamma^{(0)i}_{ki}X^k &= \frac{1}{2}\Big(\frac{\partial X^i}{\partial x^i} + \sum_k \Gamma^i_{ki}X^k\Big)\\ &\quad + \frac{1}{2}\Big(\frac{\partial X^i}{\partial x^i} + \sum_k \Gamma^{*\ i}_{ki}X^k\Big),\end{aligned}$$

which, in the virtue of (9.8.43), becomes

$$div^{(0)}X = \frac{1}{2}(divX) + \frac{1}{2}(div^*X).$$

■

**Corollary 9.8.6** *Let $(\nabla, \nabla^*)$ and $(\nabla', \nabla'')$ be two pairs of dual connections, with respect to the metric $g$. Then*

$$divX - div'X = div^*X - div''X, \qquad \forall X \in \mathcal{X}(M),$$

*i.e., the variation in the divergence of two connections is the same as the variation in the divergence of the dual connections.*

In the following we shall assume that there is a volume element $\omega$ associated with the connection $\nabla$, i.e., an $n$-form which satisfies the equation $\nabla\omega = 0$. Necessary and sufficient conditions for the existence of $\omega$ are given by the Theorem 9.5.2.

The following result states the relationship between the divergences associated with $\nabla$ and $\nabla^{(0)}$.

**Proposition 9.8.7** *If $\omega = f dv$ is the volume element associated with $\nabla$ and $X$ is a $C^1$-vector field, then we have*

$$divX = \frac{1}{f}X(f) + div^{(0)}X. \tag{9.8.44}$$

*Proof:* The parallelism of the volume element $\omega$, given by $\nabla_X\omega = 0$, becomes

$$\begin{aligned} X(\ln f) &= -X(\ln\sqrt{\det g}) + \Gamma_{ij}^j X^i \\ &= \Big( -\frac{X(\sqrt{\det g})}{\sqrt{\det g}} - \frac{\partial X^i}{\partial x^i}\Big) + \left(\frac{\partial X^i}{\partial x^i} + \Gamma_{ij}^j X^i\right) \\ &= -div^{(0)}X + divX, \end{aligned} \tag{9.8.45}$$

by Lemma 9.8.2 and formula (9.8.43). Substituting $X(\ln f) = \dfrac{X(f)}{f}$ in (9.8.45) yields the desired result. ■

In the following we shall recover Theorem 9.4.1, and also a result proved in Zhang [87].

**Proposition 9.8.8** *Let $\omega = f dv$ and $\omega^* = f^* dv$ be the volume elements parallel with respect to the connections $\nabla, \nabla^*$, respectively. Then $ff^* = C$, constant.*

*Proof:* Equation (9.8.44) applied to $\omega$ and $\omega^*$ provides

$$\begin{aligned} divX &= \frac{1}{f}X(f) + div^{(0)}X \\ div^*X &= \frac{1}{f^*}X(f^*) + div^{(0)}X. \end{aligned}$$

Adding the previous relations and using Proposition 9.8.5 yields

$$\frac{1}{f}X(f) + \frac{1}{f^*}X(f^*) = 0.$$

Multiplying by $ff^*$ and using that $X$ satisfies Leibniz rule, we obtain

$$X(ff^*) = 0, \qquad \forall X \in \mathcal{X}(M),$$

which is equivalent to $ff^* = C$, constant. ■

**Corollary 9.8.9** *For any $C^1$-vector field $X$ and any functions $f$, $f^*$ in the relation $ff^* = C$, we have*

$$\begin{aligned} divX &= \frac{1}{f}div^{(0)}(fX) \\ div^*X &= \frac{1}{f^*}div^{(0)}(f^*X) = f div^{(0)}\left(\frac{1}{f}X\right). \end{aligned}$$

*Proof:* Lemma 9.8.3 provides

$$div^{(0)}(fX) = f div^{(0)}X + X(f).$$

Dividing by $f$ yields

$$\frac{1}{f}div^{(0)}(fX) = div^{(0)}X + \frac{1}{f}X(f) = divX.$$

The second part results from $f^* = C/f$

$$div^*X = \frac{1}{f^*}div^{(0)}(f^*X) = \frac{f}{C}div^{(0)}\left(\frac{C}{f}X\right) = f div^{(0)}\left(\frac{1}{f}X\right).$$

■

The $\alpha$-divergence is the divergence taken with respect to the $\nabla^\alpha$-connection. This can be computed as in the following.

**Lemma 9.8.10** *For any $C^1$-vector field $X$, we find*

$$div^{(\alpha)}X = \frac{1+\alpha}{2}div^*X + \frac{1-\alpha}{2}divX.$$

*Proof:* Using (8.5.11), we obtain

$$g(\nabla^{\alpha}_Y X, Y) = \frac{1+\alpha}{2}g(\nabla_Y X, Y) + \frac{1-\alpha}{2}g(\nabla^*_Y X, Y).$$

Taking the trace with respect to $Y$ yields the desired result. ■

The relation between the $\alpha$-divergence and the dual volume elements is given below. The functions $f$, $f^*$ are the ones defined by the volume elements $\omega = f\,dv$ and $\omega^* = f^*\,dv$.

**Proposition 9.8.11** *For any $C^1$-vector field $X$, the $\alpha$-divergence can be expressed as*

$$div^{(\alpha)}(X) = -\alpha X(\ln f) + div^{(0)}X \qquad (9.8.46)$$

$$= \alpha X(\ln f^*) + div^{(0)}X. \qquad (9.8.47)$$

*Proof:* Combining Lemma 9.8.10 and Proposition 9.8.7, we write

$$\begin{aligned} div^{(\alpha)}(X) &= \frac{1+\alpha}{2}div^*X + \frac{1-\alpha}{2}divX \\ &= \frac{1+\alpha}{2}\frac{1}{f^*}X(f^*) + \frac{1-\alpha}{2}\frac{1}{f}X(f) + div^{(0)}X \\ &= X(\ln f^{*\frac{1+\alpha}{2}}) + X(\ln f^{\frac{1-\alpha}{2}}) + div^{(0)}X \\ &= X\big(\ln(f^{*\frac{1+\alpha}{2}} f^{\frac{1-\alpha}{2}})\big) + div^{(0)}X \\ &= X\big(\ln(Cf^{\frac{-1-\alpha}{2}} f^{\frac{1-\alpha}{2}})\big) + div^{(0)}X \\ &= -\alpha X\big(\ln f\big) + div^{(0)}X. \end{aligned}$$

The second relation is a consequence of the formula $f^* = C/f$, with $C$ constant. ■

The reason that makes divergence worthy to be studied is its geometric significance. It is used to describe the evolution (expansion or contraction) of the volume element along the integral curves of a vector field. This approach involves the study of the Lie derivative and is the subject of the next section.

## 9.9 Lie Derivative of a Volume Element

The Lie derivative is a useful tool in studying the behavior of some geometric objects evolving under the one-parameter group of diffeomorphisms generated by a vector field $X$ on a manifold $M$. If T is a covariant tensor of order $p$ on $M$, the pull-back of T under the diffeomorphism $\varphi : M \to M$ is defined by

$$(\varphi^*\mathrm{T})_x(u_1, \dots, u_p) = \mathrm{T}_{\varphi(x)}(d\varphi(u_1), \dots, d\varphi(u_p))$$

where $u_i \in T_xM$ and $d\varphi$ denotes the differential map of $\varphi$, see Sect. 7.6.

The *Lie derivative* of the tensor T, with respect to a vector field $X$, is the derivative along the integral curves of $X$, i.e.,

$$(L_X\mathrm{T})(Y_1, \dots, Y_p) = \lim_{t\to 0} \frac{1}{t}\Big((\varphi_t^*\mathrm{T})(Y_1, \dots, Y_p) - \mathrm{T}(Y_1, \dots, Y_p)\Big), \tag{9.9.48}$$

where $(\varphi_t)$ is the one-parameter group of diffeomorphisms of $X$.

The Lie derivative of the tensor $T$ of order $p$ can be expressed invariantly by the following formula

$$(L_X\mathrm{T})(Y_1, \dots, Y_p) = X\Big(\mathrm{T}(Y_1, \dots, Y_p)\Big) - \sum_{i=1}^{p} \mathrm{T}(Y_1, \dots, [X, Y_i], \dots, Y_p). \tag{9.9.49}$$

In particular, if $T = \omega$ is a 1-form, then

$$(L_X\omega)(Y) = X\omega(Y) - \omega([X, Y]).$$

If $T$ is a vector field $Y$, then $L_XY = [X, Y]$, the Lie bracket of $X$ and $Y$. If $p = 0$, then $T$ becomes a function $f$ and hence

$$L_Xf = X(f).$$

Among other properties of the Lie derivative we quote the following:

$$\begin{aligned}
L_{[X,Y]} &= L_XL_Y - L_YL_X = [L_X, L_Y], \\
L_X(f\omega) &= (L_Xf)\omega + fL_X\omega, \\
L_X(df) &= d(Xf), \quad \forall f \in \mathcal{F}(M), \\
L_{aX+bY} &= aL_X + bL_Y, \quad \forall a, b \in \mathbb{R}, X, Y \in \mathcal{X}(M).
\end{aligned}$$

The evolution of the volume element along the integral curves of $X$ is described by the Lie derivative with respect to $X$. In the case of Riemannian geometry we have the following result.

**Proposition 9.9.1** *Let $dv$ be the Riemannian volume element and $X$ be a $C^1$-vector field on $M$. Then*

$$L_X dv = (div^{(0)} X) dv. \tag{9.9.50}$$

*Proof:* It suffices to verify the formula on a basis. Applying relation (9.9.49) to the $n$-form $\mathrm{T} = dv = \sqrt{\det g}\, dx^1 \wedge \cdots \wedge dx^n$ and taking $Y_j = \partial_j$ yields

$$\begin{aligned}
& (L_X dv)(\partial_1, \dots, \partial_n) \\
= \;& X dv(\partial_1, \dots, \partial_n) - \sum_{i=1}^{n} dv(\partial_1, \dots, [X^j \partial_j, \partial_i], \dots, \partial_n) \\
= \;& X(\sqrt{\det g}) - \sum_i dv(\partial_1, \dots, -\frac{\partial X^j}{\partial x^i} \partial_j, \dots, \partial_n) \\
= \;& X(\sqrt{\det g}) + \sum_i \frac{\partial X^j}{\partial x^i} dv(\partial_1, \dots, \partial_j, \dots, \partial_n) \\
= \;& X(\sqrt{\det g}) + \frac{\partial X^i}{\partial x_i} \sqrt{\det g} \\
= \;& \sqrt{\det g} \Big( \frac{\partial X^i}{\partial x_i} + \frac{1}{\sqrt{\det g}} X(\sqrt{\det g}) \Big) \\
= \;& \sqrt{\det g}\, div^{(0)} X,
\end{aligned}$$

by Lemma 9.8.2. On the other side, we have

$$(div^{(0)} X) dv(\partial_1, \dots, \partial_n) = \sqrt{\det g}\; div^{(0)} X,$$

which proves the desired identity. ■

As a consequence of formula (9.9.48) and Proposition 9.9.1, the Riemannian volume element $dv$ expands (or contracts, or is invariant) along the integral curves of $X$ if and only if $div^{(0)} X > 0$ (or $< 0$, or $= 0$, respectively).

In the following we deal with a similar result for the volume element $\omega$ and divergence associated with an arbitrary linear connection $\nabla$.

**Proposition 9.9.2** *For any $C^1$-vector field $X$ and any volume element $\omega$, we have*

$$L_X \omega = (div X) \omega.$$

*Proof:* Let $\omega = f dv$. Using that $L_X$ acts as a derivation, we have

$$\begin{aligned} L_X\omega &= L_X(fdv) = (L_X f)dv + f(L_X dv) \\ &= X(f)dv + f(div_0 X)dv \\ &= \frac{1}{f}X(f)\omega + (div^{(0)}X)\omega \\ &= \Big(\frac{1}{f}X(f) + div^{(0)}X\Big)\omega \\ &= (divX)\omega, \end{aligned}$$

by Propositions 9.8.7 and 9.9.1. ■

We shall compute the divergence of a vector field in two important particular cases.

**Example 9.9.3** In the case of the exponential model, the $\alpha$-volume element has the coefficient $f = C(\det g)^{\frac{-\alpha}{2}}$, see Proposition 9.2.1, and hence

$$\begin{aligned} div^{(\alpha)}_{exp}X &= \frac{1}{f}X(f) + div^{(0)}X \\ &= \frac{-\alpha}{2}\frac{1}{\det g}X(\det g) + div^{(0)}X \\ &= \frac{-\alpha}{2}X(\ln(\det g)) + div^{(0)}X. \end{aligned}$$

**Example 9.9.4** The volume element in the case of the mixture model is given by Proposition 9.3.1. The $\alpha$-divergence in this case takes the following form

$$div^{(\alpha)}_{mixt}X = \frac{\alpha}{2}X(\ln(\det g)) + div^{(0)}X.$$

The next section will deal with $\alpha$-volume elements in more detail.

## 9.10 $\alpha$-Volume Elements

Let $\nabla$ and $\nabla^*$ be two dual connections, with respect to the metric $g$, and consider the $\nabla^{(\alpha)}$-connection defined by (8.5.11). Proposition 9.6.2 can be restated equivalently as:

The following conditions are equivalent:

$(i)$ The Ricci tensor associated with $\nabla^{(\alpha)}$ is symmetric;

$(ii)$ The Ricci tensor associated with $\nabla$ is symmetric;

$(iii)$ The Ricci tensor associated with $\nabla^*$ is symmetric.

Therefore, in the virtue of Theorem 9.5.2, if the volume element exists for the connection $\nabla$, then it also exists for the connections $\nabla^*$ and $\nabla^\alpha$. This was first shown by Takeuchi and Amari [79] and noted in Matsuzoe et al. [55].

In the following we shall assume that the aforementioned volume elements exist and we shall investigate their relationship using the concept of divergence. Let $\omega^{(\alpha)}$, $\omega^{(1)} = \omega^*$, and $\omega^{(-1)} = \omega$ be the volume elements corresponding to connections $\nabla^{(\alpha)}$, $\nabla^{(1)} = \nabla^*$, and $\nabla^{(-1)} = \nabla$, respectively:

$$\nabla^{(\alpha)}\omega^{(\alpha)} = \nabla^*\omega^{(1)} = \nabla\omega^{(-1)} = 0.$$

Since the previous volume elements are $n$-forms, they are proportional, so we can write

$$\omega^{(\alpha)} = f_1\omega^* = f_{-1}\omega,$$

with $f_1$ and $f_{-1}$ nonvanishing functions on $M$. The next result provides explicit formulas for the coefficients $f_1$ and $f_{-1}$.

**Theorem 9.10.1** *The coefficients $f_1$ and $f_{-1}$ can be written in terms of $f$ and $f^*$ as follows*

$$\begin{aligned} f_1 &= C(f^*)^{\alpha-1} = Cf^{1-\alpha} \\ f_{-1} &= C(f^*)^{\alpha+1} = Cf^{-(1+\alpha)}, \end{aligned}$$

*with $C$ real positive constant.*

*Proof:* Let $X$ be a $C^1$-vector field on $M$. Proposition 9.9.2 applied to the $n$-forms $\omega$, $\omega^*$ and $\omega^{(\alpha)}$ provides

$$\begin{aligned} L_X\omega &= (divX)\omega \\ L_X\omega^* &= (div^*X)\omega^* \\ L_X\omega^{(\alpha)} &= (div^{(\alpha)}X)\omega^\alpha. \end{aligned} \tag{9.10.51}$$

We shall compute both sides of (9.10.51). For the right side we use Proposition 9.8.11

$$(div^{(\alpha)}X)\omega^{\alpha} = \big(\alpha X(\ln f^*) + div^{(0)}X\big)\omega^{\alpha}. \qquad (9.10.52)$$

Using the properties of Lie derivative, the left side of (9.10.51) becomes

$$\begin{aligned} L_X\omega^{(\alpha)} &= L_X(f_1\omega^*) = X(f_1)\omega^* + f_1 L_X\omega^* \\ &= \frac{1}{f_1}X(f_1)\omega^{(\alpha)} + f_1(div^*X)\omega^* \\ &= X(\ln f_1)\omega^{(\alpha)} + (div^*X)\omega^{(\alpha)} \\ &= [X(\ln f_1) + X(\ln f^*) + div^{(0)}]\omega^{(\alpha)}, \qquad (9.10.53) \end{aligned}$$

where the last identity used the second relation of Proposition 9.8.11, with $\alpha = 1$. Equating (9.10.52) and (9.10.53) yields

$$\begin{aligned} X(\ln f_1) + X(\ln f^*) &= \alpha X(\ln f^*) \\ X(\ln f_1) &= X(\ln (f^*)^{\alpha-1}) \\ X\Big(\ln \frac{f_1}{(f^*)^{\alpha-1}}\Big) &= 0, \qquad \forall X \in \mathcal{X}(M), \end{aligned}$$

so there is a constant $c$ such that $\ln \dfrac{f_1}{(f^*)^{\alpha-1}} = c$. And then choosing $C = e^c$, we have $f_1 = C(f^*)^{\alpha-1}$. Since $ff^* =$ constant, the previous formula can be written also as $f_1 = Cf^{1-\alpha}$.

A similar computation is used to find the coefficient $f_{-1}$. In this case the right and left sides of (9.10.51) become

$$(div^{(\alpha)}X)\omega^{(\alpha)} = \big(-\alpha X(\ln f) + div^{(0)}X\big)\omega^{(\alpha)}, \qquad (9.10.54)$$

and

$$\begin{aligned} L_X\omega^{(\alpha)} &= L_X(f_{-1}\omega) = X(f_{-1})\omega + f_{-1}L_X\omega \\ &= \big(X(\ln f_{-1}) + X(\ln f) + div^{(0)}X\big)\omega^{(\alpha)}. \qquad (9.10.55) \end{aligned}$$

Equating (9.10.54) and (9.10.55) leads to

$$X\Big(\ln(f_{-1}f^{\alpha+1})\Big) = 0, \qquad (9.10.56)$$

whence $f_1 f^{\alpha+1} = C$, with $C > 0$. We note that the constants $C$ that appears in the formulas of $f_1$ and $f_{-1}$ are generic constants, and they are not necessarily equal. ■

The relationship between the coefficient functions $f_1$ and $f_{-1}$ is given in the next result.

**Corollary 9.10.2** *We have*

$$f_1^{\frac{1+\alpha}{2}} f_{-1}^{\frac{1-\alpha}{2}} = C,$$

*where $C$ is a positive constant.*

*Proof:* It is a direct computation using Theorem 9.10.1. We have

$$\begin{aligned} f_1^{\frac{1+\alpha}{2}} &= Cf^{(1-\alpha)(1+\alpha)/2} = Cf^{(1-\alpha^2)/2} \\ f_{-1}^{\frac{1-\alpha}{2}} &= Cf^{(1+\alpha)(\alpha-1)/2} = Cf^{(\alpha^2-1)/2}, \end{aligned}$$

and multiplying, we get

$$f_1^{\frac{1+\alpha}{2}} f_{-1}^{\frac{1-\alpha}{2}} = C.$$

■

It is worth to mention the particular case $\alpha = 0$, which recovers a well-known result. In this case $\omega^{(0)} = dv$ and $dv = f_1\omega^* = f_{-1}\omega$, so $f^* = 1/f_1$ and $f = 1/f_{-1}$, where $f$ and $f^*$ are defined by $\omega^* = f^*dv$ and $\omega = fdv$. Corollary 9.10.2 writes as $f_1 f_{-1} = C$, constant, which implies $f^*f = C$.

## 9.11 Problems

**9.1.** Find explicit formulas for the $\alpha$-volume elements in the case of the following distributions:

($a$) exponential; ($b$) normal; ($c$) gamma; ($d$) beta.

**9.2.** Let $\nabla$ be a linear connection on a manifold $M$, with components $\Gamma_{ij}^r$. In the following we assume Einstain summation convention.

($a$) Contract with $r = k$ in formula of $R_{ikj}^r$, and use Problem 1.14 to show the following formula for the Ricci tensor:

$$\frac{1}{2}R_{ij} = \partial_r\Gamma_{ji}^r - \partial_j\Gamma_{ri}^r + \Gamma_{\ell r}^r\Gamma_{ji}^\ell - \Gamma_{j\ell}^r\Gamma_{ri}^\ell.$$

($b$) Prove that

$$R_{ij} = R_{ji} \Leftrightarrow \partial_j(\Gamma_{ir}^r) = \partial_i(\Gamma_{jr}^r).$$

($c$) If $\nabla$ is the Levi–Civita connection, prove that $R_{ij} = R_{ji}$.

**9.3.** Let $\tau$ be a 1-form and $\nabla$, $\overline{\nabla}^*$ two torsion-free connections on a Riemannian manifold $(M, g)$ satisfying the condition

$$Xg(Y,Z) = g(\nabla_X Y, Z) + g(Y, \overline{\nabla}^*_X Z) - \tau(X)g(Y,Z) \quad (9.11.57)$$

for all $X, Y, Z \in \mathcal{X}(M)$.

(*a*) Show that

$$\partial_k g_{ij} = \Gamma_{kij} + \overline{\Gamma}^*_{kji} - \tau_k g_{ij},$$

where $\tau = \tau_k dx^k$, $\nabla_{\partial_i}\partial_j = \Gamma^k_{ij}\partial_k$, and $\overline{\nabla}^*_{\partial_i}\partial_j = \overline{\Gamma}^{*k}_{ij}\partial_k$.

(*b*) Given a 1-form $\tau$ and a linear connection $\nabla$, show that there is a unique linear connection $\overline{\nabla}^*$ satisfying relation (9.11.57).

(*c*) Assume $\nabla$ is a torsion-free, equiaffine connection and $\tau$ is exact (i.e., there is a function $f$ on $M$ such that $\tau = df$). Prove that $\overline{\nabla}^*$ is equiaffine.

(*d*) Assume the connections $\nabla$, $\overline{\nabla}^*$ are torsion-free and equiaffine. Show that the 1-form $\tau$ is exact.

**9.4.** Let $X$ be a vector field on $\mathcal{X} = \mathbb{R}^n$ and $\omega = dx^1 \wedge \cdots \wedge dx^n$ be the associated volume form. Show that the flow

$$\dot{x}(t) = X(x(t))$$

conserves the volume if $\operatorname{div} X(x) = 0$.

**9.5.** The continuity PDE

$$\frac{\partial f}{\partial t}(x,t) = -\frac{\partial}{\partial x^i}(f(x,t)X^i(x))$$

describes the dynamics induced by a probability density $f(x,t)$ on the associated phase space.

(*a*) Prove that for an incompressible flow (conservation of volume), the continuity PDE rewrites as

$$\frac{\partial f}{\partial t}(x,t) = -X^i(x)\frac{\partial f}{\partial x^i}(x,t).$$

This means that the probability density $f(x,t)$ is constant along the flow, i.e., $f(x, t+dt) = f(x - X(x)dt, t)$.

(*b*) Show that the function $\ln f(x,t)$ is also a solution of the previous continuity PDE.

(c) Show that, for any continuous function $h$, the integral

$$H(f) = \int_{\mathcal{X}} h(f(x,t))\,\omega$$

does not change in time, provided the probability density $f(x,t)$ satisfies the continuity PDE and the flow $X(x)$ conserves the phase volume.

(d) For $h(f) = -f \ln f$, the foregoing integral gives the classical Boltzmann–Gibbs–Shannon entropy functional

$$S(f) = -\int_{\mathcal{X}} f(x,t) \ln(f(x,t))\,\omega.$$

Prove that, for flows with conservation of volume, the entropy is conserved, i.e., $\dfrac{dS}{dt} = 0$.

**9.7.** Suppose the phase volume is not invariant with respect to the given flow and $f^*(x)$ is a steady-state solution (equilibrium point) of the continuity PDE, i.e.,

$$\frac{\partial}{\partial x^i}(f^* X^i)(x) = 0.$$

Check the following:

(a) in this case, instead of invariant phase volume form $\omega$, we have another invariant volume form, namely $\eta = f^*(x)\omega$;

(b) in case of volume conservation, we have

$$\frac{\partial}{\partial t}\frac{f(x,t)}{f^*(x)} = -X^i(x)\frac{\partial}{\partial x^i}\frac{f(x,t)}{f^*(x)}.$$

The function $\frac{f(x,t)}{f^*(x)}$ is constant along the flow and the measure $\eta$ is invariant.

(c) for any continuous function $h(f)$, the integral

$$H(f) = \int_{\mathcal{X}} h\left(\frac{f(x,t)}{f^*(x)}\right)\eta$$

does not change in time, if the probability density $f(x,t)$ satisfies the continuity PDE. Now we take $h(f) = -f \ln f$. We obtain the Kullback entropy functional

$$S_K(f) = -\int_{\mathcal{X}} f(x,t) \ln \frac{f(x,t)}{f^*(x)}\,\omega.$$

We note that this situation does not differ significantly from the entropy conservation in systems with conservation of volume. It is just a kind of change of variables.

**9.8.** If $f(x,t)$ is a probability density satisfying the continuity PDE, then the Boltzmann–Gibbs–Shannon entropy functional satisfies

$$\frac{d}{dt}S(f) = \int_{\mathcal{X}} f(x,t) \operatorname{div} X(x) \, \omega$$

if the left hand side exists.

(*Hint:* This *entropy production formula* can be proven for small phase drops with constant density, and then for finite sums of such distributions with positive coefficients. After that, we obtain the foregoing formula by limit transition.)

**9.9.** Show that, for a regular invariant density $f^*(x)$ (equilibrium), the entropy $S(f^*)$ exists, and for this distribution $\frac{d}{dt}S(f) = 0$ and consequently

$$\int_{\mathcal{X}} f^*(x) \operatorname{div} X(x) \, \omega = 0.$$

# Chapter 10

# Dual Laplacians

Each linear connection induces a divergence, which is used to define a Laplacian. Dual connections yield to dual Laplacians. This chapter deals with the definition and main properties of dual Laplacians and $\alpha$-Laplacians. Their relationship with Hessians, curvature vector fields, and dual volume elements is emphasized.

In this chapter $(M, g, \nabla, \nabla^*)$ is a manifold $M$ structured by a metric $g$, and endowed with a pair of dual connections $\nabla$ and $\nabla^*$.

## 10.1 Definition of Hessian

The Hessian of a function $f \in \mathcal{F}(M)$ taken with respect to the linear connection $\nabla$ is the covariant derivative of the 1-form $df$, i.e.,

$$H^f(X, Y) = (\nabla df)(X, Y). \tag{10.1.1}$$

Using the covariant differentiation formula for 1-forms

$$(\nabla \omega)(X, Y) = X\omega(Y) - \omega(\nabla_X Y),$$

then relation (10.1.1) provides the explicit formula

$$H^f(X, Y) = Xdf(Y) - df(\nabla_X Y) = XY(f) - (\nabla_X Y)f. \tag{10.1.2}$$

The local coordinates representation is $H^f(X, Y) = X^i Y^j H^f_{ij}$, where

$$H^f_{ij} = \frac{\partial^2 f}{\partial x^i \partial x^j} - \Gamma^k_{ij} \frac{\partial f}{\partial x^k}.$$

It is worth noting that $H^f$ is symmetric if and only if $\nabla$ is torsion-free, i.e., $\Gamma^k_{ij} = \Gamma^k_{ji}$ (symmetric).

O. Calin and C. Udrişte, *Geometric Modeling in Probability and Statistics*,
DOI 10.1007/978-3-319-07779-6_10,

In the following the gradient vector field of $f$ is taken with respect to the Riemannian metric $g$. This means

$$g(grad\, f, X) = X(f), \qquad \forall X \in \mathcal{X}(M),$$

which in local coordinates is equivalent to $(grad\, f)^j = g^{jk}\partial_{x^k} f$.

The relation between the Hessian and gradient is given by the next result.

**Lemma 10.1.1** *If $\nabla$, $\nabla^*$ are dual connections, then*

$$H^f(X,Y) = g\big(\nabla^*_X(grad\, f), Y\big), \qquad \forall X, Y \in \mathcal{X}(M).$$

*Proof:* Using the definition of Hessian, gradient and dual connections, we have

$$\begin{aligned} H^f(X,Y) &= X\big(Y(f)\big) - (\nabla_X Y)f \\ &= Xg(grad\, f, Y) - g(grad\, f, \nabla_X Y) \\ &= g(\nabla^*_X(grad\, f), Y) + g(grad\, f, \nabla_X Y) - g(grad\, f, \nabla_X Y) \\ &= g(\nabla^*_X(grad\, f), Y). \end{aligned}$$

■

## 10.2 Dual Hessians

Let $\nabla$, $\nabla^*$ be a pair of torsion-free dual connections on the Riemannian manifold $(M, g)$. For each smooth function $f$ on $M$, we associate a pair of Hessians $H^f$ and $H^{*f}$ given by

$$\begin{aligned} H^f(X,Y) &= (\nabla df)(X,Y) \\ H^{*f}(X,Y) &= (\nabla^* df)(X,Y). \end{aligned}$$

The dual Hessian has the components given by

$$H^{*f}_{ij} = \frac{\partial^2 f}{\partial x^i \partial x^j} - \Gamma^{*k}_{ij}\frac{\partial f}{\partial x^k},$$

and, according to Lemma 10.1.1, we can write

$$H^{*f}(X,Y) = g\big(\nabla_X(grad\, f), Y\big), \qquad \forall X, Y \in \mathcal{X}(M).$$

The relation with the difference tensor $K(X,Y)$ is given by the next result.

**Proposition 10.2.1** *For any $C^2$-function $f$, the Hessians $H^f$, $H^{*f}$ and the difference tensor $K$ are related by*

$$H^f(X,Y) - H^{*f}(X,Y) = K(X,Y)(f).$$

*Proof:* Using (10.1.2) we have

$$\begin{aligned} H^f(X,Y) - H^{*f}(X,Y) &= XY(f) - (\nabla_X Y)f \\ &\quad -\big(XY(f) - (\nabla^*_X Y)f\big) \\ &= (\nabla^*_X Y)f - (\nabla_X Y)f = K(X,Y)f. \end{aligned}$$

■

The relation with skewness tensor $C(X,Y,Z)$ is found in the next result.

**Proposition 10.2.2** *For any $C^2$-function $f$, the Hessians $H^f$, $H^{*f}$ and the skewness tensor $C$ are related by*

$$H^f(X,Y) - H^{*f}(X,Y) = C(grad\, f, X, Y).$$

*Proof:* Using Lemma 10.1.1 and Proposition 8.8.1, we obtain

$$\begin{aligned} H^f(X,Y) - H^{*f}(X,Y) &= g\big(\nabla^*_X(grad\, f), Y\big) - g\big(\nabla_X(grad\, f), Y\big) \\ &= g\big((\nabla^*_X - \nabla_X)(grad\, f), Y\big) \\ &= g\big(K(X, grad\, f), Y\big) \\ &= C(X, grad\, f, Y) = C(grad\, f, X, Y). \end{aligned}$$

The last identity follows from the symmetry of $C$. ■

## 10.3 The Laplacian

For every linear connection $\nabla$, any metric $g$ and any $C^2$-function $f$, we define the operator

$$\Delta f = div(grad\, f),$$

called *Laplacian*, where *div* is taken with respect to $\nabla$, see formula (9.8.42).

**Proposition 10.3.1** *The Laplacian is given by the trace of the dual Hessian*

$$\Delta f = Trace\big((X,Y) \to H^{*f}(X,Y)\big).$$

*Proof:* From Lemma 10.1.1

$$H^{*f}(X,Y) = g\big(\nabla_X(grad\, f), Y\big), \qquad \forall X, Y \in \mathcal{X}(M).$$

Taking the trace and using the divergence formula (9.8.42) yields

$$\begin{aligned} Trace\big((X,Y) \to H^{*f}(X,Y)\big) &= Trace\big((X,Y) \to g\big(\nabla_X(grad\, f), Y\big)\big) \\ &= div(grad\, f) = \Delta f. \end{aligned}$$

■

The previous formula can be written locally as

$$\Delta f = g^{ij} H^{*f}_{ij} = g^{ij}\Big(\frac{\partial^2 f}{\partial x^i \partial x^j} - \Gamma^{*k}_{\ ij}\frac{\partial f}{\partial x^k}\Big).$$

## 10.4 Dual Laplacians

Let $div$ and $div^*$ be divergences taken with respect to dual connections $\nabla$, $\nabla^*$ and the metric $g$. This induces a pair of dual Laplacians

$$\begin{aligned} \Delta f &= div(grad\, f) \\ \Delta^* f &= div^*(grad\, f). \end{aligned}$$

The relationship between $\Delta$ and $\Delta^*$ is given below.

**Proposition 10.4.1** *Let $K$ denote the curvature vector field associated with dual connections $\nabla$ and $\nabla^*$, see formula (8.7.22). Then*

$$\Delta^* f = \Delta f - K(f). \tag{10.4.3}$$

*Proof:* Using Propositions 10.3.1 and 10.2.1 together with formula (8.7.23) yields

$$\begin{aligned} \Delta^* f - \Delta f &= Trace H^f - Trace H^{*f} = Trace(H^f - H^{*f}) \\ &= -g^{ij} K^k_{ij}\frac{\partial f}{\partial x^k} = -K^k \frac{\partial f}{\partial x^k} = -K(f). \end{aligned}$$

■

**Corollary 10.4.2** *A function $f$ is constant along the curvature vector field, $K(f) = 0$, if and only if $\Delta f = \Delta^* f$.*

**Corollary 10.4.3** *Let $\Delta^{(0)}$ be the Laplacian with respect to the Levi–Civita connection $\nabla^{(0)}$. Then*

$$\Delta^{(0)} = \frac{1}{2}(\Delta + \Delta^*).$$

## 10.5 $\alpha$-Laplacians

Let $div^{(\alpha)}$ be the divergence with respect to the $\alpha$-connection $\nabla^{(\alpha)}$, see (9.8.42), on a Riemannian manifold $(M, g)$. For any $C^2$-function $f$, the $\alpha$-Laplacian is defined as

$$\Delta^{(\alpha)} f = div^{(\alpha)}(grad\, f).$$

Assuming the convention $\nabla^* = \nabla^{(1)}$, $\nabla = \nabla^{(-1)}$, write

$$\nabla^{(\alpha)} = \nabla^{(0)} + \frac{\alpha}{2}(\nabla^* - \nabla).$$

Taking the trace yields

$$div^{(\alpha)} = div^{(0)} + \frac{\alpha}{2}(div^* - div).$$

Applying it to $grad\, f$, we get

$$\Delta^{(\alpha)} f = \Delta^{(0)} f + \frac{\alpha}{2}(\Delta^* f - \Delta f).$$

Using (10.4.3) yields the following formula for the $\alpha$-Laplacian

$$\Delta^{(\alpha)} f = \Delta^{(0)} f - \frac{\alpha}{2} K(f). \tag{10.5.4}$$

Taking the values $\alpha = 1$ and $\alpha = -1$ yields the pair of dual Laplacians

$$\Delta^* = \Delta^{(0)} - \frac{1}{2} K \tag{10.5.5}$$

$$\Delta = \Delta^{(0)} + \frac{1}{2} K. \tag{10.5.6}$$

Consequently, we have

$$\frac{1}{2}(\Delta\Delta^* + \Delta^*\Delta) = \Delta^{(0)}\Delta^{(0)} - \frac{1}{4} KK.$$

## 10.6 Hopf's Lemma

Let us use the manifold $(M, g, \nabla)$.

**Proposition 10.6.1** *The Laplacian $\Delta$ satisfies the condition*

$$\Delta(f^2) = 2f\Delta f + 2\|grad\, f\|_g^2.$$

*Proof:* Let $X = grad\, f$ in Lemma 9.8.3 and obtain

$$\begin{aligned}\Delta(f^2) &= div(grad\, f^2) = div(2f grad\, f) \\ &= 2f div(grad\, f) + 2g(grad\, f, grad\, f) \\ &= 2f\Delta f + 2\|grad\, f\|_g^2.\end{aligned}$$

■

In the following $dv$ denotes the Riemannian volume element on $(M, g)$.

**Proposition 10.6.2** *Let $(M, g)$ be a compact Riemannian manifold, with $\partial M = \emptyset$. Consider a nonconstant $C^2$-function $f$ such that $\Delta f = 0$ on $M$. Then*

$$\int_M K(f^2)\, dv > 0. \tag{10.6.7}$$

*Proof:* Making $\Delta f = 0$ in Proposition 10.6.1 yields

$$\Delta(f^2) = 2\|grad\, f\|_g^2.$$

Using (10.5.6), we have

$$\Delta^{(0)}(f^2) + \frac{1}{2}K(f^2) = 2\|grad\, f\|_g^2.$$

Integrating we obtain

$$\int_M \Delta^{(0)}(f^2)\, dv + \frac{1}{2}\int_M K(f^2)\, dv = 2\int_M \|grad\, f\|_g^2\, dv > 0. \tag{10.6.8}$$

Since $\partial M = \emptyset$, the first integral vanishes from the divergence theorem

$$\int_M \Delta^{(0)}(f^2)\, dv = \int_M div(grad\, f^2)\, dv = 0,$$

and hence (10.6.8) yields the inequality (10.6.7). ■

It is worth noting that the condition $\Delta^* f = 0$ implies the reverse of inequality (10.6.7).

## 10.7 Laplacians and Volume Elements

This section deals with the relationship between the Laplacian and the volume element associated with the underlying connection. An explicit formula for the curvature vector field $K$ is provided as the gradient of a potential depending on the volume element.

To formulate our theory, we shall use the statistical manifold $(M, g, \nabla, \nabla^*)$.

Consider two dual connections $\nabla$, $\nabla^*$, the Levi–Civita connection $\nabla^{(0)}$, and their associated volume elements

$$\omega = \phi\, dv \qquad \omega^* = \phi^* dv \qquad \omega^{(0)} = dv.$$

**Proposition 10.7.1** *Let $f \in \mathcal{F}(M)$. Then the dual Laplacians can be written in terms of the volume element as*

$$\Delta f = \Delta^{(0)} f + g\Big(grad\, f, grad(\ln \phi)\Big) \qquad (10.7.9)$$

$$\Delta^* f = \Delta^{(0)} f - g\Big(grad\, f, grad(\ln \phi)\Big). \qquad (10.7.10)$$

*Proof:* Substituting $f = \phi$ and $X = grad\, f$ in (9.8.44) yields

$$\begin{aligned} \Delta f &= div(grad\, f) = \frac{1}{\phi}(grad\, f)(\phi) + div^{(0)}(grad\, f) \\ &= \Delta^{(0)} f + \frac{1}{\phi} g\big(grad\, f, grad\, \phi\big) \\ &= \Delta^{(0)} f + g\big(grad\, f, grad(\ln \phi)\big). \end{aligned}$$

A similar computation provides

$$\begin{aligned} \Delta^* f &= \Delta^{(0)} f + g\big(grad\, f, grad(\ln \phi^*)\big) \\ &= \Delta^{(0)} f + g\big(grad\, f, grad(\ln \frac{C}{\phi})\big) \\ &= \Delta^{(0)} f - g\big(grad\, f, grad(\ln \phi)\big). \end{aligned}$$

■

The curvature vector field, $K$, was defined in Sect. 8.7 as the trace of the difference tensor. The curvature vector appeared in formulas (10.4.3), (10.5.4)–(10.5.6) and Propositions 10.6.2 and 10.9.3. In the following we shall prove that $K$ is a gradient vector field.

**Theorem 10.7.2** *The curvature vector field is a gradient vector field given by*

$$K = grad(\ln \phi^2). \qquad (10.7.11)$$

*Proof:* Subtracting the relations (10.7.9) and (10.7.10), we obtain

$$\Delta^* f = \Delta f - 2g\big(grad\, f, grad(\ln \phi)\big).$$

Comparing with relation (10.4.3) yields

$$\begin{aligned} K(f) &= 2g\big(grad\, f, grad(\ln \phi)\big) \\ &= 2grad(\ln \phi)(f) = grad(\ln \phi^2)(f). \end{aligned}$$

Dropping the argument $f$ leads to Eq. (10.7.11). ■

**Remark 10.7.3** Consider two dual connections $\nabla, \nabla^*$ with the curvature vector field $K$. Then

$$K = 0 \Longleftrightarrow \phi = \text{constant},$$

i.e., the forms $\omega$ and $\omega^*$ are equal, up to a scaling factor, to the volume form $dv$.

## 10.8 Divergence of Tensors

The fundamental ingredient is the statistical manifold $(M, g, \nabla, \nabla^*)$. Let $T$ be a 2-covariant symmetric $C^1$-tensor. Its divergence (with respect to connection $\nabla$) is the vector field $div(T) = (div\, T)^i \frac{\partial}{\partial x^i}$, with components given by

$$(div\, T)^i = \nabla_{\partial_{x^j}} T^{ji}.$$

Two torsion-free dual connections, $\nabla$ and $\nabla^*$, induce the dual divergences $div(T)$ and $div^*(T)$. The $\alpha$-divergence is defined as the convex combination

$$div^{(\alpha)} T = \frac{1+\alpha}{2}\, div^* T + \frac{1-\alpha}{2}\, div T.$$

Then the divergence with respect to the Levi–Civita connection, $\nabla^{(0)}$, is the average of dual divergences, i.e.,

$$div^{(0)} T = \frac{1}{2}(div T + div^* T).$$

We shall investigate in the following the relation between the aforementioned divergences for the case of the metric tensor.

## 10.9 Divergence of the Metric Tensor

It is well known that the divergence of the metric tensor $g$ vanishes if the divergence is taken with respect to the Levi–Civita connection $\nabla^{(0)}$. In the following we deal with the divergence of the metric tensor $g$ with respect to a pair of dual connections $\nabla$ and $\nabla^*$, as well as to the $\alpha$-connection $\nabla^{(\alpha)}$.

First, we compute the covariant derivative

$$\begin{aligned}
(\nabla_{\partial_i} g)_{jk} &= (\nabla_{\partial_i} g)(\partial_j, \partial_k) \\
&= \partial_i g(\partial_j, \partial_k) - g(\nabla_{\partial_i}\partial_j, \partial_k) - g(\partial_j, \nabla_{\partial_i}\partial_k) \\
&= g(\nabla^*_{\partial_i}\partial_j, \partial_k) + g(\partial_j, \nabla_{\partial_i}\partial_k) - g(\nabla_{\partial_i}\partial_j, \partial_k) - g(\partial_j, \nabla_{\partial_i}\partial_k) \\
&= g(\nabla^*_{\partial_i}\partial_j - \nabla_{\partial_i}\partial_j, \partial_k) = g\big(K(\partial_i, \partial_j), \partial_k\big) \\
&= C(\partial_i, \partial_j, \partial_k) = C_{ijk}, \qquad (10.9.12)
\end{aligned}$$

which is the skewness tensor. The computation used the definitions of dual connections, difference and skewness tensors. Raising indices in (10.9.12) we obtain

$$\begin{aligned}
\big(\nabla_{\partial_i} g\big)^{lr} &= \big(\nabla_{\partial^i} g\big)_{jk} g^{jl} g^{kr} \\
&= C_{ijk} g^{jl} g^{kr} = K^l_{ik} g^{kr}.
\end{aligned}$$

Then making $r = i$, and summing over $i$ provides

$$\big(\nabla_{\partial_i} g\big)^{li} = K^l_{ik} g^{ik} = K^l,$$

which is the $p$-th component of the curvature vector field. Using the divergence definition

$$(div\, g)^p = \big(\nabla_{\partial^i} g\big)^{pi},$$

we arrive at the following result.

**Proposition 10.9.1** *The divergence of the metric tensor $g$ with respect to $\nabla$ is equal to the curvature vector field $K$, i.e.,*

$$div\, g = K^l \partial_l. \qquad (10.9.13)$$

Similar computations applied to the dual connection lead to

$$(\nabla^*_{\partial_i} g)_{jk} = -C_{ijk}. \qquad (10.9.14)$$

Following the same lines of computation as before, we arrive at the dual result.

**Proposition 10.9.2** *The divergence of the metric tensor $g$, with respect to $\nabla^*$, is equal to the negative curvature vector field $K$,*

$$div^* \, g = -K^l \partial^l. \tag{10.9.15}$$

As a consequence, the metric tensor has opposite divergences with respect two dual connections, i.e., $div^* \, g = -div \, g$.

The next result deals with the $\alpha$-divergence, which is taken with respect to the $\nabla^{(\alpha)}$-connection.

**Proposition 10.9.3** *The $\alpha$-divergence of the metric tensor is related to the curvature vector by*

$$div^{(\alpha)} \, g = -\alpha K. \tag{10.9.16}$$

*Proof:* Writing the $\alpha$-divergence as a linear combination of dual divergences, using (10.9.13) and (10.9.15) we get

$$\begin{aligned} div^{(\alpha)} \, g &= \frac{1+\alpha}{2} div^* \, g + \frac{1-\alpha}{2} div \, g \\ &= \frac{1+\alpha}{2}(-K) + \frac{1-\alpha}{2} K \\ &= -\alpha K. \end{aligned}$$

■

It is worthy to observe that for $\alpha = 0$ we recover the well-known result of Riemannian geometry $div^{(0)} \, g = 0$, quoted in the beginning of this section.

**Corollary 10.9.4** *The $\alpha$-Laplacian has the following expression*

$$\Delta^{(\alpha)} f = \Delta^{(0)} f + \frac{1}{2}(div^{(\alpha)} \, g)(f). \tag{10.9.17}$$

*Proof:* It follows from (10.5.4) and (10.9.16). We note that the difference

$$\Delta^{(\alpha)} - \frac{1}{2}(div^{(\alpha)} \, g)$$

is independent of $\alpha$. ■

## 10.10 Problems

**10.1.** Consider the statistical model $p_{\mu,\sigma}(x) = \frac{1}{\sqrt{2\pi}\sigma} e^{-\frac{(x-\mu)^2}{2\sigma^2}}$, $\sigma > 0$, $\mu \in \mathbb{R}$ and the function $f(\mu, \sigma) = \frac{1}{2}(\sigma^2 + \mu^2)$. Find the Hessians $H^f$ and $H^{*f}$.

**10.2.** Consider the statistical model $p_\xi(x) = \xi e^{-\xi x}$, $\xi > 0$, $x \in \mathbb{R}$.

(*a*) Find the dual Laplacians $\Delta$, $\Delta^*$.

(*b*) Deduct the curvature vector field $K$, and find the potential function $\phi$ such that $K = \frac{d}{d\xi}\big(\ln \phi^2(\xi)\big)$.

(*c*) Verify the relation $\Delta^{(0)} = \frac{1}{2}(\Delta + \Delta^*)$.

(*d*) Find the expression for the $\alpha$-Laplacian.

**10.3.** (*a*) Solve the equation $H^{*f}_{ij}(\xi) = 0$ in the case of the exponential family.

(*b*) Solve the equation $H^f_{ij}(\xi) = 0$ in the case of the mixture family.

**10.4.** Find the Laplacians $\Delta^0$, $\Delta$, and $\Delta^*$ in the following cases:

(*a*) exponential distribution.

(*b*) normal distribution.

**10.5.** Find the curvature vector field $K$ and the potential function $\phi$ in the case of the normal distribution.

**10.6.** Let $(M, g, \nabla, \nabla^*)$ be a statistical manifold. Define the $\alpha$-Hessian of the smooth function $f$ by $H^{(\alpha)f} = \nabla^{(\alpha)} df$.

(*a*) Show that in a local system of coordinates we have

$$H^{(\alpha)f} = \frac{\partial^2 f}{\partial x^i \partial x^j} - \frac{\partial f}{\partial x^k}\Gamma^{(\alpha)k}_{ij}.$$

(*b*) Verify the relation $\Delta^{(\alpha)} f = Trace\big(H^{(\alpha)f}\big)$.

**10.7.** Let $(M, g, \nabla^{(\alpha)})$ be a statistical manifold and $f : M \to \mathbb{R}$ a differentiable function. Consider $\gamma : (a, b) \to M$ a $\nabla^{(\alpha)}$-autoparallel curve on $M$.

(*a*) Compute $\dfrac{d}{ds} f\big(\gamma(s)\big)$.

(*b*) Show that

$$\frac{d^2}{ds^2} f\big(\gamma(s)\big) = H^{(\alpha)f}_{|\gamma(s)} \dot\gamma^i(s)\dot\gamma^j(s).$$

(*c*) Show that

$$\frac{d^3}{ds^3} f\big(\gamma(s)\big) = \frac{\partial H^{(\alpha)f}_{ij}}{\partial x^k} \dot\gamma^k(s)\dot\gamma^i(s)\dot\gamma^j(s) + 2H^{(\alpha)f}_{|\gamma(s)} \dot\gamma^i(s)\ddot\gamma^j(s),$$

with

$$\frac{\partial H^{(\alpha)f}_{ij}}{\partial x^k} = \frac{\partial^3 f}{\partial x^i \partial x^j \partial x^k} - \frac{\partial^2 f}{\partial x^k \partial x^r} \Gamma^{(\alpha)r}_{ij} - \frac{\partial f}{\partial x^r}\frac{\partial \Gamma^{(\alpha)r}_{ij}}{\partial x^k}.$$

(*d*) Write formulas (*a*)–(*c*) in the case when $M$ is the statistical model defined by an exponential family and $\alpha = 1$.

**10.8.** Consider the statistical manifold $(M, g, \nabla^{(\alpha)})$. Let $p \in M$ and $\gamma_1, \dots, \gamma_n$ be $\nabla^{(\alpha)}$-autoparallel curves on $M$ such that $\gamma_j(0) = p$, $\dot\gamma_j(0) = v_j$, with $\{v_1, \dots, v_n\}$ orthonormal system in $T_pM$. Prove that for any function $f \in \mathcal{F}(M)$ we have

$$\Delta^{(\alpha)} f = \sum_{j=1}^{n} \frac{d^2}{ds^2} f\big(\gamma_j(s)\big)_{|s=0}.$$

**10.9.** Let $(M, g)$ be a Riemannian manifold where the metric is given as a Hessian, $g_{ij}(\xi) = \partial_{\xi^i}\partial_{\xi^j}\varphi(\xi)$, with $\varphi(\xi)$ strictly convex. Show that:

(*a*) $\dfrac{H^{\varphi}_{ij}}{\partial x^k} = C_{ijk}$, where $C_{ijk}$ denotes the skewness tensor.

(*b*) Let $\gamma(s)$ be a $\nabla$-autoparallel curve on $M$. Show that

$$\frac{d^3}{ds^3}\varphi\big(\varphi(s)\big) = C(\dot\gamma(s), \dot\gamma(s), \dot\gamma(s)) + 2\langle H^{\varphi}\dot\gamma(s), \ddot\gamma(s)\rangle.$$

**10.10.** Let $(M, g, \nabla, \nabla^*)$ be a statistical manifold endowed with an equiaffine structure.

(*a*) Prove that $div\big(Ric^{(0)} - \frac{1}{2}R^{(0)}g\big) = 0$, where $R^{(0)} = Ric^{(0)}_{ij} g^{ij}$.

(*b*) Find a formula for $div\big(Ric^{(\alpha)} - \frac{1}{2}R^{(\alpha)}g\big)$, where $R^{(\alpha)} = Ric^{(\alpha)}_{ij} g^{ij}$.

# Chapter 11

# Contrast Functions Geometry

*Contrast functions*, called also divergence functions, are distance-like quantities which measure the asymmetric "proximity" of two probability density functions on a statistical manifold or statistical model $\mathcal{S}$. A contrast function, $D(p||q)$, for density functions $p, q \in \mathcal{S}$, is a smooth, non-negative function that vanishes for $p = q$. Eguchi [38, 39, 41] has shown that a contrast function $D$ induces a Riemannian metric by its second order derivatives, and a pair of dual connections by its third order derivatives.

This chapter introduces contrast functionals on statistical manifolds, which are natural extensions of Kullback–Leibler relative entropy from statistical models, and analyzes their corresponding geometric structures and how these interact with the dualistic structure of a statistical manifold. The chapter also investigates the geometry generated by a contrast functional on the space of probability distributions of a statistical model and provides examples of contrast functions.

It has been shown in Chap. 4 that Kullback–Leibler relative entropy is positive, non-degenerate, its first variation along the diagonal $\xi^0 = \xi$ vanishes, and the Hessian along the diagonal defines the Fisher metric.

O. Calin and C. Udrişte, *Geometric Modeling in Probability and Statistics*, DOI 10.1007/978-3-319-07779-6_11,

The contrast functions mimic the aforementioned properties of the Kullback–Leibler relative entropy. The only difference in the new context is that there are no density functions and no formula of expectation type can be used here.

We overcome this flaw by defining the contrast functions abstractly in two stages: (*i*) on an open set of $\mathbb{R}^k$; (*ii*) on a smooth manifold $\mathcal{S}$.

## 11.1 Contrast Functions on $\mathbb{R}^k$

Consider an open set $\mathbb{E}$ in $\mathbb{R}^k$, and let $\xi_1, \xi_2 \in \mathbb{E}$. A *contrast function* on $\mathbb{E}$ is a smooth function $D(\cdot \,||\, \cdot) : \mathbb{E} \times \mathbb{E} \to \mathbb{R}$ satisfying the following properties:

(*i*) *positive:* $D(\xi_1||\xi_2) \geq 0$, $\forall \xi_1, \xi_2 \in \mathbb{E}$;

(*ii*) *non-degenerate:* $D(\xi_1||\xi_2) = 0 \Longleftrightarrow \xi_1 = \xi_2$;

(*iii*) *the first variation along the diagonal* $\{\xi_1 = \xi_2\}$ *vanishes:*

$$\partial_{\xi_1^i} D(\xi_1||\xi_2)_{|\xi_1=\xi_2} = \partial_{\xi_2^i} D(\xi_1||\xi_2)_{|\xi_1=\xi_2} = 0;$$

(*iv*) *the Hessian along the diagonal* $\xi^0 = \xi$

$$g_{ij}(\xi_1) = \partial_{\xi_2^i} \partial_{\xi_2^j} D(\xi_1||\xi_2)_{|\xi_2=\xi_1}$$

*is strictly positive definite and smooth with respect to* $\xi_1$.

Some comments regarding the notation are worthy to make. Even if the function $D(\xi_1||\xi_2)$ is not a distance (the symmetry and the triangle inequality are not satisfied), it is a useful distance-like measure of the separation between two points $\xi_1$, $\xi_2$. The separation notation is represented by the symbol $||$.

Another observation worthy to make is the redundancy of part (*iii*) of the definition; this is a consequence of parts (*i*) and (*ii*) as follows:

$$\lim_{\epsilon \searrow 0} \frac{D(\xi_1 + \epsilon||\xi_1) - D(\xi_1||\xi_1)}{\epsilon} = \lim_{\epsilon \searrow 0} \frac{D(\xi_1 + \epsilon||\xi_1)}{\epsilon} \geq 0$$

$$\lim_{\epsilon \nearrow 0} \frac{D(\xi_1 + \epsilon||\xi_1) - D(\xi_1||\xi_1)}{\epsilon} = \lim_{\epsilon \nearrow 0} \frac{D(\xi_1 + \epsilon||\xi_1)}{\epsilon} \leq 0,$$

which implies the limit equal to 0. We assumed $\xi_1 \in \mathbb{R}$ for the sake of notation simplicity, but the result holds true in multiple dimensions.

We note two facts, which are direct consequences of the definition:

(1) *The point $\xi_0$ is a global minimum of the map $\xi \to D(\xi_0||\xi)$.*

(2) *The quadratic approximation of a contrast function is given by*

$$D(\xi_1||\xi_2) = \frac{1}{2}\sum_{i,j} g_{ij}(\xi_1)(\xi_1^i - \xi_2^i)(\xi_1^j - \xi_2^j) + o(\|\Delta(\xi_1 - \xi_2)\|^2) \tag{11.1.1}$$

*when $\xi_2 - \xi_1 \to 0$.*

Hence, for any two close enough neighbor vectors $\xi_1, \xi_2 \in \mathbb{E}$, the contrast function is approximated by half the length of their difference measured in the inner product induced by the matrix $g_{ij}$

$$D(\xi_1||\xi_2) \approx \frac{1}{2}\langle \xi_1 - \xi_2, \xi_1 - \xi_2\rangle_g = \frac{1}{2}\|\xi_1 - \xi_2\|_g^2.$$

In the following we show how a contrast function can be induced by a strictly convex function.

**Proposition 11.1.1** *Let $\varphi : \mathbb{E} \to \mathbb{R}$ be a strictly convex function. Then*

$$\begin{aligned} D(\xi_0||\xi) &= \varphi(\xi) - \varphi(\xi_0) - \sum_j \partial_j\varphi(\xi_0)(\xi^j - \xi_0^j) \qquad (11.1.2)\\ &= \varphi(\xi) - \varphi(\xi_0) - \langle \partial\varphi(\xi_0), \xi - \xi_0\rangle \end{aligned}$$

*is a contrast function on $\mathbb{E}$.*

*Proof:*

$(i)$ Positivity: since the graph of the strictly convex function $\varphi$ is above the tangent plane at each point, we have

$$\varphi(\xi) \geq \varphi(\xi_0) + \sum_j \partial_j\varphi(\xi_0)(\xi^j - \xi_0^j). \tag{11.1.3}$$

This implies $D(\xi_0||\xi) \geq 0$.

$(ii)$ Non-degenerate: Since the equality in (11.1.3) occurs only for $\xi = \xi_0$, it follows that $D(\xi_0||\xi) = 0$ implies $\xi = \xi_0$.

$(iii)$ Differentiating with respect to $\xi_i$ yields

$$\partial_{\xi_i} D(\xi_0||\xi) = \partial_{\xi_i}\varphi(\xi) - \partial_{\xi_i}\varphi(\xi_0),$$

and hence $\partial_{\xi_i} D(\xi_0||\xi)_{|\xi=\xi_0} = 0$.

(*iv*) Since the function $\varphi$ is strictly convex, and

$$\partial_{\xi_i}\partial_{\xi_j}D(\xi_0||\xi) = \partial_{\xi_i}\partial_{\xi_j}\varphi(\xi) \tag{11.1.4}$$

it follows that $\partial_{\xi_i}\partial_{\xi_j}D(\xi_0||\xi)$ is strictly positive definite. Hence $D(\xi_0||\xi)$ satisfies the properties of a contrast function.

∎

We shall discuss in the following a few particular cases.

**Example 11.1.2 (Exponential Model)** Consider the convex function $\varphi(\xi) = -\ln\xi$, with $\xi > 0$. The induced contrast function is given by

$$D(\xi_0||\xi) = \frac{\xi}{\xi_0} - \ln\frac{\xi}{\xi_0} - 1,$$

which is exactly the Kullback–Leibler relative entropy for the exponential distribution. It is worth noting that the convex function $\varphi(\xi) = \xi - \ln\xi$ induces the same contrast function. Hence, there is no one-to-one correspondence between convex functions and contrast functions.

**Example 11.1.3** The convex function $\varphi(\xi) = \xi^2 - \ln\xi$, with $\xi > 0$, induces the contrast function

$$D(\xi_0||\xi) = (\xi - \xi_0)^2 + \frac{\xi}{\xi_0} - \ln\frac{\xi}{\xi_0} - 1.$$

**Example 11.1.4** If consider $\varphi(\xi) = \xi^2$, with $\xi > 0$, the induced contrast function is

$$D(\xi_0||\xi) = (\xi - \xi_0)^2.$$

Not all contrast functions are induced by strictly convex functions. For instance, one can show that

$$D(\xi_0||\xi) = \frac{(\xi - \xi_0)^2}{\xi_0\xi^2}$$

is a contrast function on $(0,\infty)^2$, which cannot be written in the form of formula (11.1.2). We make the note that this contrast function is related to the problem of minimum chi-squared estimator, as described in Kass and Vos [49], p.244. There are many other contrast functions that are not in the form (11.1.2), for instance most

$f$-divergences, see Sect. 12.2. It can be shown that a contrast function derived from a strictly convex function by formula (11.1.2) is a dually flat contrast function.

It is worth noting that the definition of the contrast function adopted by Kass and Vos [49], p.240, is slightly modified, replacing condition $(iv)$ by the following condition:

$(iv')$ *the matrix*

$$g_{ij}(\xi_1) = \partial_{\xi_1^i}\partial_{\xi_1^j} D(\xi_1||\xi_2)$$

*is positive definite and a smooth function of* $\xi_1$ *alone.*

The contrast function given by formula (11.1.2) is sometimes called *Bregman divergence*, see Bregman [20], and it is widely used in convex optimization, see Bauschke [14], Bauschke and Combettes [16], and Bauschke et al. [15].

The term of "contrast function" has been defined slightly different by other authors, and under different names (divergence, yoke, etc.) see Eguchi [40], Rao [72] and Barndorff-Nielsen [11].

## 11.2 Contrast Functions on a Manifold

Let $\mathcal{S}$ be a smooth manifold. A *contrast function* on $\mathcal{S}$ is a smooth mapping $D_{\mathcal{S}}(\,\cdot\,\|\,\cdot\,) : \mathcal{S} \times \mathcal{S} \to \mathbb{R}$, such that any parametrization $\phi : \mathbb{E} \to \mathcal{S}$ makes

$$D(\xi_1||\xi_2) = D_{\mathcal{S}}\big(\phi(\xi_1)||\phi(\xi_2)\big)$$

a contrast function on $\mathbb{E}$. This definition was given for the first time in Amari [5].

We note the local character of a contrast function on a manifold. If $p_1, p_2 \in \mathcal{S}$ belong to the same coordinate chart, there are $\xi_1, \xi_2 \in \mathbb{E}$ such that $\phi(p_i) = \xi_i$ and then we have $D(\xi_1||\xi_2) = D_{\mathcal{S}}\big(p_1||p_2\big)$. Since there might be no coordinate charts to include both points $p_1, p_2$, then the contrast function $D_{\mathcal{S}}(\,\cdot\,\|\,\cdot\,)$ makes sense only locally. In general, there might be no global defined contrast functions on a manifold $\mathcal{S}$.

The invariance of the contrast function with respect to charts is given in the following result.

**Theorem 11.2.1** *Consider two local parametrizations* $\phi : \mathbb{E}_\xi \to U$, $\varphi : \mathbb{E}_\eta \to V$ *on the manifold* $\mathcal{S}$. *If*

$$D(\xi_1||\xi_2) = D_{\mathcal{S}}\big(\phi(\xi_1)||\phi(\xi_2)\big)$$

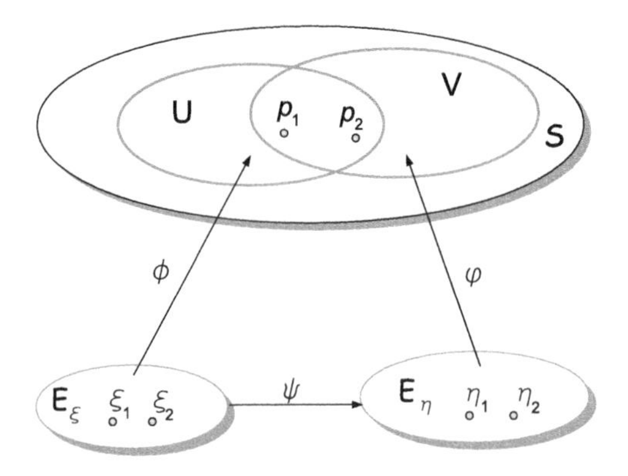


Figure 11.1: The parameterizations $\phi$ and $\varphi$ on a manifold $\mathcal{S}$

*is a contrast function on the parameter space* $\mathbb{E}_\xi$*, then*

$$D(\eta_1||\eta_2) = D_{\mathcal{S}}\big(\varphi(\eta_1)||\varphi(\eta_2)\big)$$

*is also a contrast function on the parameter space* $\mathbb{E}_\eta$*.*

*Proof:* For any two points $p_1, p_2 \in U \cap V \subset \mathcal{S}$ denote $p_1 = \phi(\xi_1) = \varphi(\eta_1)$, $p_2 = \phi(\xi_2) = \varphi(\eta_2)$. Let $\psi : \mathbb{E}_\xi \to \mathbb{E}_\eta$, $\psi(\xi) = \eta$ be the change of parametrization map, which is invertible as a composition of invertible maps $\psi = \varphi^{-1} \circ \phi$, see Fig. 11.1.

(*i*) The positivity follows obviously from

$$D(\eta_1||\eta_2) = D_{\mathcal{S}}\big(p_1||p_2\big) = D(\xi_1||\xi_2) \geq 0.$$

(*ii*) To check the non-degeneracy we note that $D(\eta_1||\eta_2) = 0$ implies $D(\xi_1||\xi_2) = 0$, and hence $\xi_1 = \xi_2$, or $\psi^{-1}(\eta_1) = \psi^{-1}(\eta_2)$. Since $\psi^{-1}$ is one-to-one, we obtain $\eta_1 = \eta_2$.

(*iii*) The fact that the first variation along the diagonal $\{\eta_1 = \eta_2\}$ vanishes is a consequence of (*i*) and (*ii*).

(*iv*) We investigate first how does $g_{ij}$ change when changing the parameter $\xi$ into $\eta$

$$\begin{aligned} g_{ij}(\xi) &= g(\partial_{\xi^i}, \partial_{\xi^j}) = g\Big(\frac{\partial \eta^r}{\partial \xi^i}\partial_{\eta^r}, \frac{\partial \eta^k}{\partial \xi^j}\partial_{\eta^k}\Big) \\ &= \frac{\partial \eta^r}{\partial \xi^i}\frac{\partial \eta^k}{\partial \xi^j} g(\partial_{\eta^r}, \partial_{\eta^k}) = \frac{\partial \eta^r}{\partial \xi^i}\frac{\partial \eta^k}{\partial \xi^j}\bar{g}_{rk}(\eta), \end{aligned}$$

and hence

$$g_{ij}(\xi) = \frac{\partial \eta^r}{\partial \xi^i} \frac{\partial \eta^k}{\partial \xi^j} \bar{g}_{rk}(\eta). \qquad (11.2.5)$$

Consider the points $p_1$ and $p_2$ infinitesimally close. Then writing the quadratic approximation formula (11.1.1) in differential form for $D(\xi_1||\xi_2)$ and $D(\eta_1||\eta_2)$ and combining with (11.2.5) and the chain rule yields

$$\begin{aligned} D(\xi_1||\xi_2) &= \frac{1}{2}\sum_{i,j} g_{ij}(\xi_1) d\xi^i d\xi^j \\ &= \frac{1}{2}\sum_{i,j}\sum_{r,k} \bar{g}_{rk}(\eta_1) \frac{\partial \eta^r}{\partial \xi^i} \frac{\partial \eta^k}{\partial \xi^j} d\xi^i d\xi^j \qquad (11.2.6) \end{aligned}$$

$$\begin{aligned} D(\eta_1||\eta_2) &= \frac{1}{2}\sum_{r,k} h_{rk}(\eta_1) d\eta^r d\eta^k \\ &= \frac{1}{2}\sum_{i,j}\sum_{r,k} h_{rk}(\eta_1) \frac{\partial \eta^r}{\partial \xi^i} \frac{\partial \eta^k}{\partial \xi^j} d\xi^i d\xi^j. \qquad (11.2.7) \end{aligned}$$

Comparing (11.2.6) and (11.2.7) yields $\bar{g}_{rk}(\eta) = h_{rk}(\eta)$. Since $\bar{g}_{rk}(\eta)$ is strictly positive definite, then $h_{rk}(\eta)$ is the same. Hence $D(\eta_1, \eta_2)$ verifies all the conditions of a contrast function.

■

**Corollary 11.2.2** *The diagonal part of the Hessians*

$$g_{ij}(\xi_1) = \partial_{\xi_2^i} \partial_{\xi_2^j} D(\xi_1||\xi_2)_{|\xi_2=\xi_1}$$

$$h_{ij}(\eta_1) = \partial_{\eta_2^i} \partial_{\eta_2^j} D(\eta_1||\eta_2)_{|\eta_2=\eta_1}$$

*are related by the following relation*

$$g_{ij}(\xi_1) = \frac{\partial \eta^r}{\partial \xi^i} \frac{\partial \eta^k}{\partial \xi^j} h_{rk}(\eta_1). \qquad (11.2.8)$$

## 11.3 Induced Riemannian Metric

One of the useful consequences of the invariance property given by Theorem 11.2.1 is that a contrast function provides a unique Riemannian metric on the manifold $\mathcal{S}$. This metric is the inner product $g_p : T_p\mathcal{S} \times T_p\mathcal{S} \to \mathbb{R}$ defined in a particular chart as

$$g_p(\partial_i, \partial_j) = \partial_{\xi_2^i}\partial_{\xi_2^j} D(\xi_1||\xi_2)_{|\xi_2=\xi_1}, \tag{11.3.9}$$

for any coordinate vector fields $\partial_i, \partial_j$ on $\mathcal{S}$ about $p$.

In the following we shall develop two formulas equivalent with (11.3.9). Consider the notation $\rho(\xi_1, \xi_2) = D(\xi_1||\xi_2)$. By $(ii)$ we have

$$\begin{aligned} \partial_{\xi_1^i}\rho(\xi_1,\xi_2)_{|\xi_1=\xi_2=\xi} &= \partial_{\xi_1^i}\rho(\xi,\xi) = 0 \\ \partial_{\xi_2^i}\rho(\xi_1,\xi_2)_{|\xi_1=\xi_2=\xi} &= \partial_{\xi_2^i}\rho(\xi,\xi) = 0. \end{aligned}$$

Denote $\partial_j = \frac{\partial}{\partial \xi^j}$. Differentiating the function $\varphi(\xi) = \partial_{\xi_1^i}\rho(\xi,\xi)$ with respect to $\partial_j$ we get

$$0 = \partial_j\varphi(\xi) = \partial_{\xi_1^j}\partial_{\xi_1^i}\rho(\xi,\xi) + \partial_{\xi_2^j}\partial_{\xi_1^i}\rho(\xi,\xi),$$

which implies

$$\partial_{\xi_1^j}\partial_{\xi_1^i}\rho(\xi,\xi) = -\partial_{\xi_2^j}\partial_{\xi_1^i}\rho(\xi,\xi). \tag{11.3.10}$$

Differentiating the function $\phi(\xi) = \partial_{\xi_2^i}\rho(\xi,\xi)$ with respect to $\partial_j$ we obtain

$$0 = \partial_j\phi(\xi) = \partial_{\xi_1^j}\partial_{\xi_2^i}\rho(\xi,\xi) + \partial_{\xi_2^j}\partial_{\xi_2^i}\rho(\xi,\xi),$$

which implies

$$\partial_{\xi_2^j}\partial_{\xi_2^i}\rho(\xi,\xi) = -\partial_{\xi_1^j}\partial_{\xi_2^i}\rho(\xi,\xi). \tag{11.3.11}$$

Assuming $\rho(\cdot\,,\,\cdot)$ smooth enough, the partial derivatives commute and using (11.3.10) and (11.3.11) we arrive at the following equivalent local formulas for the induced Riemannian metric:

$$\begin{aligned} g_{ij}(\xi) &= \partial_{\xi_1^i}\partial_{\xi_1^j} D(\xi_1||\xi_2)_{|\xi_2=\xi_1} && (11.3.12) \\ &= \partial_{\xi_2^i}\partial_{\xi_2^j} D(\xi_1||\xi_2)_{|\xi_2=\xi_1} && (11.3.13) \\ &= -\partial_{\xi_1^i}\partial_{\xi_2^j} D(\xi_1||\xi_2)_{|\xi_2=\xi_1} && (11.3.14) \\ &= -\partial_{\xi_1^j}\partial_{\xi_2^i} D(\xi_1||\xi_2)_{|\xi_2=\xi_1}. && (11.3.15) \end{aligned}$$

Another relation which will be useful in a later section is obtained by differentiating with respect to $\partial_k (= \frac{\partial}{\partial \xi^k})$ in relation (11.3.11) and applying the chain rule

$$\begin{aligned} \partial_k\partial_{\xi_2^j}\partial_{\xi_2^i}\rho(\xi,\xi) &= -\partial_k\partial_{\xi_1^j}\partial_{\xi_2^i}\rho(\xi,\xi) \Longleftrightarrow \\ \partial_{\xi_1^k}\partial_{\xi_2^j}\partial_{\xi_2^i}\rho(\xi,\xi) + \partial_{\xi_2^k}\partial_{\xi_2^j}\partial_{\xi_2^i}\rho(\xi,\xi) &= -\partial_{\xi_1^k}\partial_{\xi_1^j}\partial_{\xi_2^i}\rho(\xi,\xi) \\ &\quad -\partial_{\xi_2^k}\partial_{\xi_1^j}\partial_{\xi_2^i}\rho(\xi,\xi). \end{aligned} \tag{11.3.16}$$

The following notation is adopted for the representation of a vector field $X$ on $\mathcal{S}$ with respect to two local coordinate systems $(\xi_1^i)$ and $(\xi_2^i)$

$$X_{(\xi_1)} = \sum_i X^i(\xi_1)\partial_{\xi_1^i}, \quad X_{(\xi_2)} = \sum_i X^i(\xi_2)\partial_{\xi_2^i}.$$

We note that for any vector field $X$ we have

$$X_{(\xi_1)}D(\xi_1||\xi_2)_{|\xi_1=\xi_2} = X_{(\xi_2)}D(\xi_1||\xi_2)_{|\xi_1=\xi_2} \quad = \quad 0.$$

Next we provide the global definition of the induced Riemannian metric.

**Proposition 11.3.1** *The inner product of two vector fields is given by the following equivalent formulas*

$$\begin{aligned} g(X,Y) &= X_{(\xi_1)}Y_{(\xi_1)}D(\xi_1||\xi_2)_{|\xi_1=\xi_2} \\ &= X_{(\xi_2)}Y_{(\xi_2)}D(\xi_1||\xi_2)_{|\xi_1=\xi_2} \\ &= -X_{(\xi_1)}Y_{(\xi_2)}D(\xi_1||\xi_2)_{|\xi_1=\xi_2} \\ &= -X_{(\xi_2)}Y_{(\xi_1)}D(\xi_1||\xi_2)_{|\xi_1=\xi_2}. \end{aligned}$$

*Proof:* The proof follows from the bilinearity of $g$ and an application of relations (11.3.12)–(11.3.15). For instance, the first relation can be shown as

$$\begin{aligned} g(X,Y) &= \sum_{i,j} X^iY^j g(\partial_i, \partial_j) \\ &= \sum_{i,j} X^iY^j \partial_{\xi_1^i}\partial_{\xi_1^j} D(\xi_1||\xi_2)_{|\xi_1=\xi_2} \\ &= X_{(\xi_1)}Y_{(\xi_1)}D(\xi_1||\xi_2)_{|\xi_1=\xi_2}. \end{aligned}$$

■

## 11.4 Dual Contrast Function

If $D$ is a contrast function on $\mathbb{R}^k$, then the associated dual contrast function is defined by

$$D^*(\xi_1||\xi_2) = D(\xi_2||\xi_1).$$

The fact that $D^*$ satisfies properties $(i)$–$(iv)$ from the definition of a contrast function follows obviously from the fact that $D$ satisfies the same properties. Similarly, we can define the dual contrast function on a manifold by

$$D^*_{\mathcal{S}}(p||q) = D_{\mathcal{S}}(q||p), \qquad \forall p, q \in \mathcal{S}.$$

It is worthy to note that the contrast functions $D$ and $D^*$ induce the same Riemannian metric on the manifold $\mathcal{S}$. However, the connections induced by $D$ and $D^*$ play a central role in the geometry of contrast functions, as we shall see in the next couple of sections.

## 11.5 Induced Primal Connection

Let $g$ be the Riemannian metric on $\mathcal{S}$ induced by the contrast function $D_{\mathcal{S}}$. Consider the operator $\nabla^{(D)}$ given by

$$g(\nabla^{(D)}_X Y, Z) \;\; = \;\; -X_{(\xi_1)} Y_{(\xi_1)} Z_{(\xi_2)} D(\xi_1||\xi_2)_{|\xi_1=\xi_2}, \quad (11.5.17)$$

for any vector fields $X, Y, Z$ defined on the overlap of the chart neighborhoods associated with the coordinate systems $(\xi^i_1)$ and $(\xi^i_2)$. We shall check that $\nabla^{(D)}$ satisfies the properties of a connection. The $\mathbb{R}$-bilinearity is obvious. Let $f \in \mathcal{F}(\mathcal{S})$ be an arbitrary smooth function. Then

$$g(\nabla^{(D)}_{fX} Y, Z) \;\; = \;\; -f X_{(\xi_1)} Y_{(\xi_1)} Z_{(\xi_2)} D(\xi_1||\xi_2)_{|\xi_1=\xi_2} = g(f\nabla^{(D)}_X Y, Z),$$

and dropping the $Z$-argument implies $\nabla^{(D)}_{fX} Y = f\nabla^{(D)}_X Y$. Next we check Leibniz rule in the second argument

$$\begin{aligned} g(\nabla^{(D)}_X fY, Z) &= -X_{(\xi_1)}(fY_{(\xi_1)}) Z_{(\xi_2)} D(\xi_1||\xi_2)_{|\xi_1=\xi_2} \\ &= -f X_{(\xi_1)} Y_{(\xi_1)} Z_{(\xi_2)} D(\xi_1||\xi_2)_{|\xi_1=\xi_2} \\ &\quad -X_{(\xi_1)}(f)\, Y_{(\xi_1)} Z_{(\xi_2)} D(\xi_1||\xi_2)_{|\xi_1=\xi_2} \\ &= f g(\nabla^{(D)}_X fY, Z) + X_{(\xi_1)}(f) g(Y, Z) \\ &= g(f\nabla^{(D)}_X fY + X(f)Y, Z), \end{aligned}$$

so $\nabla^{(D)}_X fY = f\nabla^{(D)}_X fY + X(f)Y$.

Writing formula (11.5.17) in local coordinates we obtain the components of the linear connection $\nabla^{(D)}$ as in the following

$$\Gamma^{(D)}_{ij,k} = g(\nabla^{(D)}_{\partial_i} \partial_j, \partial_k) = -\partial_{\xi^i_1} \partial_{\xi^j_1} \partial_{\xi^k_2} D(\xi_1||\xi_2)_{|\xi_1=\xi_2}. \qquad (11.5.18)$$

The commutativity of the partial derivatives imply $\Gamma^{(D)}_{ij,k} = \Gamma^{(D)}_{ji,k}$, and hence the connection $\nabla^{(D)}$ has zero torsion. We can arrive to the same result in the following equivalent way. Starting from the global definition of the connection and Riemannian metric we write

$$\begin{aligned} g(\nabla^{(D)}_X Y - \nabla^{(D)}_Y X, Z) &= -X_{(\xi_1)}Y_{(\xi_1)}Z_{(\xi_2)}D(\xi_1||\xi_2)_{\xi_1=\xi_2} \\ &\quad +Y_{(\xi_1)}X_{(\xi_1)}Z_{(\xi_2)}D(\xi_1||\xi_2)_{\xi_1=\xi_2} \\ &= -[X,Y]_{(\xi_1)}Z_{(\xi_2)}D(\xi_1||\xi_2)_{\xi_1=\xi_2} \\ &= g([X,Y],Z). \end{aligned}$$

Dropping the $Z$-argument implies $\nabla^{(D)}_X Y - \nabla^{(D)}_Y X = [X,Y]$, i.e., the torsion of connection $\nabla^{(D)}$ is zero.

## 11.6 Induced Dual Connection

The dual connection $\nabla^{(D^*)}$ is the connection induced by the dual contrast function $D^*$, i.e., it is given by

$$\begin{aligned} g(\nabla^{(D^*)}_X Y, Z) &= -X_{(\xi_2)}Y_{(\xi_2)}Z_{(\xi_1)}D^*(\xi_2||\xi_1)_{|\xi_1=\xi_2} \\ &= -X_{(\xi_2)}Y_{(\xi_2)}Z_{(\xi_1)}D(\xi_1||\xi_2)_{|\xi_1=\xi_2}, \end{aligned}$$

for any vector fields $X, Y, Z$. This can be written locally as

$$\Gamma^{(D^*)}_{ij,k} = g(\nabla^{(D^*)}_{\partial_i}\partial_j, \partial_k) = -\partial_{\xi_2^i}\partial_{\xi_2^j}\partial_{\xi_1^k}D(\xi_1||\xi_2)_{|\xi_1=\xi_2}.$$

**Theorem 11.6.1** *The connections $\nabla^{(D)}$ and $\nabla^{(D^*)}$ are torsion-less dual connections.*

*Proof:* The fact that the torsions vanish follows from the symmetry in the first two indices of the connection components $\Gamma^{(D)}_{ij,k} = \Gamma^{(D)}_{ji,k}$ and $\Gamma^{(D^*)}_{ij,k} = \Gamma^{(D^*)}_{ji,k}$. The duality relation will be shown in local coordinates. Differentiating with respect to $\partial_k = \partial_{\xi^k}$ in relation $g_{ij}(\xi) = -\partial_{\xi_1^i}\partial_{\xi_2^j}D(\xi||\xi)$ we obtain

$$\begin{aligned} \partial_k g_{ij} &= -\partial_{\xi_1^k}\partial_{\xi_1^i}\partial_{\xi_2^j}D(\xi||\xi) \\ &\quad -\partial_{\xi_2^k}\partial_{\xi_1^i}\partial_{\xi_2^j}D(\xi||\xi) \\ &= \Gamma^{(D)}_{ki,j} + \Gamma^{(D^*)}_{kj,i}, \end{aligned}$$

which is equivalent with the duality of $D$ and $D^*$. ■

Therefore, a contrast function $D$ on a manifold $\mathcal{S}$ induces a statistical structure $(g, \nabla^{(D)}, \nabla^{(D^*)})$. Hence, $(\mathcal{S}, g, \nabla^{(D)}, \nabla^{(D^*)})$ becomes the statistical manifold induced by the contrast function $D$.

**Proposition 11.6.2** *The Levi–Civita connection of the Riemannian space $(\mathcal{S}, g)$ is given by*

$$\nabla^{(0)} = \frac{1}{2}\big(\nabla^{(D)} + \nabla^{(D^*)}\big).$$

*Proof:* Since $\nabla^{(D)}$ and $\nabla^{(D^*)}$ have zero torsion, the same applies to $\nabla^{(0)}$. Using the duality relation we show that $\nabla^{(0)}$ is a metrical connection

$$\begin{aligned}
Xg(Y,Z) &= \frac{1}{2}Xg(Y,Z) + \frac{1}{2}Xg(Y,Z)\\
&= \frac{1}{2}\Big\{g(\nabla_X^{(D)}Y, Z) + g(Y, \nabla_X^{(D^*)}Z)\Big\}\\
&= \frac{1}{2}\Big\{g(\nabla_X^{(D^*)}Y, Z) + g(Y, \nabla_X^{(D)}Z)\Big\}\\
&= g\Big(\frac{\nabla_X^{(D)}Y + \nabla_X^{(D^*)}Y}{2}, Z\Big) + g\Big(Y, \frac{\nabla_X^{(D)}Z + \nabla_X^{(D^*)}Z}{2}\Big)\\
&= g(\nabla_X^{(0)}Y, Z) + g(Y, \nabla_X^{(0)}Z).
\end{aligned}$$

■

## 11.7 Skewness Tensor

Besides a Riemannian metric $g$ and a pair of dual connections $\nabla^{(D)}$, $\nabla^{(D^*)}$, a contrast function $D$ also induces the skewness tensor by

$$\begin{aligned}
C^{(D)}(X,Y,Z) &= g\big(\nabla_X^{(D^*)}Y - \nabla_X^{(D)}Y, Z\big)\\
&= \Big(X_{(\xi_1)}Y_{(\xi_1)}Z_{(\xi_2)} - X_{(\xi_2)}Y_{(\xi_2)}Z_{(\xi_1)}\Big)D(\xi_1||\xi_2)_{|\xi_1=\xi_2}.
\end{aligned}$$

In local coordinates this becomes

$$\begin{aligned}
C_{ijk}^{(D)} &= \Gamma_{ij,k}^{(D^*)} - \Gamma_{ij,k}^{(D)}\\
&= \partial_{\xi_1^i}\partial_{\xi_1^j}\partial_{\xi_2^k}D(\xi_1||\xi_2)_{|\xi_1=\xi_2} - \partial_{\xi_2^i}\partial_{\xi_2^j}\partial_{\xi_1^k}D(\xi_1||\xi_2)_{|\xi_1=\xi_2}.
\end{aligned}$$

In the virtue of identities (11.3.12)–(11.3.15), the tensor $C_{ijk}^{(D)}$ becomes completely symmetric.

## 11.8 Third Order Approximation of $D(p||\cdot)$

This section will present the third order approximation of a contrast function $D_{\mathcal{S}}$ on a manifold $\mathcal{S}$. Let $p, q \in \mathcal{S}$ be two points in the same chart with coordinates $\xi_1 = \phi^{-1}(p)$ and $\xi_2 = \phi^{-1}(q)$. Denote $\Delta\xi^i = \xi_2^i - \xi_1^i$. The third order approximation of $D_{\mathcal{S}}(p||\cdot)$ about $p$ is given by

$$\begin{aligned} D_{\mathcal{S}}(p||q) &= D_{\mathcal{S}}(p||p) + \partial_{\xi_2^i} D(\xi_1||\xi_2)_{|\xi_1=\xi_2=\xi}\,\Delta\xi^i \\ &\quad + \frac{1}{2}\partial_{\xi_2^i}\partial_{\xi_2^j} D(\xi_1||\xi_2)_{|\xi_1=\xi_2=\xi}\,\Delta\xi^i\Delta\xi^j \\ &\quad + \frac{1}{6}\partial_{\xi_2^i}\partial_{\xi_2^j}\partial_{\xi_2^k} D(\xi_1||\xi_2)_{|\xi_1=\xi_2=\xi}\,\Delta\xi^i\Delta\xi^j\Delta\xi^k + o(\|\Delta\xi\|^2), \end{aligned}$$

where $o(\|\Delta\xi\|^2)$ is a term which converges to 0 faster than $\|\Delta\xi\|^2$ does, as $p \to q$. Since from the definition of a contrast function the first two terms are zero, then

$$D_{\mathcal{S}}(p||q) = \frac{1}{2}g_{ij}(\xi_1)\Delta\xi^i\Delta\xi^j + \frac{1}{6}h_{ijk}(\xi_1)\Delta\xi^i\Delta\xi^j\Delta\xi^k + o(\|\Delta\xi\|^2),$$

where $g_{ij}$ is the induced Riemannian metric. It suffices to compute the coefficients

$$h_{ijk}(\xi_1) = \partial_{\xi_2^i}\partial_{\xi_2^j}\partial_{\xi_2^k} D(\xi_1||\xi_2)_{|\xi_1=\xi_2=\xi}.$$

Writing relation (11.3.16) in terms of the induced connections components, see formula (11.5.18), we have

$$-\Gamma^*_{ij,k} + h_{ijk} = \Gamma_{jk,i} + \Gamma^*_{ik,j}$$

from where

$$\begin{aligned} h_{ijk} &= \Gamma^*_{ij,k} + \Gamma_{jk,i} + \Gamma^*_{ik,j} \\ &= \partial_j g_{ik} + \Gamma^*_{ik,j} \\ &= \partial_k g_{ij} + \Gamma^*_{ij,k}. \end{aligned}$$

The last two identities follow from formula (8.1.2). A similar argument can be used to show also the relation

$$h_{ijk} = \partial_i g_{kj} + \Gamma^*_{jk,i}.$$

This relations imply the total symmetry of $h_{ijk}$

$$h_{ijk} = h_{ikj} = h_{kji} = h_{jik}.$$

It is worthy to mention that if $D(\cdot || \cdot)$ induces a dually flat statistical manifold (i.e., $\Gamma = \Gamma^* = 0$), then $h_{ijk} = 0$.

We have seen that any contrast function induces a dualistic structure $(g^{(D)}, \nabla^{(D)}, \nabla^{(D^*)})$ on $\mathcal{S}$. Next we consider the converse implication, which states that any triple $(g, \nabla, \nabla^*)$, which consists in a metric and two dual torsion-free connections, is induced from a divergence. The divergence can be given locally by

$$D(p||q) = \frac{1}{2}g_{ij}(p)\Delta\xi^i\Delta\xi^j + \frac{1}{6}h_{ijk}(p)\Delta\xi^i\Delta\xi^j\Delta\xi^k, \tag{11.8.19}$$

where $\Delta\xi^i = \xi^i(q) - \xi^i(p)$ and $h_{ijk} = \partial_i g_{kj} + \Gamma^*_{jk,i}$. The existence of a globally defined contrast function is proved in Matumoto [56].

However, the contrast function is not unique. An alternative construction for (11.8.19) is

$$D(p||q) = \frac{1}{2}g_{ij}(p)\Delta\xi^i\Delta\xi^j - \frac{1}{6}h^*_{ijk}(p)\Delta\xi^i\Delta\xi^j\Delta\xi^k,$$

where $h^*_{ijk} = \partial_i g_{jk} + \Gamma^*_{jk,i}$.

## 11.9 Hessian Geometry

Assume now that there is a local coordinate chart with respect to which the contrast function $D_{\mathcal{S}}$ is induced locally by a convex function $\varphi$ via formula (11.1.2). We make the remark that it is not necessarily true that there is always a local system of coordinates in which the contrast function is induced by a convex function. However, when this occurs, it defines a dually flat structure of statistical manifold, as we shall see next. This type of contrast function is sometimes called *Bregman divergence*, see Bregman [20], and it is widely used in convex optimization, see Bauschke [14–16]. For a generalization of this contrast function to an $\alpha$-family, see Zhang [86].

Using (11.1.4) we obtain that the metric is given by the Hessian of the strictly convex potential function $\varphi$

$$g_{ij}(\xi) = \partial_{\xi^i}\partial_{\xi^j}\varphi(\xi). \tag{11.9.20}$$

A straightforward computation shows that the components of the induced dual connections $\nabla^{(D)}$ and $\nabla^{(D^*)}$ are given by

$$\Gamma^{(D)}_{ij,k}(\xi) = 0, \qquad \Gamma^{(D^*)}_{ij,k}(\xi) = \partial_{\xi^i}\partial_{\xi^j}\partial_{\xi^k}\varphi(\xi). \tag{11.9.21}$$

A further computation shows that the Riemann curvature tensors are $R = R^* = 0$, i.e., the connections are dually flat.

It is worth noting that there are topological obstructions to the existence of dually flat structures. Ay and Tuschmann [10] proved that if $(\mathcal{S}, g, \nabla, \nabla^*)$ is dually flat and $\mathcal{S}$ is compact, then the first fundamental group $\pi_1(\mathcal{S})$ must be finite.

The skewness tensor is given by the third order derivatives as

$$C_{ijk}^{(D)} = \partial_{\xi^i}\partial_{\xi^j}\partial_{\xi^k}\varphi(\xi).$$

This geometry is commonly referred to in the literature as the *Hessian geometry*. Some authors considered weaker conditions than strictly convexity for the potential function $\varphi$, see Shima [74] and Shima and Yagi [75]. For more details on hessian metrics, the reader is referred to Bercu [17] and Corcodel [29].

## 11.10 Problems

**11.1.** Let $\gamma : (a, b) \to (M, g)$ be a regular curve, i.e., $\dot{\gamma} \neq 0$. Define

$$D(s||t) = \int_s^t (t - u)|\dot{\gamma}(u)|_g^2\, du.$$

Show that $D(\cdot\, ||\, \cdot)$ is a contrast function on $(a, b)$.

**11.2.** Let $\mathcal{S}$ be a statistical model and consider two distributions $p_0, p_1 \in \mathcal{S}$. Define the following curves in $\mathcal{S}$

$$p_t^{(m)} = (1 - t)p_0 + tp_1, \quad p_t^{(e)} = C_t p_0^{1-t} p_1^t, \quad 0 \leq t \leq 1,$$

where $C_t$ is a normalization function. Denote by $g^{(m)}(t)$ and $g^{(e)}(t)$ the Fisher metrics along the aforementioned curves. Let

$$D^{(m)}(p_1||p_0) = \int_0^1 (1 - s)g^{(m)}(s)\, ds,$$

$$D^{(e)}(p_1||p_0) = \int_0^1 (1 - s)g^{(e)}(s)\, ds.$$

(*a*) Prove that $D^{(m)}(\cdot\, ||\, \cdot)$ and $D^{(e)}(\cdot\, ||\, \cdot)$ are contrast functions on $\mathcal{S}$.

(*b*) What is the relationship between $D^{(m)}(\cdot\, ||\, \cdot)$ and $D^{(e)}(\cdot\, ||\, \cdot)$?

**11.3.** Let $(M, g, \nabla, \nabla^*)$ be a dually flat statistical manifold and $(x^i)$ and $(\zeta_\alpha)$ a pair of dual coordinate systems associated with potentials $\varphi$ and $\psi$ (i.e., $x^i = \partial_{\zeta_i}\varphi(\zeta)$, $\zeta_j = \partial_{x^j}\psi(x)$). Define $D : M \times M \to \mathbb{R}$ as

$$D(p||q) = \psi\big(x(p)\big) + \varphi\big(\zeta(q)\big) - x^i(p)\zeta_i(q).$$

(*a*) Prove that $D(\cdot\,||\,\cdot)$ is a contrast function (called the **canonical divergence** of $(M, g, \nabla, \nabla^*)$).

(*b*) Find the dual contrast function $D^*(\cdot\,||\,\cdot)$.

(*c*) Show that for any $p, q, r \in M$ the following relation holds

$$D(p||q) + D(q||r) = D(p||r) - \big(x^i(q) - x^i(p)\big)\big(\zeta_i(q) - \zeta_i(q)\big).$$

(*d*) Let $\theta$ be the angle made at $q$ by the $\nabla$-geodesic joining $p$ and $q$, $\gamma_{pq}$, and the $\nabla^*$-geodesic joining $q$ and $r$, $\gamma^*_{qr}$. Show that

$$D(p||q) + D(q||r) = D(p||r) - \|\dot{\gamma}_{pq}\| \cdot \|\dot{\gamma}^*_{qr}\| \cos(\pi - \theta).$$

(*e*) If $\theta = \dfrac{\pi}{2}$ show the following Pythagorean relation:

$$D(p||r) = D(p||q) + D(q||r).$$

(*f*) Find the skewness tensor associated with $D(\cdot\,||\,\cdot)$.

**11.4.** Consider the Euclidean space $(M, g) = (\mathbb{R}^n, \delta_{ij})$, with $\nabla = \nabla^*$ given by $\nabla_U V = U(V^j)e_j$, for any $U, V \in \mathcal{X}(M)$.

(*a*) Show that the Euclidean coordinates system is self-dual, i.e., $x^i = \zeta_i$.

(*b*) Show that in this case the potential functions are given by

$$\psi(x) = \frac{1}{2}\sum_i (x^i)^2, \qquad \phi(x) = \frac{1}{2}\sum_i (\zeta_i)^2.$$

(*c*) Prove that the canonical divergence is given by $D(p||q) = \frac{1}{2}d_E^2(p, q)$, where $d_E(p, q)$ denotes the Euclidean distance between $p$ and $q$.

**11.5.** How many of the previous requirements still hold on a Riemannian manifold $(M, g, \nabla)$ with a flat Levi–Civita connection $\nabla$?

**11.6.** Let $(M, g, \nabla, \nabla^*)$ be a dually flat statistical manifold, and denote by $D(\cdot \| \cdot)$ the associated canonical divergence. Consider the $D$-sphere centered at $p \in M$ of radius $\rho$, defined by

$$S^{(D)} = \{q \in M; D(p\|q) = \rho\}.$$

Show that every $\nabla$-geodesic starting at the center $p$ intersects $S^{(D)}$ orthogonally.

**11.7.** Consider the exponential family $p(x;\xi) = e^{C(x)+\xi^i F_i(x)-\psi(\xi)}$, $x \in \mathcal{X}$, with $\{F_i(x)\}$ linearly independent on $\mathcal{X}$. Define $\eta_j = E_\xi[F_j]$, $1 \le j \le n$.

(*a*) Show that $\eta_j = \partial_j \psi(\xi)$.

(*b*) Prove that $(\xi^i)$ and $(\eta_j)$ are dual systems of coordinates.

(*c*) Verify that $(\xi^i)$ is a 1-affine coordinate system and $(\eta_j)$ is a $(-1)$-affine coordinate system.

(*d*) Let $\varphi(\eta)$ be the potential associated with $\xi$, i.e., $\xi^j = \partial_{\eta^j}\varphi(\eta)$. Show that $\varphi(\eta) = E_\xi[\ln p_\xi(x) - C(x)]$.

(*e*) Let $H(p)$ be the entropy of distribution $p$. Validate the relation

$$H(p_\xi) = -\varphi(\xi) - E_\xi[C(x)].$$

(*f*) Let $\hat{\eta}_j = F_j(x)$. Show that $\hat{\eta}$ is an unbiased estimator for $\eta$, and that the covariance matrix provides the Fisher metric, i.e., $V_\eta(\hat{\eta}) = g_{ij}$.

(*g*) Find the contrast function given by the canonical divergence associated with the dual system of coordinates $(\xi^i)$, $(\eta_i)$. What is its relationship with the Kullback–Leibler relative entropy?

**11.8.** Consider the statistical model given by the Poisson distribution $p(x;\xi) = e^{-\xi}\frac{\xi^x}{x!}$, $x \in \{0, 1, 2, \dots\}$, $\xi > 0$. Consider $\eta = \xi$ and $\theta = \ln \xi$.

(*a*) Prove that $\eta$ and $\theta$ are dual coordinates.

(*b*) Find the canonical divergence associated with the above dual coordinates.

**11.9.** Consider the statistical model given by the normal family

$$p(x;\xi) = \frac{1}{\sqrt{2\pi}\sigma} e^{-\frac{(x-\mu)^2}{2\sigma^2}}, \quad \mu \in \mathbb{R}, \sigma > 0.$$

Show that $(\theta^i)$ are $(\eta_i)$ are dual systems of coordinates, where

$$\eta_1 = \mu, \qquad \eta_2 = \mu^2 + \sigma^2$$

$$\frac{\theta^1}{2\theta^2} = -\mu, \qquad \frac{(\theta^1)^2 - 2\theta^2}{4(\theta^2)^2} = \mu^2 + \sigma^2.$$

**11.10.** Consider the statistical model given by the exponential distribution $p(x; \xi) = \xi e^{-\xi x}$, $x \geq 0$, $\xi > 0$.

$(a)$ Find a pair of dual coordinates on the above statistical model.

$(b)$ Find the potentials $\psi$ and $\varphi$ associated with the dual coordinates obtained at $(a)$.

$(c)$ Deduct the expression for the Fisher metric.

$(d)$ Find the canonical divergence associated with the dual coordinates obtained at $(a)$.

# Chapter 12

# Contrast Functions on Statistical Models

This chapter deals with some important examples of contrast functions on a space of density functions, such as: Bregman divergence, Kullback–Leibler relative entropy, $f$-divergence, Hellinger distance, Chernoff information, Jefferey distance, Kagan divergence, and exponential contrast function. The relation with the skewness tensor and $\alpha$-connection is made. The goal of this chapter is to produce hands-on examples for the theoretical concepts introduced in Chap. 11.

## 12.1 A First Example

We start with a suggestive example of Bregman divergence. We show that the Kullback–Leibler relative entropy on a statistical model is a particular example of Bregman divergence.

Let $\mathcal{S} = \mathcal{P}(\mathcal{X})$, where $\mathcal{X} = \{x^1, \dots, x^{n+1}\}$ and consider the global chart $\phi : \mathcal{S} \to \mathbb{E} \subset \mathbb{R}^n$

$$\phi(p) = (\ln p_1, \dots, \ln p_n) = (\xi^1, \dots, \xi^n),$$

with the parameter space

$$\mathbb{E} = \{(\xi^1, \dots, \xi^n); \xi^k > 0, \sum_{k=1}^{n} \xi^k < 1\}.$$

O. Calin and C. Udrişte, *Geometric Modeling in Probability and Statistics*,
DOI 10.1007/978-3-319-07779-6_12, 

The contrast function on $\mathcal{S}$ is then given by

$$\begin{aligned} D_{\mathcal{S}}(p||q) &= D_{\mathcal{S}}\big(\phi^{-1}(p)||\phi^{-1}(q)\big) \\ &= D(\xi_1||\xi_2), \end{aligned}$$

where $D(\cdot\,||\,\cdot)$ is the Bregman divergence on $\mathbb{E}$ induced by the convex function $\varphi(\xi) = \sum_{i=1}^n e^{\xi_i}$, i.e.,

$$D(\xi_1||\xi_2) = \varphi(\xi_2) - \varphi(\xi_1) - \sum_{i=1}^n \partial_i\varphi(\xi_1)(\xi_2^i - \xi_1^i).$$

Therefore

$$\begin{aligned} D_{\mathcal{S}}(p||q) &= D(\xi_1||\xi_2) \\ &= \sum_i e^{\xi_2^i} - \sum_i e^{\xi_1^i} - \sum_i e^{\xi_1^i}(\xi_2^i - \xi_1^i) \\ &= \sum_i p_i - \sum_i q_i - \sum_i p_i \ln\frac{q_i}{p_i} \\ &= \sum_i p_i \ln\frac{p_i}{q_i} = D_{KL}(p||q). \end{aligned}$$

Hence, the induced contrast function $D_{\mathcal{S}}$ on $\mathcal{P}(\mathcal{X})$ in this case is the Kullback–Leibler relative entropy.

## 12.2 $f$-Divergence

An important class of contrast functions on statistical models was introduced by Csiszár [31, 32]. Let $f : (0, \infty) \to \mathbb{R}$ be a function satisfying the following conditions

(*a*) $f$ is convex;

(*b*) $f(1) = 0$;

(*c*) $f''(1) = 1$.

For each probability distributions $p$, $q$, consider

$$D_f(p||q) = E_p\Big[f\Big(\frac{q(x)}{p(x)}\Big)\Big] = \int_{\mathcal{X}} p(x) f\Big(\frac{q(x)}{p(x)}\Big)\,dx. \tag{12.2.1}$$

We shall assume that the previous integral converges and we can differentiate under the integral sign.

**Proposition 12.2.1** *The operator $D_f(\cdot\,||\,\cdot)$ is a contrast function on the statistical model $\mathcal{S} = \{p_\xi\}$.*

*Proof:* We check the properties of a contrast function.

$(i)$ positive: Jensen's inequality applied to the convex function $f$ provides

$$\begin{aligned} D_f(p||q) &= E_p\Big[f\Big(\frac{q(x)}{p(x)}\Big)\Big] \geq f\Big(E_p\Big[\frac{q(x)}{p(x)}\Big]\Big) \\ &= f\Big(\int_{\mathcal{X}} p(x)\frac{q(x)}{p(x)}\,dx\Big) = f(1) = 0. \end{aligned}$$

$(ii)$ non-degenerate: Let $p \neq q$. Since $f$ is strictly convex at 1, then

$$D_f(p||q) = E_p\Big[f\Big(\frac{q(x)}{p(x)}\Big)\Big] > f\Big(E_p\Big[\frac{q(x)}{p(x)}\Big]\Big) = f(1) = 0,$$

and hence $D(p||q) \neq 0$, which implies the non-degenerateness.

$(iii)$ The vanishing property of the first variation along the diagonal $\{\xi_1 = \xi_2\}$ is a consequence of $(i)$ and $(ii)$.

$(iv)$ Let $p = p_{\xi_0}$ and $q = p_\xi$. We shall compute the Hessian of

$$D_f(p_{\xi_0}||p_\xi) = \int_{\mathcal{X}} p_{\xi_0}(x) f\Big(\frac{p_\xi(x)}{p_{\xi_0}(x)}\Big)\,dx \qquad (12.2.2)$$

along the diagonal $\xi^0 = \xi$. Differentiating we have

$$\begin{aligned} \partial_{\xi^j} f\Big(\frac{p_\xi}{p_{\xi_0}}\Big) &= f'\Big(\frac{p_\xi}{p_{\xi_0}}\Big)\frac{1}{p_{\xi_0}}\partial_{\xi^j} p_\xi \\ \partial_{\xi^i}\partial_{\xi^j} f\Big(\frac{p_\xi}{p_{\xi_0}}\Big) &= f''\Big(\frac{p_\xi}{p_{\xi_0}}\Big)\Big(\frac{p_\xi}{p_{\xi_0}}\Big)^2 \partial_{\xi^i}(\ln p_\xi)\partial_{\xi^j}(\ln p_\xi) \\ &\quad + f'\Big(\frac{p_\xi}{p_{\xi_0}}\Big)\frac{1}{p_\xi}\partial_{\xi^i}\partial_{\xi^j} p_\xi. \end{aligned}$$

Differentiating under the integral we get

$$\begin{aligned}\partial_{\xi^i}\partial_{\xi^j} D_f(p_{\xi_0}||p_\xi)_{|\xi=\xi_0} &= f''(1)\int p_{\xi_0}\partial_{\xi^i}\ln p_\xi\,\partial_{\xi^j}\ln p_\xi\,dx_{|\xi=\xi_0}\\ &\quad +f'(1)\partial_{\xi^i}\partial_{\xi^j}\int p_\xi(x)\,dx\\ &= f''(1)E_\xi[\partial_{\xi^i}\ell(\xi)\partial_{\xi^j}\ell(\xi)]\\ &= E_\xi[(\partial_{\xi^i}\ell)(\partial_{\xi^j}\ell)] = g_{ij}(\xi),\end{aligned}$$

which is strictly positive definite, since it is the Fisher–Riemann information matrix. Hence $D_f(\cdot\,||\,\cdot)$ is a contrast function.

■

**Theorem 12.2.2** *The Riemannian metric induced by the contrast function $D_f(\cdot\,||\,\cdot)$ on the statistical model $\mathcal{S} = \{p_\xi\}$ is the Fisher–Riemann information matrix*

$$g_{ij}(\xi) = \partial_{\xi^i}\partial_{\xi^j} D_f(p_{\xi_0}||p_\xi)_{|\xi=\xi_0}.$$

*Proof:* It follows from the calculation performed in the part *(iv)* above. ■

Let $f^*(u) = uf\Big(\frac{1}{u}\Big)$. Since

$$\begin{aligned}f^*(1) &= f(1) = 0\\ f^{*\prime\prime}(u) &= \frac{1}{u^3}f''\Big(\frac{1}{u}\Big) \geq 0\\ f^{*\prime\prime}(1) &= f''(1) = 1,\end{aligned}$$

then $f^*$ satisfies properties $(a)$–$(c)$, and hence $D_{f^*}(\cdot\,||\,\cdot)$ is a contrast function, which defines the same Riemannian metric as $D_f(\cdot\,||\,\cdot)$.

**Proposition 12.2.3** *The contrast function $D_{f^*}(\cdot\,||\,\cdot)$ is the dual of $D_f(\cdot\,||\,\cdot)$.*

*Proof:* Consider the dual $D_f^*(p||q) = D_f(q||p)$. Then we have

$$\begin{aligned}D_{f^*}(p||q) &= \int_{\mathcal{X}} p(x)f^*\Big(\frac{q(x)}{p(x)}\Big)\,dx\\ &= \int_{\mathcal{X}} p(x)\frac{q(x)}{p(x)}f\Big(\frac{p(x)}{q(x)}\Big)\,dx\\ &= \int_{\mathcal{X}} q(x)f\Big(\frac{p(x)}{q(x)}\Big)\,dx\\ &= D_f(q||p) = D_f^*(p||q), \qquad \forall p,q\in\mathcal{S}.\end{aligned}$$

Therefore $D_{f^*} = D_f^*$. ■

In the following we shall find the induced connections. Let $\nabla^{(f)}$ be the linear connection induced by the contrast function $D_f(\cdot\,||\,\cdot)$, and denote by $\Gamma^{(f)}_{ij,k}$ its components on a local basis.

**Proposition 12.2.4** *We have*

$$\Gamma^{(f)}_{ij,k}(\xi) = E_\xi\Big[\big(\partial_i\partial_j\ell - (f'''(1)+1)\partial_i\partial_j\ell\big)\partial_k\ell\Big]. \tag{12.2.3}$$

*Proof:* From formula (11.5.18) we find

$$\Gamma^{(f)}_{ij,k}(\xi) = -\partial_{\xi_0^i}\partial_{\xi_0^j}\partial_{\xi^k} D_f(p_{\xi_0}||p_\xi)_{|\xi=\xi_0}. \tag{12.2.4}$$

We shall compute the derivatives on the right side. Differentiating in (12.2.2) yields

$$\partial_{\xi^k} D_f(p_{\xi_0}||p_\xi) = \int_{\mathcal{X}} f'\Big(\frac{p_\xi}{p_{\xi_0}}\Big) p_\xi \partial_{\xi^k}\ell(\xi)\,dx. \tag{12.2.5}$$

Before continuing the computation we note that

$$\begin{aligned}
\partial_{\xi_0^j} f'\Big(\frac{p_\xi}{p_{\xi_0}}\Big) &= f''\Big(\frac{p_\xi}{p_{\xi_0}}\Big)\Big(\frac{-p_\xi}{p_{\xi_0}}\Big)\partial_{\xi_0^j}\ell(\xi_0)\\
\partial_{\xi_0^i}\partial_{\xi_0^j} f'\Big(\frac{p_\xi}{p_{\xi_0}}\Big) &= f'''\Big(\frac{p_\xi}{p_{\xi_0}}\Big)\frac{p_\xi^2}{p_{\xi_0}^2}\partial_{\xi_0^i}\ell(\xi_0)\partial_{\xi_0^j}\ell(\xi_0)\\
&\quad +f''\Big(\frac{p_\xi}{p_{\xi_0}}\Big)\frac{p_\xi}{p_{\xi_0}}\partial_{\xi_0^i}\ell(\xi_0)\partial_{\xi_0^j}\ell(\xi_0)\\
&\quad -f''\Big(\frac{p_\xi}{p_{\xi_0}}\Big)\frac{p_\xi}{p_{\xi_0}}\partial_{\xi_0^i}\partial_{\xi_0^j}\ell(\xi_0).
\end{aligned}$$

Applying now $\partial_{\xi_0^i}\partial_{\xi_0^j}$ to (12.2.5), using the foregoing formulas, and taking $\xi_0 = \xi$, yields

$$\begin{aligned}
\partial_{\xi_0^i}\partial_{\xi_0^j}\partial_{\xi^k} D_f(p_{\xi_0}||p_\xi)_{|\xi=\xi_0} &= \int_{\mathcal{X}} \Big[(f'''(1)+f''(1))p_\xi(\partial_{\xi^i}\ell)(\partial_{\xi^j}\ell)(\partial_{\xi^k}\ell)\\
&\quad -f''(1)(\partial_{\xi^i}\partial_{\xi^j}\ell)(\partial_{\xi^k}\ell)\Big]\,dx\\
&= E_\xi\Big[\big(\partial_{\xi^i}\partial_{\xi^j}\ell - (f'''(1)+1)\partial_{\xi^i}\partial_{\xi^j}\ell\big)\partial_{\xi^k}\ell\Big].
\end{aligned}$$

Applying (12.2.4) we arrive at (12.2.3). ∎

The relation with the geometry of $\alpha$-connections is given below.

**Theorem 12.2.5** *The connection induced by $D_f(\cdot\,||\,\cdot)$ is an $\alpha$-connection*

$$\nabla^{(f)} = \nabla^{(\alpha)},$$

*with $\alpha = 2f'''(1) + 3$.*

*Proof:* It suffices to show the identity in local coordinates. Recall first the components of the $\alpha$-connection given by (1.11.34)

$$\Gamma^{(\alpha)}_{ij,k} = E_\xi\Big[\Big(\partial_i\partial_j\ell + \frac{1-\alpha}{2}\partial_i\ell\partial_j\ell\Big)\partial_k\ell\Big]. \tag{12.2.6}$$

Comparing with (12.2.3) we see that $\Gamma^{(f)}_{ij,k} = \Gamma^{(\alpha)}_{ij,k}$ if and only if $\alpha = 2f'''(1) + 3$. ■

We make the remark that $\nabla^{(f^*)} = \nabla^{(-\alpha)}$, which follows from the properties of dual connections induced by contrast functions. We shall show shortly that for any $\alpha$ there is a function $f$ satisfying $(a)$–$(c)$ and solving the equation $\alpha = 2f'''(1) + 3$.

**Proposition 12.2.6** *The skewness tensor induced by the contrast function $D_f(\cdot\,||\,\cdot)$ is given in local coordinates by*

$$T^{(f)}_{ijk} = (2f'''(1) + 3)E_\xi[(\partial_i\ell)(\partial_j\ell)(\partial_k\ell)].$$

*Proof:* Using Theorem 12.2.5, formula (12.2.6) and the aforementioned remarks, we have

$$\begin{aligned} T^{(f)}_{ijk} &= \Gamma^{(f^*)}_{ijk} - \Gamma^{(f)}_{ijk} = \Gamma^{(-\alpha)}_{ijk} - \Gamma^{(\alpha)}_{ijk} \\ &= E_\xi\Big[\Big(\partial_i\partial_j\ell + \frac{1+\alpha}{2}\partial_i\ell\partial_j\ell\Big)\partial_k\ell\Big] \\ &\quad - E_\xi\Big[\Big(\partial_i\partial_j\ell + \frac{1-\alpha}{2}\partial_i\ell\partial_j\ell\Big)\partial_k\ell\Big] \\ &= \alpha E_\xi[(\partial_i\ell)(\partial_j\ell)(\partial_k\ell)] \\ &= (2f'''(1) + 3)E_\xi[(\partial_i\ell)(\partial_j\ell)(\partial_k\ell)]. \end{aligned}$$

■

## 12.3 Particular Cases

This section presents a few classical examples of contrast functions as particular examples of $D_f(\cdot\,||\cdot\,)$. These are constructed by choosing several examples of functions $f$ that satisfy conditions $(a)$–$(c)$ and verify the equation $\alpha = 2f'''(1) + 3$. We make the remark that if $f$ is such a function, then $f_c(u) = f(u) + c(u-1)$, $c \in \mathbb{R}$, is also a function that induces the same contrast function, $D_{f_c} = D_f$. Therefore, the correspondence between functions $f$ and contrast functions is not one-to-one.

### 12.3.1 Hellinger Distance

Consider $f(u) = 4(1 - \sqrt{u})$ and the associated contrast function

$$\begin{aligned}
D_f(p||q) &= 4\int_{\mathcal{X}} p(x)\Big(1 - \sqrt{\frac{q(x)}{p(x)}}\Big)dx = 4\Big(1 - \int_{\mathcal{X}} \sqrt{p(x)q(x)}\,dx\Big) \\
&= 2\Big(2 - \int_{\mathcal{X}} 2\sqrt{p(x)q(x)}\,dx\Big) \\
&= 2\int_{\mathcal{X}} \Big(p(x) - 2\sqrt{p(x)q(x)} + q(x)\Big)\,dx \\
&= 2\int_{\mathcal{X}} \big(\sqrt{p(x)} - \sqrt{q(x)}\big)^2\,dx \\
&= H^2(p,q).
\end{aligned}$$

$H(p,q)$ is called the *Hellinger distance*, and is a true distance on the statistical model $\mathcal{S} = \{p_\xi\}$. Since in this case $\alpha = 2f'''(1) + 3 = 0$, the linear connection induced by $H^2(p,q)$ is exactly the Levi–Civita connection, $\nabla^{(0)}$, on the Riemannian manifold $(\mathcal{S}, g)$.

**Example 12.3.1** Consider two exponential distributions, $p(x) = \alpha e^{-\alpha x}$ and $q(x) = \beta e^{-\beta x}$, $x \geq 0$, $\alpha, \beta > 0$. Then

$$\begin{aligned}
H^2(p,q) &= 4 - 4\int_0^\infty \sqrt{p(x)q(x)}\,dx \\
&= 4 - 4\sqrt{\alpha\beta}\int_0^\infty e^{-\frac{\alpha+\beta}{2}x}\,dx \\
&= 4 - \frac{8\sqrt{\alpha\beta}}{\alpha+\beta},
\end{aligned}$$

hence the Hellinger distance is $H(p,q) = 2\sqrt{1 - \dfrac{2\sqrt{\alpha\beta}}{\alpha+\beta}}$.

The Hellinger distance can also be defined between two discrete distributions $p = (p_k)$ and $q = (q_k)$, replacing the integral by a sum

$$H(p,q) = 2\Big(1 - \sum_{k\geq 0}\sqrt{p_k q_k}\Big)^{1/2} = \Big(2\sum_{k\geq 0}\big(\sqrt{p_k} - \sqrt{q_k}\big)^2\Big)^{1/2}.$$

**Example 12.3.2** Consider two Poisson distributions, $p_k = \frac{\alpha^k}{k!}e^{-\alpha}$ and $q_k = \frac{\beta^k}{k!}e^{-\beta}$ , $k \geq 0$. Then

$$\begin{aligned}\sum_{k\geq 0}\sqrt{p_k q_k} &= \sum_{k\geq 0}\frac{(\sqrt{\alpha\beta})^k}{k!}e^{-\frac{\alpha+\beta}{2}} \\ &= e^{-\frac{\alpha+\beta}{2}}e^{\sqrt{\alpha\beta}}\sum_{k\geq 0}\frac{(\sqrt{\alpha\beta})^k}{k!}e^{-\sqrt{\alpha\beta}} \\ &= e^{\sqrt{\alpha\beta}-\frac{\alpha+\beta}{2}}.\end{aligned}$$

Hence, the Hellinger distance becomes

$$H(p,q) = 2\Big(1 - \sum_{k\geq 0}\sqrt{p_k q_k}\Big)^{1/2} = 2\sqrt{1 - e^{\sqrt{\alpha\beta}-\frac{\alpha+\beta}{2}}}.$$

### 12.3.2 Kullback–Leibler Relative Entropy

The contrast function associated with function $f(u) = -\ln u$ is given by

$$D_f(p||q) = \int_{\mathcal{X}} p(x)\ln\frac{p(x)}{q(x)}\,dx = D_{KL}(p||q),$$

which is the *Kullback–Leibler information* or the *relative entropy*. In this case $\alpha = 2f'''(1)+3 = -1$, so the associated connection is $\nabla^{(-1)}$.

It is worthy to note that the convex function $f(u) = u\ln u$ induces the contrast function

$$D_f(p||q) = \int_{\mathcal{X}} q(x)\ln\frac{q(x)}{p(x)}\,dx = D_{KL}(q||p) = D^*_{KL}(p||q),$$

which is the dual of the *Kullback–Leibler information*, see [51, 53]. Since $\alpha = 2f'''(1)+3 = 1$, the induced connection is $\nabla^{(1)}$.

### 12.3.3 Chernoff Information of Order $\alpha$

The convex function

$$f^{(\alpha)} = \frac{1}{1-\alpha^2}(1 - u^{\frac{1+\alpha}{2}}), \qquad \alpha \neq \pm 1$$

induces the contrast function

$$D^{(\alpha)}(p||q) = \frac{4}{1-\alpha^2}\Big\{1 - \int_{\mathcal{X}} p(x)^{\frac{1-\alpha}{2}} q(x)^{\frac{1+\alpha}{2}}\, dx\Big\},$$

see Chernoff [27]. For the computation of $D^{(\alpha)}$ in the case of exponential, normal and Poisson distributions, see Problems 12.9., 12.10. and 12.11. We note that for $\alpha = 0$ we retrieve the squared Hellinger distance, $D^{(0)}(p||q) = H^2(p, q)$.

### 12.3.4 Jeffrey Distance

The function $f(u) = \frac{1}{2}(u-1)\ln u$ induces the contrast function

$$\begin{aligned} J(p,q) &= D_f(p||q) = \frac{1}{2}\int_{\mathcal{X}} p(x)\Big(1 - \frac{q(x)}{p(x)}\Big)\ln\frac{p(x)}{q(x)}\, dx \\ &= \frac{1}{2}\int_{\mathcal{X}} \big(p(x) - q(x)\big)\big(\ln p(x) - \ln q(x)\big)\, dx, \end{aligned}$$

see Jeffrey [47]. A computation shows that $\alpha = 0$, so the induced connection is the Levi–Civita connection $\nabla^{(0)}$. In fact, the Jeffrey contrast function is the same as the symmetric Kullback–Leibler relative entropy

$$\begin{aligned} J(p,q) &= \frac{1}{2}\int_{\mathcal{X}} p(x)\Big(1 - \frac{q(x)}{p(x)}\Big)\ln\frac{p(x)}{q(x)}\, dx \\ &= \frac{1}{2}\int_{\mathcal{X}} p(x)\ln\frac{p(x)}{q(x)} + \frac{1}{2}\int_{\mathcal{X}} q(x)\ln\frac{q(x)}{p(x)}\, dx \\ &= \frac{1}{2}\Big(D_{KL}(p||q) + D_{KL}(q||p)\Big). \end{aligned}$$

### 12.3.5 Kagan Divergence

Choosing $f(u) = \frac{1}{2}(1-u)^2$ yields

$$\begin{aligned} D_{\chi^2}(p||q) &= D_f(p||q) = \frac{1}{2}\int_{\mathcal{X}} p(x)\Big(1 - \frac{q(x)}{p(x)}\Big)\, dx \\ &= \frac{1}{2}\int_{\mathcal{X}} \frac{(p(x) - q(x))^2}{q(x)}\, dx, \end{aligned}$$

called the *Kagan contrast function*, see Kagan [48]. In this case $\alpha = 2f'''(1) + 3 = 3$, and therefore the induced connection is $\nabla^{(3)}$. It is worth noting the relation with the minimum chi-squared estimation in the discrete case, see Kass and Vos [49], p.243. In this case the Kagan divergence becomes

$$D_{\chi^2}(p, q) = \frac{1}{2} \sum_{i=1}^{n} \frac{(p_i - q_i)^2}{q_i}.$$

### 12.3.6 Exponential Contrast Function

The contrast function associated with the convex function $f(u) = \frac{1}{2}(\ln u)^2$ is

$$\mathcal{E}(p||q) = D_f(p||q) = \frac{1}{2} \int_{\mathcal{X}} p(x) \big( \ln p(x) - \ln q(x) \big)^2 \, dx.$$

The induced connection in this case is $\nabla^{(-3)}$.

We note that all function candidates of the form $f(u) = K(\ln u)^{2k}$ are convex, but the condition $f''(1) = 1$ is verified only for $k = 1, 2$ (with appropriate constants $K$).

### 12.3.7 Product Contrast Function with $(\alpha, \beta)$-Index

The following 2-parameter family of contrast functions is introduced and studied in Eguchi [40]

$$D_{\alpha,\beta}(p||q) = \frac{2}{(1-\alpha)(1-\beta)} \int \Big\{ 1 - \Big( \frac{p(x)}{q(x)} \Big)^{\frac{1-\alpha}{2}} \Big\} \Big\{ 1 - \Big( \frac{p(x)}{q(x)} \Big)^{\frac{1-\beta}{2}} \Big\} \, dx,$$

and is induced by the function

$$f_{\alpha,\beta}(u) = \frac{2}{(1-\alpha)(1-\beta)} (1 - u^{\frac{1-\alpha}{2}})(1 - u^{\frac{1-\beta}{2}}).$$

This connects to the previous contrast functions, see Problem 12.3.

It is worthy to note that the contrast function $D_{\alpha,\beta}(\cdot \,||\, \cdot)$ can be written as the following convex combination of Chernoff informations, see Problem 12.3, part (*e*).

We end this section with a few suggestive examples. The computations are left as exercises to the reader.

**Example 12.3.1** Consider the statistical model $\mathcal{S} = \{p_\mu; \mu \in \mathbb{R}^k\}$, where

$$p_\mu(x) = (2\pi)^{-k/2} e^{-\frac{\|x-\mu\|^2}{2}}, \qquad x \in \mathbb{R}^k$$

is a $k$-dimensional Gaussian density with $\sigma = 1$. Problem 12.4 provides exact formulas for the aforementioned contrast functions in terms of the Euclidean distance $\| \cdot \|$.

**Example 12.3.2 (Exponential Model)** Let $\mathcal{S} = \{p_\xi\}$, where

$$p_\xi = \xi e^{-\xi x}, \qquad \xi > 0, x > 0.$$

A computation shows

$$\begin{aligned}
D_{KL}(p_\xi || p_{\xi'}) &= \frac{\xi'}{\xi} - \ln \frac{\xi'}{\xi} - 1 \\
J(p_\xi, p_{\xi'}) &= \frac{(\xi' - \xi)^2}{2\xi\xi'} \\
H^2(p_\xi, p_{\xi'}) &= \frac{4(\sqrt{\xi} - \sqrt{\xi'})^2}{\xi + \xi'} \\
D^{(\alpha)}(p_\xi || p_{\xi'}) &= \frac{4}{1 - \alpha^2}\left\{1 - \frac{\xi^{\frac{1-\alpha}{2}} {\xi'}^{\frac{1+\alpha}{2}}}{\frac{1+\alpha}{2}\xi' + \frac{1-\alpha}{2}\xi}\right\} \\
D_{\chi^2}(p_\xi || p_{\xi'}) &= \frac{1}{2}\left[\frac{1}{\left(2 - \frac{\xi}{\xi'}\right)\frac{\xi}{\xi'}} - 1\right] \\
\mathcal{E}(p_\xi || p_{\xi'}) &= \frac{1}{2}\left\{\frac{\xi'}{\xi} - \ln \frac{\xi'}{\xi} - 1\right\}.
\end{aligned}$$

It is worthy to note that all these contrast functions provide the same Riemannian metric on $\mathcal{S}$ given by $g_{11} = \frac{1}{\xi^2}$, which is the Fisher information. The induced distance between $p_\xi$ and $p_{\xi'}$ is a hyperbolic distance, i.e., $dist(p_\xi, p_{\xi'}) = |\ln \frac{\xi}{\xi'}|$.

## 12.4 Problems

**12.1.** Consider the exponential family

$$p(x; \xi) = e^{C(x) + \xi^i F_i(x) - \psi(\xi)}, \quad i = 1, \cdots, n,$$

with $\psi(\xi)$ convex function, and define

$$D(\xi_0 || \xi) = \psi(\xi) - \psi(\xi_0) - \langle \partial\psi(\xi_0), \xi - \xi_0 \rangle.$$

(*a*) Prove that $D(\cdot||\cdot)$ is a contrast function;

(*b*) Find the dual contrast function $D^*(\cdot||\cdot)$;

(*c*) Prove that the Riemann metric induced by the contrast function $D(\cdot||\cdot)$ is the Fisher–Riemann metric of the exponential family. Find a formula for it using the function $\psi(\xi)$;

(*d*) Find the components of the dual connections $\nabla^{(D)}$ and $\nabla^{(D^*)}$ induced by the contrast function $D(\cdot||\cdot)$;

(*e*) Show that the skewness tensor induced by the contrast function $D(\cdot||\cdot)$ is $T_{ijk}(\xi) = \partial_i\partial_j\partial_k\psi(\xi)$.

**12.2.** Prove that the Hellinger distance

$$H(p,q) = \sqrt{2\int_{\mathcal{X}} (\sqrt{p(x)} - \sqrt{q(x)})^2\, dx}$$

satisfies the distance axioms.

**12.3.** Consider the Eguchi contrast function

$$D_{\alpha,\beta}(p||q) = \frac{2}{(1-\alpha)(1-\beta)} \int \Big\{1-\Big(\frac{p}{q}\Big)^{\frac{1-\alpha}{2}}\Big\}\Big\{1-\Big(\frac{p}{q}\Big)^{\frac{1-\beta}{2}}\Big\}\, dx.$$

Let $H(\cdot,\cdot)$, $D^{(\alpha)}(\cdot||\cdot)$, $J(\cdot,\cdot)$, $\mathcal{E}(\cdot||\cdot)$ be the Hellinger distance, the Chernoff information of order $\alpha$, the Jefferey distance, and the exponential contrast function, respectively. Prove the following relations:

(*a*) $D_{0,0}(p||q) = H^2(p,q)$

(*b*) $D_{-\alpha,\alpha}(p||q) = \frac{1}{2}\big(D^{(\alpha)}(p||q) + D^{(-\alpha)}(p||q)\big)$

(*c*) $\lim_{\alpha\to 1} D_{-\alpha,\alpha}(p||q) = J(p,q)$

(*d*) $\lim_{\alpha\to -1} D_{\alpha,\alpha}(p||q) = \mathcal{E}(p||q)$

(*e*) $D_{\alpha,\beta}(p||q) = \lambda_1 D^{(-\alpha)} + \lambda_2 D^{(-\beta)} + \lambda_3 D^{(\frac{1-\alpha-\beta}{2})}$,

where

$$\lambda_1 = \frac{1+\alpha}{2(1-\beta)},\quad \lambda_2 = \frac{1+\beta}{2(1-\alpha)},\quad \lambda_3 = -\frac{(\alpha+\beta)(2-\alpha-\beta)}{2(1-\alpha)(1-\beta)},$$

and show that $\lambda_1 + \lambda_2 + \lambda_3 = 1$.

**12.4.** Consider the statistical model defined by the $k$-dimensional Gaussian family, $\mathcal{S} = \{p_\mu; \mu \in \mathbb{R}^k\}$,

$$p_\mu(x) = (2\pi)^{-k/2} e^{-\frac{\|x-\mu\|^2}{2}}, \qquad x \in \mathbb{R}^k.$$

Prove the following relations:

$$(a)\quad D_{KL}(p_\mu||p_{\mu'}) = \frac{1}{2}\|\mu - \mu'\|^2$$
$$(b)\quad J(p_\mu, p_{\mu'}) = \frac{1}{2}\|\mu - \mu'\|^2$$
$$(c)\quad H^2(p_\mu, p_{\mu'}) = 4\Big[1 - e^{-\frac{\|\mu-\mu'\|^2}{8}}\Big]$$
$$(d)\quad D^{(\alpha)}(p_\mu||p_{\mu'}) = \frac{4}{1-\alpha^2}\Big[1 - e^{-\frac{1-\alpha^2}{8}\|\mu-\mu'\|^2}\Big]$$
$$(e)\quad \mathcal{E}(p_\mu||p_{\mu'}) = \frac{1}{2}\|\mu - \mu'\|^2\Big[1 + \frac{1}{4}\|\mu - \mu'\|^2\Big],$$

where $\|\cdot\|$ denotes the Euclidean norm on $\mathbb{R}^k$.

**12.5.** Let $D_f(\cdot||\cdot)$ be the $f$-divergence. Prove the following convexity property

$$D_f\Big(\lambda p_1 + (1-\lambda)p_2||\lambda q_1 + (1-\lambda)q_2\Big) \leq \lambda D_f(p_1||q_1) + (1-\lambda)D_f(p_2||q_2),$$

$\forall \lambda \in [0,1]$ and $p_1, p_2, q_1, q_2$ distribution functions.

**12.6.** Prove the formulas for the contrast function in the case of the exponential distribution presented by Example 12.3.2.

**12.7.** Consider the normal distributions $p(x) = \frac{1}{\sqrt{2\pi}\sigma_1}e^{-\frac{(x-\mu_1)^2}{2\sigma_1^2}}$ and $q(x) = \frac{1}{\sqrt{2\pi}\sigma_2}e^{-\frac{(x-\mu_2)^2}{2\sigma_2^2}}$.

$(a)$ Show that

$$\int_{-\infty}^{\infty}\sqrt{p(x)q(x)}\,dx = \sqrt{\frac{2\sigma_1\sigma_2}{\sigma_1^2+\sigma_2^2}e^{A-B}},$$

where

$$A = \frac{\Big(\frac{\mu_1}{2\sigma_1^2} + \frac{\mu_2}{2\sigma_2^2}\Big)^2}{\frac{1}{\sigma_1^2} + \frac{1}{\sigma_2^2}}, \qquad B = \frac{\mu_1^2}{4\sigma_1^2} + \frac{\mu_2^2}{4\sigma_2^2}.$$

$(b)$ Find the Hellinger distance $H(p,q)$.

**12.8.** Find the Hellinger distance between two gamma distributions.

**12.9.** Consider two exponential distributions, $p(x) = ae^{-ax}$ and $q(x) = be^{-bx}$, $x \geq 0$. Show that the Chernoff information of order $\alpha$ is

$$D^{\alpha}(p||q) = \frac{4}{1-\alpha^2}\Big\{1 - \frac{2a^{\frac{1-\alpha}{2}}b^{\frac{1+\alpha}{2}}}{a(1-\alpha)+b(1+\alpha)}\Big\}, \quad \alpha \neq \pm 1.$$

**12.10.** Consider the normal distributions $p(x) = \frac{1}{\sqrt{2\pi}\sigma_1}e^{-\frac{(x-\mu_1)^2}{2\sigma_1^2}}$ and $q(x) = \frac{1}{\sqrt{2\pi}\sigma_2}e^{-\frac{(x-\mu_2)^2}{2\sigma_2^2}}$. Show that the Chernoff information of order $\alpha$ is

$$D^{\alpha}(p||q) = \frac{4}{1-\alpha^2}\Big\{1 - A\sqrt{\frac{\pi}{a}}e^{\frac{b^2}{4a}-c}\Big\}, \qquad |\alpha| < 1,$$

where

$$a = \frac{1-\alpha}{4\sigma_1^2} + \frac{1+\alpha}{4\sigma_2^2}$$

$$b = \frac{\mu_1(1-\alpha)}{2\sigma_1^2} + \frac{\mu_2(1+\alpha)}{2\sigma_2^2}$$

$$c = \frac{\mu_1^2(1-\alpha)}{4\sigma_1^2} + \frac{\mu_2^2(1+\alpha)}{4\sigma_2^2}.$$

**12.11.** The Chernoff information of order $\alpha$ for discrete distributions $(p_n)$ and $(q_n)$ is given by

$$D^{(\alpha)}(p||q) = \frac{4}{1-\alpha^2}\Big\{1 - \sum_{n\geq 0} p_n^{\frac{1-\alpha}{2}} q_n^{\frac{1+\alpha}{2}}\Big\}.$$

Let $p_n = \frac{\lambda_1^n}{n!}e^{-\lambda_1}$ and $q_n = \frac{\lambda_2^n}{n!}e^{-\lambda_2}$ be two Poisson distributions.

(*a*) Show that

$$D^{(\alpha)}(p||q) = \frac{4}{1-\alpha^2}\Big\{1 - e^{\lambda_1^{(1-\alpha)/2}\lambda_2^{(1+\alpha)/2} - \lambda_1(1-\alpha)/2 - \lambda_2(1+\alpha)/2}\Big\}.$$

(*b*) Show that the square of the Hellinger distance is given by

$$H^2(p,q) = 4\{1 - e^{\sqrt{\lambda_1\lambda_2} - \frac{\lambda_1+\lambda_2}{2}}\}.$$

# Chapter 13

# Statistical Submanifolds

This chapter studies the geometric structure induced on a submanifold by the dualistic structure of a statistical manifold. This includes the study of the first and second fundamental forms, curvatures, mean curvatures, and the relations among them.

This material adapts the well-known theory of submanifolds to the statistical manifolds framework and consists mainly in the contributions of the authors.

## 13.1 First Fundamental Form

Let $\mathcal{M}$ and $\mathcal{S}$ be two manifolds, and consider the immersion $\iota : \mathcal{M} \to \mathcal{S}$, i.e., a one-to-one map, onto on its image, with $\iota_*$ isomorphism of $T_p\mathcal{M}$ into the image $\iota_*(T_p\mathcal{M}) \subset T_{\iota(p)}\mathcal{S}$. This implies $\dim \mathcal{M} \leq \dim \mathcal{S}$. Then the manifold $\mathcal{M}$ is called an *immersed submanifold* of $\mathcal{S}$. In the case when $\iota$ is a homeomorphism onto its image in the induced topology, then $\mathcal{M}$ is called an *imbedded submanifold* of $\mathcal{S}$.

If $(\mathcal{S}, g)$ is a Riemannian manifold, then the immersion $\iota : \mathcal{M} \to \mathcal{S}$ induces a Riemannian metric $h$ on the submanifold $\mathcal{M}$ by $h = \iota^*(g)$, i.e.,

$$h(X, Y) = g(\iota_* X, \iota_* Y), \qquad \forall X, Y \in \mathcal{X}(\mathcal{M}). \tag{13.1.1}$$

For the sake of simplicity, it is useful to consider $\mathcal{M}$ as a subset of $\mathcal{S}$, and the imbedding $\iota$ as the canonical inclusion. This assumption simplifies notations. For instance, both the vector $X \in T_p\mathcal{M}$ and its image $\iota_*(X)_p \in T_{\iota(p)}\mathcal{S}$ will be denoted by $X$; whether $X$ is considered

O. Calin and C. Udrişte, *Geometric Modeling in Probability and Statistics*,
DOI 10.1007/978-3-319-07779-6_13,

tangent to $\mathcal{M}$ or to $\mathcal{S}$ is to be understood from the context. Therefore, in the virtue of the previous assumption, relation (13.1.1) becomes

$$h(X,Y) = g(X,Y), \qquad \forall X, Y \in \mathcal{X}(\mathcal{M}).$$

The tensor $h$ is called the *first fundamental form* on $\mathcal{M}$. The pair $(\mathcal{M}, h)$ becomes a Riemannian manifold with the metric and topological structure induced by $(\mathcal{S}, g)$.

## 13.2 Induced Dual Connections

Let $(\mathcal{M}, h)$ be a submanifold of $(\mathcal{S}, g, \nabla, \nabla^*)$, where $\nabla$, $\nabla^*$ are dual, torsion-free connections on $\mathcal{S}$. Each tangent space of $\mathcal{S}$ has the orthogonal decomposition

$$T_p\mathcal{S} = T_p\mathcal{M} \oplus N_p,$$

where $N_p = \{Y;\ g(Y,X) = 0, \forall X \in T_p\mathcal{M}\}$. This way, any vector $Z \in T_p\mathcal{S}$ can be written in a unique way as $Z = Z^T + Z^N$, with $Z^T \in T_p\mathcal{M}$ and $Z^N \in N_p$.

Applying the previous decomposition, for any $X, Y \in \mathcal{X}(\mathcal{M})$ we have

$$\nabla_X Y = D_X Y + L(X,Y) \tag{13.2.2}$$
$$\nabla^*_X Y = D^*_X Y + L^*(X,Y), \tag{13.2.3}$$

where

$$D_X Y = (\nabla_X Y)^T, \qquad D^*_X Y = (\nabla^*_X Y)^T \tag{13.2.4}$$
$$L(X,Y) = (\nabla_X Y)^N, \quad L^*(X,Y) = (\nabla^*_X Y)^N. \tag{13.2.5}$$

These formulas define the maps $D$, $D^*$, $L$, and $L^*$.

**Theorem 13.2.1** *The maps $D, D^* : \mathcal{X}(\mathcal{M}) \times \mathcal{X}(\mathcal{M}) \to \mathcal{X}(\mathcal{M})$ are torsion-free dual connections on $(\mathcal{M}, g)$.*

*Proof:* First, we note that $D$ and $D^*$ are well defined and $\mathbb{R}$-bilinear. Then we check the $\mathcal{F}(\mathcal{M})$-linearity in the first argument and the Leibnitz' rule in the second as follows

$$D_{fX}Y = (\nabla_{fX}Y)^T = (f\nabla_X Y)^T = f(\nabla_X Y)^T = fD_X Y$$
$$D_X(fY) = (\nabla_X fY)^T = (X(f) + f\nabla_X Y)^T = X(f) + fD_X Y,$$

$\forall X, Y \in \mathcal{X}(\mathcal{M})$, so $D$ is a linear connection on $\mathcal{M}$. Since $\nabla$ is torsion-free and $[X, Y]$ is tangent to $\mathcal{M}$, we find

$$D_X Y - D_Y X = (\nabla_X Y - \nabla_Y X)^T = [X, Y]^T = [X, Y],$$

so $D$ is torsion-free (symmetric). A similar approach shows that $D^*$ is a torsion-free connection. For any $X, Y, Z \in \mathcal{X}(\mathcal{M})$, we have

$$\begin{aligned} Zh(X, Y) = Zg(X, Y) &= g(\nabla_Z X, Y) + g(X, \nabla_Z^* Y) \\ &= g\big((\nabla_Z X)^T, Y\big) + g\big(X, (\nabla_Z^* Y)^T\big) \\ &= g\big(D_Z X, Y\big) + g\big(X, D_Z^* Y\big), \end{aligned}$$

which shows that $D$ and $D^*$ are dual connections on $(\mathcal{M}, h)$. ■

Therefore, a dualistic structure on $\mathcal{S}$, i.e., a quadruple $(\mathcal{S}, g, \nabla, \nabla^*)$, induces a natural dualistic structure $(\mathcal{M}, h, D, D^*)$ on $\mathcal{M}$, which is called a *statistical submanifold* of $(\mathcal{S}, g, \nabla, \nabla^*)$.

The $\nabla^{(\alpha)}$-connection of $\mathcal{S}$

$$\nabla^{(\alpha)} = \frac{1 - \alpha}{2} \nabla + \frac{1 + \alpha}{2} \nabla^*$$

induces a $D^{(\alpha)}$-connection on $\mathcal{M}$ as in the following

$$\begin{aligned} D_X^{(\alpha)} Y &= (\nabla_X^{(\alpha)} Y)^T = \frac{1 - \alpha}{2} (\nabla_X Y)^T + \frac{1 + \alpha}{2} (\nabla_X^* Y)^T \\ &= \frac{1 - \alpha}{2} D_X Y + \frac{1 + \alpha}{2} D_X^* Y, \quad \forall X, Y \in \mathcal{X}(\mathcal{M}). \end{aligned}$$

Similarly, we can define the $L^{(\alpha)}$ operator by taking the normal part of the $\alpha$-connection

$$\begin{aligned} L^{(\alpha)}(X, Y) &= (\nabla_X^{(\alpha)} Y)^N = \frac{1 - \alpha}{2} (\nabla_X Y)^N + \frac{1 + \alpha}{2} (\nabla_X^* Y)^N \\ &= \frac{1 - \alpha}{2} L(X, Y) + \frac{1 + \alpha}{2} L^*(X, Y), \quad \forall X, Y \in \mathcal{X}(\mathcal{M}). \end{aligned}$$

## 13.3 Dual Second Fundamental Forms

The operators $L$ and $L^*$ defined by (13.2.5) have tensor-like properties, even if they are not tensors on $\mathcal{M}$ (their image is not a vector field on $\mathcal{M}$).

**Theorem 13.3.1** *The mappings* $L, L^* : \mathcal{X}(\mathcal{M}) \times \mathcal{X}(\mathcal{M}) \to \mathcal{X}(\mathcal{S})$ *are symmetric, i.e.,*

$$L(X,Y) = L(Y,X), \qquad L^*(X,Y) = L^*(Y,X),$$

*and* $\mathcal{F}(\mathcal{M})$*-bilinear. In particular* $\forall f_1, f_2 \in \mathcal{F}(\mathcal{M})$*, we have*

$$L(f_1X, f_2Y) = f_1f_2L(X,Y), \quad L^*(f_1X, f_2Y) = f_1f_2L(X,Y).$$

*Proof:* Using the torsion-free property of $\nabla$ and $D$, we get

$$\begin{aligned} L(X,Y) &= \nabla_XY - D_XY = \nabla_YX + [X,Y] - (D_YX + [X,Y]) \\ &= \nabla_XY - D_XY = L(Y,X). \end{aligned}$$

Using the symmetry, we find

$$\begin{aligned} L(f_1X,Y) &= \nabla_{f_1X}Y - D_{f_1X}Y = f_1(\nabla_XY - D_XY) = f_1L(X,Y) \\ L(X,f_2Y) &= L(f_2Y,X) = f_2L(Y,X) = f_2L(X,Y), \end{aligned}$$

whence $L(f_1X, f_2Y) = f_1f_2L(X,Y)$. A similar approach applies for $L^*$. ■

**Corollary 13.3.2** *The operator*

$$L^{(\alpha)}(X,Y) = \frac{1-\alpha}{2}L(X,Y) + \frac{1+\alpha}{2}L^*(X,Y), \quad \forall X,Y \in \mathcal{X}(\mathcal{M})$$

*is symmetric and* $\mathcal{F}(\mathcal{M})$*-bilinear.*

The operators $L$ and $L^*$ are called *dual second fundamental forms* of $(\mathcal{M}, h, D, D^*)$ with respect to $(\mathcal{S}, g, \nabla, \nabla^*)$. Sometimes, they are also referred to as *embedding curvatures* (see Amari [8] p.23), since they describe how the submanifold $\mathcal{M}$ is curved inside of $\mathcal{S}$ with respect to the dual connections $\nabla, \nabla^*$.

The submanifold $\mathcal{M}$ is called $\nabla$*-autoparallel* if $L(X,Y) = 0$ for all $X,Y \in \mathcal{X}(\mathcal{M})$. The submanifold $\mathcal{M}$ is called *dual-autoparallel* if it is both $\nabla$- and $\nabla^*$-autoparallel, i.e., if $L(X,Y) = L^*(X,Y) = 0$ for all $X,Y \in \mathcal{X}(\mathcal{M})$. Finally, the submanifold $\mathcal{M}$ is called *geodesic* if it is $\nabla^{(0)}$-autoparallel.

Let $(\mathcal{M}, h, D, D^*)$ be a statistical submanifold of $(\mathcal{S}, g, \nabla, \nabla^*)$. The following statements:

($a$) $\mathcal{M}$ is $\nabla$-autoparallel in $\mathcal{S}$;

($b$) $\mathcal{M}$ is $\nabla^*$-autoparallel in $\mathcal{S}$;

(*c*) $\mathcal{S}$ has $R = 0$;

(*d*) $\mathcal{S}$ has $R^* = 0$;

(*e*) $\mathcal{M}$ has $R_{\mathcal{M}} = 0$;

(*f*) $\mathcal{M}$ has $R^*_{\mathcal{M}} = 0$.

(*g*) $R_{\mathcal{M}}(X, Y, Z) = R(X, Y, Z)$, $\forall X, Y, Z \in \mathcal{X}(\mathcal{M})$;

(*h*) $R^*_{\mathcal{M}}(X, Y, Z) = R^*(X, Y, Z)$, $\forall X, Y, Z \in \mathcal{X}(\mathcal{M})$

are related as in the next lemma.

**Lemma 13.3.3** *The following relations hold:*

$$(a) \Rightarrow (g),\ (b) \Rightarrow (h),\ (c) \Leftrightarrow (d),\ \text{and } (e) \Leftrightarrow (f).$$

*Proof:* Assume $(a)$ holds true. Then $L = 0$, and hence $\nabla_X Y = D_X Y$ for any tangent vector fields $X, Y$ to $\mathcal{M}$. This implies

$$\begin{aligned} R_{\mathcal{M}}(X, Y, Z) &= [D_X, D_Y]Z - D_{[X,Y]}Z \\ &= [\nabla_X, \nabla_Y]Z - \nabla_{[X,Y]}Z \\ &= R(X, Y, Z), \qquad \forall X, Y, Z \in \mathcal{X}(\mathcal{M}), \end{aligned}$$

and hence $R_{\mathcal{M}} = R$, which is $(g)$. The implication $(b) \Rightarrow (h)$ can be proved similarly. The equivalences $(c) \Leftrightarrow (d)$ and $(e) \Leftrightarrow (f)$ follow from Proposition 8.1.4. ■

**Lemma 13.3.4** *Let $\mathcal{S}$ be $\nabla$-flat. Then any $\nabla$-autoparallel submanifold $\mathcal{M}$ of $\mathcal{S}$ is $D$-flat.*

*Proof:* Since $\mathcal{M}$ is $\nabla$-autoparallel we have $L(\partial_i, \partial_j) = 0$ and then $\nabla_{\partial_i}\partial_j = D_{\partial_i}\partial_j$ for $\partial_i, \partial_j$ tangent vector fields to $\mathcal{M}$. Using that $\mathcal{S}$ is $\nabla$-flat, we find

$$h(D_{\partial_i}\partial_j, \partial_k) = g(D_{\partial_i}\partial_j, \partial_k) = g(\nabla_{\partial_i}\partial_j, \partial_k) = 0,$$

which implies that $\mathcal{M}$ is $D$-flat. ■

The following result can be found in Amari [8], p.58.

**Theorem 13.3.5** *Let $(\mathcal{S}, g, \nabla, \nabla^*)$ be a dually flat statistical manifold and $(\mathcal{M}, h, D, D^*)$ be a submanifold. If $(\mathcal{M}, h, D, D^*)$ is either $\nabla$-autoparallel or $\nabla^*$-autoparallel, then $(\mathcal{M}, h, D, D^*)$ is a dually flat statistical manifold.*

*Proof:* Let $\{\partial_i\}_{1\leq i\leq m}$ be local coordinate vector fields on $\mathcal{M}$. Since $\nabla$ and $\nabla^*$ are dually flat connections, we have

$$\partial_k g(\partial_i, \partial_j) = g(\nabla_{\partial_k}\partial_i, \partial_j) + g(\partial_i, \nabla^*_{\partial_k}\partial_j) = 0.$$

It follows that $\partial_k h(\partial_i, \partial_j) = \partial_k g(\partial_i, \partial_j) = 0$. Combining with the duality of $D$ and $D^*$ yields

$$0 = \partial_k h(\partial_i, \partial_j) = g(D_{\partial_k}\partial_i, \partial_j) + g(\partial_i, D^*_{\partial_k}\partial_j). \tag{13.3.6}$$

Assume $\mathcal{M}$ is $\nabla$-autoparallel. Then Lemma 13.3.4 implies that $\mathcal{M}$ is $D$-flat, i.e., $g(D_{\partial_k}\partial_i, \partial_j) = 0$. Substituting in (13.3.6) implies that $g(\partial_i, D^*_{\partial_k}\partial_j) = 0$, i.e. $\mathcal{M}$ is also $D^*$-flat. ■

We formulate next a partial converse of the previous result.

**Theorem 13.3.6** *Let $(\mathcal{M}, h, D, D^*)$ be a dually flat statistical submanifold of $(\mathcal{S}, g, \nabla, \nabla^*)$. If $\mathcal{S}$ is either $\nabla$-flat or $\nabla^*$-flat, then $\mathcal{S}$ is dually flat.*

*Proof:* Let $\{\partial_1, \dots, \partial_m\}$ be local coordinate vector fields on $\mathcal{M}$. Since $\partial_k g(\partial_i, \partial_j) = \partial_k h(\partial_i, \partial_j)$, then the duality relations imply

$$g(\nabla_{\partial_k}\partial_i, \partial_j) + g(\partial_i, \nabla^*_{\partial_k}\partial_j) = g(D_{\partial_k}\partial_i, \partial_j) + g(\partial_i, D^*_{\partial_k}\partial_j).$$

Since $\mathcal{M}$ is dually flat we have $g(D_{\partial_k}\partial_i, \partial_j) = 0$, $g(\partial_i, D^*_{\partial_k}\partial_j) = 0$. Substituting in the previous relation yields

$$g(\nabla_{\partial_k}\partial_i, \partial_j) + g(\partial_i, \nabla^*_{\partial_k}\partial_j) = 0.$$

If $\mathcal{S}$ is $\nabla$-flat, then $g(\nabla_{\partial_k}\partial_i, \partial_j) = 0$, and using the previous relation we get $g(\partial_i, \nabla^*_{\partial_k}\partial_j) = 0$, i.e., $\mathcal{S}$ is $\nabla^*$-flat. The same argument applies if assume first that $\mathcal{S}$ is $\nabla^*$-flat. ■

## 13.4 Generalized Shape Operator

Let $(\mathcal{M}, h, D, D^*)$ be a statistical submanifold of $(\mathcal{S}, g, \nabla, \nabla^*)$. Consider

$$\begin{aligned}
\nabla^{(\alpha)} &= \frac{1-\alpha}{2}\nabla + \frac{1+\alpha}{2}\nabla^* \\
L^{(\beta)} &= \frac{1-\beta}{2}L + \frac{1+\beta}{2}L^*,
\end{aligned}$$

and define the generalized shape operator

$$S^{(\alpha,\beta)}(X,Y,Z,W) = g\big(\nabla_X^{(\alpha)} L^{(\beta)}(Y,Z), W\big), \quad \forall X,Y,Z,W \in \mathcal{X}(\mathcal{M}). \tag{13.4.7}$$

The tensor $S^{(\alpha,\beta)}$ measures the projection of the covariant derivative of the $\beta$-second fundamental form, which is normal to $\mathcal{M}$, with respect to the $\alpha$-connection. The following result shows that $S^{(\alpha,\beta)}$ is a $(0,4)$-type tensor on $\mathcal{M}$ that measures the angle made by two fundamental forms, i.e., describes the shape of the submanifold $\mathcal{M}$.

**Proposition 13.4.1** *For any real numbers $\alpha, \beta$ and any $X,Y,Z,W \in \mathcal{X}(\mathcal{M})$, we find*

$$S^{(\alpha,\beta)}(X,Y,Z,W) = -g\big(L^{(\beta)}(Y,Z), L^{(-\alpha)}(X,W)\big). \tag{13.4.8}$$

*Proof:* Using that $\nabla^{(\alpha)}$ and $\nabla^{(-\alpha)}$ are dual connections, we have

$$\begin{aligned} S^{(\alpha,\beta)}(X,Y,Z,W) &= g\big(\nabla_X^{(\alpha)} L^{(\beta)}(Y,Z), W\big) \\ &= X\,\underbrace{g\big(L^{(\beta)}(Y,Z), W\big)}_{=0} - g\big(L^{(\beta)}(Y,Z), \nabla_X^{(-\alpha)} W\big) \\ &= -g\big(L^{(\beta)}(Y,Z), D_X^{(-\alpha)} W + L^{(-\alpha)}(X,W)\big) \\ &= -g\big(L^{(\beta)}(Y,Z), L^{(-\alpha)}(X,W)\big), \end{aligned}$$

where we used the orthogonality of $L^{(\beta)}(Y,Z)$ and $D_X^{(-\alpha)} W$. ■

We note that the $\mathcal{F}(\mathcal{M})$-bilinearity of $g$ and $L$ implies that $S^{(\alpha,\beta)}$ is $\mathcal{F}(\mathcal{M})$-multilinear, and hence it is a tensor on $\mathcal{M}$. The symmetry of the second fundamental form implies that $S^{(\alpha,\beta)}$ is symmetric in the inner and exterior pairs

$$S^{(\alpha,\beta)}(X,Y,Z,W) = S^{(\alpha,\beta)}(X,Z,Y,W) \tag{13.4.9}$$

$$S^{(\alpha,\beta)}(X,Y,Z,W) = S^{(\alpha,\beta)}(W,Y,Z,X). \tag{13.4.10}$$

It is worth noting that $S^{(\alpha,\beta)}$ is also symmetric under the inversion order of arguments

$$S^{(\alpha,\beta)}(X,Y,Z,W) = S^{(\alpha,\beta)}(W,Z,Y,X) \tag{13.4.11}$$

**Proposition 13.4.2** *The following index symmetry holds*

$$S^{(\alpha,\beta)}(X,Y,Z,W) = S^{(-\beta,-\alpha)}(Z,W,X,Y), \quad \forall X,Y,Z,W \in \mathcal{X}(\mathcal{M}). \tag{13.4.12}$$

*Proof:* Proposition 13.4.1, the symmetry of $g$ and $L$ infer

$$\begin{aligned} S^{(\alpha,\beta)}(X,Y,Z,W) &= -g\big(L^{(\beta)}(Y,Z),L^{(-\alpha)}(X,W)\big) \\ &= -g\big(L^{(\beta)}(Z,Y),L^{(-\alpha)}(W,X)\big) \\ &= -g\big(L^{(-\alpha)}(W,X),L^{(\beta)}(Z,Y)\big) \\ &= S^{(-\beta,-\alpha)}(Z,W,X,Y). \end{aligned}$$

■

Different particular cases of $\alpha$ and $\beta$ lead to the following formulas of the shape operator:

**Corollary 13.4.3** *We have*

$$\begin{aligned} S^{(-1,-1)}(X,Y,Z,W) &= -g\big(L(Y,Z),L^*(X,W)\big) \\ S^{(1,-1)}(X,Y,Z,W) &= -g\big(L(Y,Z),L(X,W)\big) \\ S^{(-1,1)}(X,Y,Z,W) &= -g\big(L^*(Y,Z),L^*(X,W)\big) \\ S^{(1,1)}(X,Y,Z,W) &= -g\big(L^*(Y,Z),L(X,W)\big). \end{aligned}$$

Denote by $L^{(0)}$ the second fundamental form associated with the Levi–Civita connection, i.e.,

$$L^{(0)}(X,Y) = (\nabla^{(0)}_X Y)^N, \qquad \forall X,Y \in \mathcal{X}(\mathcal{M}).$$

Using

$$L^{(0)} = \frac{1}{2}(L+L^*) = \frac{1}{2}(L^{(\beta)} + L^{(-\beta)}),$$

Proposition 13.4.1 and the linearity leads to another particular value for the shape operator

$$S^{(\alpha,0)} = \frac{1}{2}(S^{(\alpha,\beta)} + S^{(\alpha,-\beta)}). \tag{13.4.13}$$

## 13.5 Mean Curvature Vector Fields

Consider $(\mathcal{M},h,D,D^*)$ a statistical submanifold of $(\mathcal{S},g,\nabla,\nabla^*)$. Let $\{E_1,\dots,E_m\}$ be an orthonormal basis in $T_p\mathcal{M}$, with $m = \dim\mathcal{M}$. The *mean curvature vector fields* with respect to $\nabla$ and $\nabla^*$ are defined as

$$\begin{aligned} H_p &= \frac{1}{m}\delta^{ij}L(E_i,E_j) \\ H_p^* &= \frac{1}{m}\delta^{ij}L^*(E_i,E_j), \end{aligned}$$

with summation in the repeated indices. Here $\delta^{ij}$ denotes the Kronecker's delta symbol. The manifold $\mathcal{M}$ is called $\nabla$-minimal (respectively, $\nabla^*$-minimal) if $H_p = 0$ (respectively $H_p^* = 0$) at every point $p \in \mathcal{M}$. The vector fields $H$ and $H^*$ are normal to the manifold $\mathcal{M}$.

The relation between $L$ and $L^*$ is given by following result.

**Proposition 13.5.1** *The fundamental forms $L$ and $L^*$ are related by*

$$g\big(\nabla_X L(Y,Z), W\big) = g\big(\nabla_Y^* L^*(X,W), Z\big), \quad \forall X, Y, Z, W \in \mathcal{X}(\mathcal{M}). \tag{13.5.14}$$

*Proof:* From the definition of the shape tensor and relations (13.4.12) and (13.4.11), we have

$$\begin{aligned} g\big(\nabla_X L(Y,Z), W\big) &= S^{(-1,-1)}(X,Y,Z,W) = S^{(1,1)}(Z,W,X,Y) \\ &= S^{(1,1)}(Y,X,W,Z) = g\big(\nabla_Y^* L^*(X,W), Z\big). \end{aligned}$$

■

A similar argument can be used to prove the more general relation

$$g\big(\nabla_X^{(\alpha)} L^{(\beta)}(Y,Z), W\big) = g\big(\nabla_Y^{(-\beta)} L^{(-\alpha)}(X,W), Z\big),$$

$\forall\, X, Y, Z, W$, tangent vector fields to $\mathcal{M}$.

Proposition 13.5.1 has a couple of interesting applications.

**Proposition 13.5.2** *Let $(\mathcal{M}, h, D, D^*)$ be a submanifold of the statistical manifold $(\mathcal{S}, g, \nabla, \nabla^*)$.*

(*i*) *If $\mathcal{M}$ is $\nabla^*$-autoparallel, then $\nabla_X L(Y,Z) = 0$, $\forall X, Y, Z, W \in \mathcal{X}(\mathcal{M})$, i.e., $L(Y,Z)$ is parallel with respect to $\nabla$;*

(*ii*) *If $\mathcal{M}$ is $\nabla$-autoparallel, then $\nabla_X^* L^*(Y,Z) = 0$, $\forall X, Y, Z, W \in \mathcal{X}(\mathcal{M})$, i.e., $L^*(Y,Z)$ is parallel with respect to $\nabla^*$.*

*Proof:*

(*i*) Assume that $\mathcal{M}$ is $\nabla^*$-autoparallel, so $L = 0$. Substituting in (13.5.14) yields

$$g\big(\nabla_X L(Y,Z), W\big) = 0, \quad \forall X, Y, Z, W \in \mathcal{X}(\mathcal{M}),$$

so $\nabla_X L(Y,Z) = 0$, i.e., $L(Y,Z)$ is parallel with respect to $\nabla$. Part (*ii*) has a similar proof.

■

**Proposition 13.5.3** *Let $(\mathcal{M}, h, D, D^*)$ be a submanifold of the statistical manifold $(\mathcal{S}, g, \nabla, \nabla^*)$ and denote by $H$ and $H^*$ the dual mean curvature vector fields.*

*(i) $divL(Y,Z) = 0$, $\forall Y, Z \in \mathcal{X}(\mathcal{M})$ if and only if $H^*$ is parallel with respect to $\nabla^*$;*

*(ii) $div^*L^*(Y,Z) = 0$, $\forall Y, Z \in \mathcal{X}(\mathcal{M})$ if and only if $H$ is parallel with respect to $\nabla$.*

*Proof:*

(i) Let $\{E_1, \ldots, E_m\}$ be an orthonormal basis in $T_p\mathcal{M}$. Making $X = W = E_i$ in relation (13.5.14) yields

$$g\big(\nabla_{E_i} L(Y_p, Z_p), E_i\big) = g\big(\nabla^*_{Y_p} L^*(E_i, E_i), Z_p\big), \quad \forall Y_p, Z_p \in T_p(\mathcal{M}). \tag{13.5.15}$$

Summing over $i$, we obtain

$$divL(Y,Z) = mg\big(\nabla^*_Y H^*, Z\big), \quad Y, Z \in \mathcal{X}(\mathcal{M}).$$

Hence $divL(Y,Z) = 0$ if and only if $\nabla^*_Y H^* = 0$, $Y, Z \in \mathcal{X}(\mathcal{M})$.

(ii) Changing the roles of $\nabla$ and $\nabla^*$ and applying a similar argument leads to the desired result.

■

We deal next with the $\nabla$- and $\nabla^*$-minimality of the submanifold $\mathcal{M}$.

**Corollary 13.5.4** *We have:*

*(i) If $H^* = 0$, then $divL(Y,Z) = 0$, $\forall Y, Z \in \mathcal{X}(\mathcal{M})$;*

*(ii) If $H = 0$, then $div^*L^*(Y,Z) = 0$, $\forall Y, Z \in \mathcal{X}(\mathcal{M})$.*

*Proof:* It follows from Proposition 13.5.3 by considering either $H = 0$, or $H^* = 0$. ■

The relation between divergences of dual mean curvature vector fields is given by the next result.

**Lemma 13.5.5** *For any submanifold $\mathcal{M}$ of $\mathcal{S}$, we have*

$$divH = div^*H^*.$$

*Proof:* Let $\{E_1, \dots, E_m\}$ be an orthonormal basis in $T_p\mathcal{M}$. Then relation (13.5.14) can be written as

$$g\big(\nabla_{E_i} L(E_j, E_j), E_i\big) = g\big(\nabla^*_{E_j} L^*(E_i, E_i), E_j\big). \tag{13.5.16}$$

Summing over $i$ and $j$ yields $divH = div^* H^*$.

■

**Lemma 13.5.6** *Let $\{E_1, \dots, E_m\}$ be an orthonormal basis in $T_p\mathcal{M}$. Then*

$$\begin{aligned} g(H_p, H_p^*) &= -\frac{1}{m^2}\sum_{i,j} S^{(-1,-1)}(E_j, E_i, E_i, E_j) \\ &= -\frac{1}{m^2}\sum_{i,j} S^{(1,1)}(E_j, E_i, E_i, E_j). \end{aligned}$$

*Proof:* From the definition of mean curvature vector field and Proposition 13.4.1, we get

$$\begin{aligned} g(H_p, H_p^*) &= \frac{1}{m^2} g\Big(\sum_i L(E_i, E_i), \sum_j L^*(E_j, E_j)\Big) \\ &= \frac{1}{m^2}\sum_{i,j} g\big(L(E_i, E_i), L^*(E_j, E_j)\big) \\ &= -\frac{1}{m^2}\sum_{i,j} S^{(-1,-1)}(E_j, E_i, E_i, E_j). \end{aligned}$$

The second identity follows from relation (13.4.12). ■

The next result contains an unexpected relation.

**Theorem 13.5.7** *The divergences of the mean curvature vector fields are related to their inner product by*

$$divH = div^* H^* = -m\, g(H, H^*). \tag{13.5.17}$$

*Proof:* Assume $\{E_1, \dots, E_m\}$ is an orthonormal basis in $T_p\mathcal{M}$. The divergence of $H_p$ with respect to the $\nabla$-connection is given by

$$\begin{aligned} divH_p &= \sum_i g(\nabla_{E_i} H, E_i) = \frac{1}{m}\sum_{i,j} g\big(\nabla_{E_i} L(E_j, E_j), E_i\big) \\ &= \frac{1}{m}\sum_{i,j} S^{(-1,-1)}(E_i, E_j, E_j, E_i), \end{aligned}$$

and applying Lemma 13.5.6 yields $divH_p = -m\, g(H_p, H_p^*)$. Using Lemma 13.5.5, we get $div^* H_p^* = -m\, g(H_p, H_p^*)$. ■

**Corollary 13.5.8** *The following statements are equivalent:*

(*i*) *$H$ and $H^*$ are orthogonal vector fields;*

(*ii*) *$divH = 0$;*

(*iii*) *$div^* H^* = 0$.*

## 13.6 Gauss–Codazzi Equations

Let $(\mathcal{M}, h, D, D^*)$ be a statistical submanifold of $(\mathcal{S}, g, \nabla, \nabla^*)$. Denote by $R_{\mathcal{S}}$ and $R^*_{\mathcal{S}}$ the curvature tensors on $\mathcal{S}$ with respect to connections $\nabla$ and $\nabla^*$. Similarly, $R_{\mathcal{M}}$ and $R^*_{\mathcal{M}}$ denote the curvature tensors on $\mathcal{M}$ with respect to connections $D$ and $D^*$. The equations of Gauss and Codazzi will be deducted from the following lemma.

**Lemma 13.6.1** *For any $X, Y, Z \in \mathcal{X}(\mathcal{M})$, we find*

$$\begin{aligned} R_{\mathcal{S}}(X,Y,Z) &= R_{\mathcal{M}}(X,Y,Z) + L(X, D_Y Z) - L(Y, D_X Z) \\ &\quad + L([X,Y], Z) + \nabla_X L(Y,Z) - \nabla_Y L(X,Z). \end{aligned} \tag{13.6.18}$$

$$\begin{aligned} R^*_{\mathcal{S}}(X,Y,Z) &= R^*_{\mathcal{M}}(X,Y,Z) + L^*(X, D_Y Z) - L^*(Y, D_X Z) \\ &\quad + L^*([X,Y], Z) + \nabla^*_X L^*(Y,Z) - \nabla^*_Y L^*(X,Z). \end{aligned} \tag{13.6.19}$$

*Proof:* It suffices to prove the first relation, the second resulting by duality. Using the decomposition of a connection into its tangent and normal part, we have

$$\begin{aligned} \nabla_X \nabla_Y Z &= \nabla_X \big(D_Y Z + L(Y,Z)\big) = \nabla_X D_Y Z + \nabla_X L(Y,Z) \\ &= D_X D_Y Z + L(X, D_Y Z) + \nabla_X L(Y,Z) \\ \nabla_Y \nabla_X Z &= D_Y D_X Z + L(Y, D_X Z) + \nabla_Y L(X,Z) \\ \nabla_{[X,Y]} Z &= D_{[X,Y]} Z - L\big([X,Y], Z\big). \end{aligned}$$

Subtracting the second and third relation from the first one provides

$$\begin{aligned} R_{\mathcal{S}}(X,Y,Z) &= R_{\mathcal{M}}(X,Y,Z) + L(X, D_Y Z) - L(Y, D_X Z) \\ &\quad + L([X,Y], Z) + \nabla_X L(Y,Z) - \nabla_Y L(X,Z), \end{aligned}$$

which is (13.6.18). ■

**Corollary 13.6.2** *(i) If $\mathcal{M}$ is a $\nabla$-autoparallel submanifold of $\mathcal{S}$, then*

$$R_{\mathcal{S}}(X,Y,Z) = R_{\mathcal{M}}(X,Y,Z), \quad \forall X,Y,Z \in \mathcal{X}(\mathcal{M}).$$

*(ii) If $\mathcal{M}$ is a $\nabla^*$-autoparallel submanifold of $\mathcal{S}$, then*

$$R^*_{\mathcal{S}}(X,Y,Z) = R^*_{\mathcal{M}}(X,Y,Z), \quad \forall X,Y,Z \in \mathcal{X}(\mathcal{M}).$$

*Proof:* It follows from (13.6.18) and (13.6.19) by making $L = 0$ and $L^* = 0$. ■

**Theorem 13.6.3 (Gauss' Equation)** *For any $X,Y,Z,W \in \mathcal{X}(\mathcal{M})$, we have*

$$\begin{aligned} R_{\mathcal{S}}(X,Y,Z,W) &= R_{\mathcal{M}}(X,Y,Z,W) + g\big(L(X,Z), L^*(Y,W)\big) \\ &\quad -g\big(L(Y,Z), L^*(X,W)\big) \end{aligned} \tag{13.6.20}$$

$$\begin{aligned} R^*_{\mathcal{S}}(X,Y,Z,W) &= R^*_{\mathcal{M}}(X,Y,Z,W) + g\big(L^*(X,Z), L(Y,W)\big) \\ &\quad -g\big(L^*(Y,Z), L(X,W)\big). \end{aligned} \tag{13.6.21}$$

*Proof:* Taking the scalar product with respect to $W$ in (13.6.18) and using that $L$ is normal to the submanifold $\mathcal{M}$, we have

$$\begin{aligned} R_{\mathcal{S}}(X,Y,Z,W) &= g\big(R_{\mathcal{S}}(X,Y,Z), W\big) = g\big(R_{\mathcal{M}}(X,Y,Z), W\big) \\ &\quad +g\big(\nabla_X L(Y,Z), W\big) - g\big(\nabla_Y L(X,Z), W\big) \\ &= g\big(R_{\mathcal{M}}(X,Y,Z), W\big) + S^{(-1,-1)}(X,Y,Z,W) \\ &\quad -S^{(-1,-1)}(Y,X,Z,W) \\ &= R_{\mathcal{M}}(X,Y,Z,W) + g\big(L(X,Z), L^*(Y,W)\big) \\ &\quad -g\big(L(Y,Z), L^*(X,W)\big), \end{aligned}$$

where in the last identity we used Corollary 13.4.3. The second relation is proved using a similar argument applied to (13.6.19). ■

We deal next with a few straightforward applications of Gauss' formula.

**Proposition 13.6.4** *(i) If $\mathcal{S}$ is $\nabla$-flat, then*

$$R_{\mathcal{M}}(X,Y,Z,W) = g\big(L(Y,Z), L^*(X,W)\big) - g\big(L(X,Z), L^*(Y,W)\big).$$

*(ii) If $\mathcal{S}$ is $\nabla^*$-flat, then*

$$R^*_{\mathcal{M}}(X,Y,Z,W) = g\big(L^*(Y,Z), L(X,W)\big) - g\big(L^*(X,Z), L(Y,W)\big).$$

*(iii) If $\mathcal{S}$ is dually flat, then*

$$R_{\mathcal{M}}(X,Y,Z,W) = -R^*_{\mathcal{M}}(X,Y,W,Z).$$

*Proof:*

(*i*) It follow from making $R_{\mathcal{S}} = 0$ in formula (13.6.20).

(*ii*) Similarly, put $R^*_{\mathcal{S}} = 0$ in formula (13.6.21).

(*iii*) It follows from (*i*) and (*ii*). It is worth noting the equivalence $R = 0 \Longleftrightarrow R^* = 0$.

■

**Proposition 13.6.5** *Assume the submanifold $\mathcal{M}$ has the property that $L$ and $L^*$ are perpendicular, i.e.,*

$$g\big(L(X,Y), L^*(Z,W)\big) = 0, \ \forall X,Y,Z,W \in \mathcal{X}(\mathcal{M}).$$

*Then for any $X,Y,Z,W \in \mathcal{X}(\mathcal{M})$, we get*

$$R_{\mathcal{S}}(X,Y,Z,W) = R_{\mathcal{M}}(X,Y,Z,W); \qquad (13.6.22)$$

$$R^*_{\mathcal{S}}(X,Y,Z,W) = R^*_{\mathcal{M}}(X,Y,Z,W). \qquad (13.6.23)$$

*Proof:* It is a direct consequence of Gauss' equation. ■

The same argument applied to the $\alpha$-connection yields the following $\alpha$-version of Gauss' equation:

$$\begin{aligned} R^{(\alpha)}_{\mathcal{S}}(X,Y,Z,W) &= R^{(\alpha)}_{\mathcal{M}}(X,Y,Z,W) + g\big(L^{(\alpha)}(X,Z), L^{(-\alpha)}(Y,W)\big) \\ &\quad - g\big(L^{(\alpha)}(Y,Z), L^{(-\alpha)}(X,W)\big), \qquad (13.6.24) \end{aligned}$$

for any $X,Y,Z,W \in \mathcal{X}(\mathcal{M})$.

The following formula regarding the coefficients of the second fundamental form and their covariant derivatives is known under the name of Codazzi equation.

**Theorem 13.6.6 (Codazzi Equation)** *Assume $\mathcal{S}$ is $\nabla$-flat.*

(*i*) *Then for any* $X, Y, Z \in \mathcal{X}(\mathcal{M})$,

$$\big(\nabla_Y L(X,Z)\big)^N - \big(\nabla_X^N L(Y,Z)\big)^N = L(X, D_Y Z) - L(Y, D_X Z) + L([X,Y], Z).$$

(*ii*) *Assume* $\mathcal{S}$ *is* $\nabla^{(\alpha)}$*-flat. Then*

$$\begin{aligned} & \big(\nabla_Y^{(\alpha)} L^{(\alpha)}(X,Z)\big)^N - \big(\nabla_X^{(\alpha)} L^{(\alpha)}(Y,Z)\big)^N \\ = \; & L^{(\alpha)}(X, D_Y^{(\alpha)} Z) - L^{(\alpha)}(Y, D_X^{(\alpha)} Z) + L^{(\alpha)}([X,Y], Z). \end{aligned}$$

*Proof:*

(*i*) Make $R_{\mathcal{S}} = 0$ in (13.6.18) and then consider the equation given by the normal components.

(*ii*) A similar argument applies.

■

**Proposition 13.6.7** *Assume* $\mathcal{S}$ *is* $\nabla$*-flat. If* $\mathcal{M}$ *is a* $\nabla^*$*-autoparallel submanifold of* $\mathcal{S}$*, then*

$$\begin{aligned} L(X, D_Y Z) - L(D_Y X, Z) &= L(Y, D_X Z) - L(D_X Y, Z); \\ L(X, D_Y Z) - L(D_Y^* X, Z) &= L(Y, D_X Z) - L(D_X^* Y, Z). \end{aligned}$$

*Proof:* From Proposition 13.5.2, we get $\nabla_X L(Y,Z) = 0$. In this case the Codazzi equation becomes

$$L(X, D_Y Z) - L(Y, D_X Z) + L([X,Y], Z) = 0.$$

Using that $D$ is symmetric, we substitute $[X,Y] = D_X Y - D_Y X$ in the previous equation and obtain the first desired relation. Using the symmetry of $D^*$, $[X,Y] = D_X^* Y - D_Y^* X$, leads to the second relation.

■

## 13.7 Induced Skewness Tensor

Let $(\mathcal{M}, h, D, D^*)$ be a statistical submanifold of $(\mathcal{S}, g, \nabla, \nabla^*)$. For any tangent vector fields $X$, $Y$ on $\mathcal{M}$, consider the difference tensors on $\mathcal{S}$ and $\mathcal{M}$, respectively,

$$\begin{aligned} K(X,Y) &= \nabla_X^* Y - \nabla_X Y \\ \mathcal{K}(X,Y) &= D_X^* Y - D_X Y. \end{aligned}$$

It follows that

$$K(X,Y) = \mathcal{K}(X,Y) + L^*(X,Y) - L(X,Y). \tag{13.7.25}$$

More precisely,

$$K(X,Y)^T = \mathcal{K}(X,Y), \qquad K(X,Y)^N = L^*(X,Y) - L(X,Y).$$

Denote by $C$ and $\mathcal{C}$ the skewness tensors on manifolds $(\mathcal{S}, g, \nabla, \nabla^*)$ and, respectively, on $(\mathcal{M}, h, D, D^*)$. The next result states that the submanifold and the manifold share the same skewness tensor.

**Proposition 13.7.1** *The restriction of skewness tensor $C$ to $\mathcal{M}$ is the tensor $\mathcal{C}$, i.e.,*

$$\mathcal{C}(X,Y,Z) = C(X,Y,Z), \qquad \forall X,Y,Z \in \mathcal{X}(\mathcal{M}).$$

*Proof:* The proof is a consequence of Proposition 8.8.1 and relation (13.7.25). For any $X, Y, Z$ tangent vector fields to $\mathcal{M}$, we find

$$\begin{aligned} C(X,Y,Z) &= g(K(X,Y),Z) = g(\mathcal{K}(X,Y),Z) + g(L^*(X,Y),Z) \\ &\quad -g(L(X,Y),Z) \\ &= g(\mathcal{K}(X,Y),Z) = h(\mathcal{K}(X,Y),Z) \\ &= \mathcal{C}(X,Y,Z), \end{aligned}$$

where we used that both $L(X,Y)$ and $L^*(X,Y)$ are orthogonal to $\mathcal{M}$. ■

In general, the difference tensors $K$ and $\mathcal{K}$ are distinct, but in the following particular case they coincide.

**Proposition 13.7.2** *If the submanifold $\mathcal{M}$ is dual-autoparallel, then $K(X,Y) = \mathcal{K}(X,Y)$, for any $X, Y \in \mathcal{X}(\mathcal{M})$.*

*Proof:* Since $\mathcal{M}$ is dual-autoparallel, then $L = L^* = 0$. Substituting in (13.7.25) yields the desired result. ■

The next result shows that the restriction of $\nabla^{(\alpha)}g$ to $\mathcal{M}$ is $D^{(\alpha)}h$.

**Proposition 13.7.3** *For any vector fields $X, Y, Z \in \mathcal{X}(\mathcal{M})$, we have*

$$\begin{aligned} (\nabla g)(X,Y,Z) &= (Dh)(X,Y,Z) \\ (\nabla^* g)(X,Y,Z) &= (D^* h)(X,Y,Z) \\ (\nabla^{(\alpha)} g)(X,Y,Z) &= (D^{(\alpha)} h)(X,Y,Z). \end{aligned}$$

*Proof:* We shall prove the first relation, the others being similar. We follow, step by step, the computation

$$\begin{aligned}(\nabla g)(X,Y,Z) &= Xg(Y,Z) - g(\nabla_X Y, Z) - g(Y, \nabla_X Z)\\ &= Xh(Y,Z) - g(D_X Y, Z) - g(Y, D_X Z)\\ &= Xh(Y,Z) - h(D_X Y, Z) - h(Y, D_X Z)\\ &= (Dh)(X,Y,Z).\end{aligned}$$

■

The skewness can be recovered from one of the submanifold connections and the Riemannian metric $h$.

**Corollary 13.7.4** *The skewness tensor $\mathcal{C}$ on $\mathcal{M}$ is given by any of the following relations*

$$Dh = \mathcal{C}, \qquad D^*h = -\mathcal{C}, \qquad D^{(\alpha)}h = -\alpha\mathcal{C}.$$

*Proof:* From Propositions 8.8.1 and 13.7.3, we obtain

$$C(X,Y,Z) = (\nabla g)(X,Y,Z) = (Dh)(X,Y,Z), \quad \forall X,Y,Z \in \mathcal{X}(\mathcal{M}).$$

Using Proposition 13.7.1, the previous relations imply $\mathcal{C}(X,Y,Z) = (Dh)(X,Y,Z)$. The other two relations are proved by a similar argument. ■

We end this section with a relation between the skewness tensor on $\mathcal{S}$ and the shape tensor.

**Proposition 13.7.5** *For any $X,Y,Z,W \in \mathcal{X}(\mathcal{M})$ we have*

$$\begin{aligned}C\big(X, L(Y,Z), W\big) &= S^{(1,-1)}(X,Y,Z,W) - S^{(-1,-1)}(X,Y,Z,W)\\ C\big(X, L^*(Y,Z), W\big) &= S^{(1,1)}(X,Y,Z,W) - S^{(-1,1)}(X,Y,Z,W).\end{aligned}$$

*Proof:* Using the definitions of skewness, difference, and shape tensors we have

$$\begin{aligned}C\big(X, L(Y,Z), W\big) &= g\big(K(X, L(Y,Z)), W\big)\\ &= g\big(\nabla^*_X L(Y,Z), W\big) - g\big(\nabla_X L(Y,Z), W\big)\\ &= S^{(1,-1)}(X,Y,Z,W) - S^{(-1,-1)}(X,Y,Z,W).\end{aligned}$$

The second relation has a similar proof. ■

## 13.8 Curve Kinematics

Let $\gamma(s)$ be a curve contained in $(\mathcal{M}, h, D, D^*)$, which is a statistical submanifold of $(\mathcal{S}, g, \nabla, \nabla^*)$. The acceleration of $\gamma(s)$ in the manifold $\mathcal{S}$ can be measured with respect to both connections. This is the sum between the acceleration in $\mathcal{M}$ and the associated normal curvature term

$$\begin{aligned} \nabla_{\dot\gamma}\dot\gamma &= D_{\dot\gamma}\dot\gamma + L(\dot\gamma, \dot\gamma) && (13.8.26)\\ \nabla^*_{\dot\gamma}\dot\gamma &= D^*_{\dot\gamma}\dot\gamma + L^*(\dot\gamma, \dot\gamma). && (13.8.27) \end{aligned}$$

The "angle" between the accelerations, measured with respect to connections $\nabla$ and $\nabla^*$ on $\mathcal{S}$, is

$$\Omega(\dot\gamma) = g(\nabla_{\dot\gamma}\dot\gamma, \nabla^*_{\dot\gamma}\dot\gamma),$$

while a similar measure with respect to connections $D$ and $D^*$ on $\mathcal{M}$ is given by

$$\omega(\dot\gamma) = h(D_{\dot\gamma}\dot\gamma, D^*_{\dot\gamma}\dot\gamma).$$

It is easy to see that if the curve $\gamma$ is either $\nabla$- or $\nabla^*$-autoparallel, then $\Omega(\dot\gamma) = 0$. Similarly, if $\gamma$ is either $D$- or $D^*$-autoparallel, then $\omega(\dot\gamma) = 0$. The relation between these two measures is computed using the normal decompositions (13.8.26)–(13.8.27):

$$\begin{aligned} \Omega(\dot\gamma) &= g(\nabla_{\dot\gamma}\dot\gamma, \nabla^*_{\dot\gamma}\dot\gamma) = g\big(D_{\dot\gamma}\dot\gamma + L(\dot\gamma, \dot\gamma), D^*_{\dot\gamma}\dot\gamma + L^*(\dot\gamma, \dot\gamma)\big)\\ &= h\big(D_{\dot\gamma}\dot\gamma, D^*_{\dot\gamma}\dot\gamma\big) + g\big(L(\dot\gamma, \dot\gamma), L^*(\dot\gamma, \dot\gamma)\big)\\ &= \omega(\dot\gamma) + g\big(L(\dot\gamma, \dot\gamma), L^*(\dot\gamma, \dot\gamma)\big), \end{aligned}$$

and hence

$$\Omega(\dot\gamma) - \omega(\dot\gamma) = g\big(L(\dot\gamma, \dot\gamma), L^*(\dot\gamma, \dot\gamma)\big). \qquad (13.8.28)$$

**Proposition 13.8.1** *Let $(\mathcal{S}, g, \nabla, \nabla^*)$ be a dually flat manifold.*

(*i*) *Then the magnitude of the acceleration of a curve $\gamma$ contained in the submanifold $\mathcal{M}$ is given by*

$$\|\ddot\gamma\|_g^2 = \omega(\dot\gamma) + g\big(L(\dot\gamma, \dot\gamma), L^*(\dot\gamma, \dot\gamma)\big).$$

(*ii*) *If $\gamma$ is either $D$- or $D^*$-autoparallel, then*

$$\|\ddot\gamma\|_g^2 = g\big(L(\dot\gamma, \dot\gamma), L^*(\dot\gamma, \dot\gamma)\big).$$

(*iii*) *If $\gamma$ is either $D$- or $D^*$-autoparallel and has one of the normal curvature equal to zero (either $L(\dot\gamma, \dot\gamma) = 0$ or $L^*(\dot\gamma, \dot\gamma) = 0$), then*

$$\gamma^k(s) = \alpha^k s + \beta^k, \qquad 1 \le k \le \dim \mathcal{S},$$

*with $\alpha^k$ and $\beta^k$ constants.*

*Proof:*

(*i*) If $\mathcal{S}$ is dually flat, then $\Gamma_{ij}^k = \Gamma_{ij}^{*k} = 0$, and hence $\nabla_{\dot\gamma}\dot\gamma = \nabla^*_{\dot\gamma}\dot\gamma = \ddot\gamma^k\partial_k$. Therefore $\Omega(\dot\gamma) = g(\ddot\gamma^k\partial_k, \ddot\gamma^l\partial_l) = g(\ddot\gamma, \ddot\gamma) = \|\ddot\gamma\|_g^2$. Substituting in (13.8.28) leads to the required relation.

(*ii*) If $\gamma$ is either $D$- or $D^*$-autoparallel, then $\omega(\dot\gamma) = 0$. Then an application of the relation from part (*i*) yields the claimed relation.

(*iii*) Applying (*ii*) yields $\|\ddot\gamma\|_g = 0$. This implies $\ddot\gamma^k\partial_k = 0$ and hence $\ddot\gamma(s) = 0$, which implies the linearity of component functions.

■

## 13.9 Problems

**13.1.** Prove that $L^\alpha$ is symmetric, i.e., $L^\alpha(X, Y) = L^\alpha(Y, X)$, for all $X, Y \in \mathcal{X}(M)$.

**13.2.** Let $(\mathcal{M}, h, D, D^*)$ be a dual autoparallel submanifold of $(\mathcal{S}, g, \nabla, \nabla^*)$.

(*a*) Show that $L^{(\alpha)} \equiv 0$.

(*b*) Verify that $S^{(\alpha,\beta)} \equiv 0$ for all $\alpha, \beta$.

**13.3.** (*a*) Show that the shape operator satisfies the equation

$$S^{(\alpha,0)} = \frac{1}{2}\Big(S^{(\alpha,\beta)} + S^{(\alpha,-\beta)}\Big).$$

(*b*) Formulate and prove a similar relation for $S^{(0,\beta)}$.

**13.4.** Let $(\mathcal{S}, g, \nabla, \nabla^*)$ be a statistical manifold, where

$$\mathcal{S} = \{p(x; \mu, \sigma) = \frac{1}{\sqrt{2\pi}\sigma} e^{-\frac{(x-\mu)^2}{2\sigma^2}}; x \in \mathbb{R}, \mu \in \mathbb{R}, \sigma > 0\},$$

$g$ is the Fisher–Riemann metric, and $\nabla = \nabla^{(-1)}$, $\nabla^* = \nabla^{(1)}$. Consider the submanifold

$$\mathcal{M} = \{p(x; \mu, \sigma) \in \mathcal{S}; \mu = 0, \sigma > 0\}.$$

(*a*) Find the Fisher–Riemann metric $h$ on $\mathcal{M}$ induced from $(\mathcal{S}, g)$.

(*b*) Find an expression for the induced dual connections $D$, $D^*$ on $(\mathcal{M}, h)$.

(*c*) Find an expression for the connection $D^{(\alpha)}$.

(*d*) Compute the second fundamental forms $L$, $L^*$, and $L^{(\alpha)}$.

(*e*) Find the Riemann curvature tensors $R_{\mathcal{M}}$ and $R^*_{\mathcal{M}}$. Is $\mathcal{M}$ dually flat?

(*f*) Compute the mean curvature vectors $H_p$ and $H^*_p$. Is $(M, h)$ a minimal submanifold?

(*g*) Is there any value of $\alpha$ for which $(M, h)$ is $\alpha$-autoparallel?

**13.5.** Consider the statistical manifold $(\mathcal{S}, g, \nabla, \nabla^*)$, where

$$\mathcal{S} = \{p(x; \alpha, \beta) = \frac{1}{\beta^\alpha \Gamma(\alpha)} x^{\alpha-1} e^{-x/\beta}; x \in [0, \infty), \alpha, \beta > 0\},$$

$g$ is the Fisher–Riemann metric, and $\nabla = \nabla^{(-1)}$, $\nabla^* = \nabla^{(1)}$. Let

$$\mathcal{M} = \{p(x; \alpha, \beta) \in \mathcal{S}; \alpha = 1, \beta > 0\}.$$

(*a*) Find the Fisher–Riemann metric $h$ on $\mathcal{M}$ induced from $(\mathcal{S}, g)$.

(*b*) Is $(\mathcal{M}, h)$ a flat submanifold?

(*c*) Is there any value of $\alpha$ for which $(M, h)$ is $\alpha$-autoparallel?

**13.6.** Define the $\alpha$-mean curvature vector of a submanifold $(\mathcal{M}, h, D, D^*)$ of the statistical manifold $(\mathcal{S}, g, \nabla, \nabla^*)$ as

$$H_p^{(\alpha)} = \frac{1}{m} Trace L_p^{(\alpha)}, \quad \forall p \in \mathcal{M}.$$

(*a*) Express $H^{(\alpha)}$ in terms of $H$ and $H^*$.

(*b*) Find a relation between the equation $div^{(\alpha)} L^{(\alpha)} = 0$ and the fact that the vector field $H^{(\alpha)}$ is parallel with respect to $\nabla^{(\alpha)}$.

(c) Prove or disprove: $div^{(\alpha)} H^{(\alpha)} = div^{(-\alpha)} H^{(-\alpha)}$, $\forall \alpha$.

(d) Find a formula for $g\big(H_p^{(\alpha)}, H_p^{(\beta)}\big)$ in terms of the generalized shape operator $S^{(\alpha,\beta)}$.

**13.7.** Prove the following $\alpha$-version of Gauss' equation:

$$R_{\mathcal{S}}^{(\alpha)}(X,Y,Z,W) = R_{\mathcal{M}}^{(\alpha)}(X,Y,Z,W) + g\big(L^{(\alpha)}(X,Z), L^{(-\alpha)}(Y,W)\big) - g\big(L^{(\alpha)}(Y,Z), L^{(-\alpha)}(X,W)\big),$$

for any $X,Y,Z,W \in \mathcal{X}(\mathcal{M})$.

**13.8.** Let $\mathcal{S}$ be an exponential family and $\mathcal{M}$ a statistical submanifold of $\mathcal{S}$. Show that the following statements are equivalent:

(a) $\mathcal{M}$ is an exponential family.

(b) $\mathcal{M}$ is $\nabla^{(1)}$-autoparallel in $\mathcal{S}$.

**13.9.** Let $\mathcal{S}$ be a mixture family and $\mathcal{M}$ a statistical submanifold of $\mathcal{S}$. Show that the following statements are equivalent:

(a) $\mathcal{M}$ is an mixture family.

(b) $\mathcal{M}$ is $\nabla^{(-1)}$-autoparallel in $\mathcal{S}$.

# Appendix A Information Geometry Calculator

The book comes with a software companion, which is an Information Geometry calculator. The software is written in $C\#$ and runs on any PC computer endowed with .NET Framework. The use of this software does not involve having installed any other softwares such as Maple, Mathematica, Matlab, or Excel.

The following instructions apply to the software version 1.1. To open the program double click on the **IGS** icon, and then enter the password **Springer 2014**. The software computes the following information geometry measures for the most used probability distributions:

- Entropy
- Informational Energy
- Cross Entropy
- Contrast Functions.

O. Calin and C. Udrişte, *Geometric Modeling in Probability and Statistics*,
DOI 10.1007/978-3-319-07779-6,

## A.1 Entropy

Select from the menu the **Entropy** tab and choose one of the following probability distributions:

- Normal Distribution
- Exponential Distribution
- Lognormal Distribution
- Gamma Distribution
- Beta Distribution
- Poisson Distribution

The pop-up window can graph the probability distribution and compute the entropy. There are two ways of computing the entropy:

1. Fill in manually the editable fields, which correspond to the parameters of the distribution. For instance, in the case of the normal distribution, there are two parameters: the mean $\mu$ and the standard deviation $\sigma$. Enter the desired values in the editable boxes. Then click the **Entropy** button and obtain the entropy in the nearby box. If the chosen values do not belong to the MIN/MAX range, adjust the range before clicking the **Entropy** button. Otherwise, the same effect is obtained if the **Entropy** button is clicked twice.

2. Using sliders is a more dynamical way of watching how the entropy changes while the graph of the probability density deforms. Each parameter value can be changed by a slider. The minima and maxima values of the slider are editable. The vertical slider modifies the scale. If the graph is out of range, then a smaller scale will make the graph to get in the desired range. Similarly, if the graph is too small, then the scale slider can be used to enlarge the graph.

The variant 1 is useful when the parameters values are either too large or too small. In this case the graph might not shown well on the canvas.

The **Help** button provides a window with explanations about the window's functions. The **Close Window** button closes the window, while the **Print** button prints the graph of the distribution, parameter values and entropy.

## A.2 Informational Energy

The use of this tab is similar with the **Entropy** tab. It can be used to compute the informational energy for the following distributions: normal, exponential, lognormal, gamma, beta, and Poisson.

The use of sliders as well as **Help**, **Close Window**, and **Print** buttons are similar as in the case of the **Entropy** tab.

## A.3 Cross Entropy

This computes the cross entropy between the following pairs of distributions: normal, exponential, Poisson, and beta. The sliders corresponding to the first distribution are pink while the sliders for the second are blue. The corresponding graphs of the distributions are also pink and blue.

A new feature here is the level bar. In order to display it, click on the **Show Level** button. This bar provides a visual measure of the cross entropy between the distributions. By modifying the sliders one can see how the level of the cross entropy increases or decreases. There is an editable box for the maximum of the level bar, while the minimum is left to 0 by default. Whenever modifying the maximum level field, clicking the **Cross Entropy** button adjusts the level bar to the new level.

All the other functional features (**Help**, **Close Window**, and **Print** buttons) are similar with the ones of the previous tabs (**Entropy**, **Informational Energy**).

## A.4 Contrast Functions

The software computes the following three types of contrast functions:

- Kullback–Leibler divergence
- Hellinger Distance
- Chernoff Information of order $\alpha$.

The functional behavior for each of these features is similar with the **Cross Entropy**'s functions. The only difference is that in the case of the Chernoff Information there is one more editable field, the order $\alpha$, which takes values between $-1$ and 1.

## A.5 Examples

This section presents a few examples which show how the main features of the software work.

1. *Find the entropy of a Poisson distribution with parameter* $\xi = 14.432$.
   From the **Entropy** tab choose **Poisson Distribution**. In the pop-out window type the value 14.432 in the text-box next to "$\xi =$" and then click the **Entropy** button. The value of the entropy is displayed in the box next to the entropy button: 2.748.

2. *Find the value of the parameter* $\xi$ *of a Poisson distribution which has the entropy equal to* 3.5.
   Open the same pop-up window as in 1. Then move the slider until the entropy value is 3.5. Since the MAX value is 60 by default, and the corresponding entropy for this value is just 3.465, we need to adjust the MAX value. Type the value 70 in the text-box under MAX. Then move the slider until the entropy value becomes 3.500. The corresponding value of $\xi$ is 64.343.

3. *An exponential distribution* $p$ *has* $\xi = 0.03$. *Find all exponential distributions* $q$ *with the cross entropy* $S(p, q) < 5$.
   From the **Cross Entropy** tab choose $p$, $q$ **Exponential Distributions**. In the pop-out window type the value 0.03 in the text-box next to "$\xi^1 =$" and then click the **Cross Entropy** button. Click the **Show level** button to see the cross entropy level. Move the second slider left and right to realize that the cross entropy increases when the slider values decreases. When the value of parameter $\xi^0$ is 0.019 the cross entropy value is 5. Therefore, the distributions $q$ with $S(p, q) < 5$ are the exponential distributions with parameter $\xi^0 > 0.019$.

4. *Consider a Poisson distribution with parameter* $\xi^1 = 30$. *Find all Poisson distributions* $q$ *with the informational energy* $I(q) < 0.045$ *such that* $D_{KL}(p||q) < 1.5$
   From the **Informational Energy** tab choose **Poisson Distribution**. In the pop-out window click the **Show level** button and move the slider to realize that $I(q) < 0.045$ for $\lambda > 38.7$.

   From the **Contrast Functions** tab choose **Kullback–Leibler Divergence > p,q Poisson Distributions**. In the pop-out window enter the value 30 in the box next to "$\lambda^1 =$". Then click the **Show level** button and move the blue slider to realize that $D_{KL}(p||q) < 1.5$ for $21.027 < \lambda^0 < 39.975$. The sharp bounds are obtained in the following way. Consider, for instance, the lower bound. At a first approximation we have something like $21.99 < \lambda^0$. Then "zoom in" by typing 20 in the MIN box and 22 in the MAX box of the blue slider. Moving the slider we have $D_{KL}(p||q) < 1.500$ for $21.027 < \lambda^0$.

Intersecting the intervals we obtain the solution interval

$$(38.7, +\infty) \cap (21.027, 39.975) = (38.7, 39.975).$$

Hence, for all Poisson distributions $q$ with $38.7 < \lambda < 39.975$ we have $I(q) < 0.045$ and $D_{KL}(p||q) < 1.5$.

# Bibliography

[1] R. Abraham, J.E. Marsden, *Foundations of Mechanics* (Addison-Wesley, Reading, 1994)

[2] H. Akaike, in *On Entropy Maximization Principle*, ed. by P.R. Krishnaiah, Applications of statistics (North-Holland, Amsterdam, 1977), pp. 27-41

[3] E. Akin, *The Geometry of Population Genetics*. Lecture Notes in Biomathematics, vol. 31 (Springer, New York, 1979)

[4] C. Akinson, A.F. Mit, *Rao's Distance Measure*. Sankhya Indian J. Stat. A **43**, 345–365 (1981)

[5] S. Amari, *Differential-Geometric Methods in Statistics (Lecture Notes in Statistics, 28)* (Springer, New York, 1885). Reprinted in 1990

[6] S. Amari, *Theory of information spaces - a geometrical foundation of statistics*. POST RAAG Report 106, 1980

[7] S. Amari, Differential geometry of curved exponential families - curvature and information loss. Ann. Stat. **10**, 375–385 (1982)

[8] S. Amari, H. Nagaoka, *Methods of Information Geometry*. AMS monographs, vol. 191 (Oxford University Press, Oxford, 2000)

[9] L. Auslander, R.E. MacKenzie, *Introduction to Differentiable Manifolds* (Dover Publications, New York, 1977)

[10] N. Ay, W. Tuschmann, Dually flat manifolds and global information geometry. Open Syst. Inform. Dynam. **9**(2), 195–200 (2002)

[11] O.E. Barndorff-Nielsen, Differential geometry and statistics: Some mathematical aspects. Indian J. Math. **29**, 335–350 (1987)

O. Calin and C. Udrişte, *Geometric Modeling in Probability and Statistics*,
DOI 10.1007/978-3-319-07779-6,

[12] O.E. Barndorff-Nielsen, On some differential geometric concepts and their relations to statistics, in *Geometrization of Statistical Theory*, ed. by C.T.J Dodson. Proceedings of the GST Workshop. Department of Mathematics, University of Lancaster (ULDM Publications, UK, 1987), pp. 53–90

[13] O.E. Barndorff-Nielsen, *Parametric Statistical Models and Likelihood.* Lecture Notes in Statistics, vol. 50 (Springer, New York, 1988)

[14] H.H. Baüschke, Duality for Bregman projections onto translated cones and affine subspaces. J. Approx. Theory **121**, 1–12 (2003)

[15] H.H. Bauschke, J.M. Borwein, P.L. Combettes, Bregman monotone optimization algorithms. SIAM J. Control Optim. **42**, 596–636 (2003)

[16] H.H. Bauschke, P.L. Combettes, Iterating Bregman retractions. SIAM J. Optim. **13**, 1159–1173 (2003)

[17] G. Bercu, Newton method on differentiable manifolds. Ph.D. Thesis, University Polytehnica of Bucharest, 2006

[18] P. Blæsid, Yokes: Elemental properties with statistical applications, in *Geometrization of Statistical Theory*, ed. by C.T.J Dodson, Proceedings of the GST Workshop. Department of Mathematics, University of Lancaster (ULDM Publications, UK, 1987), pp. 193–198

[19] P. Blæsid, Yokes and tensors derived from yokes. Ann. Inst. Stat. Math. **43**, 95–113 (1991)

[20] L.M. Bregman, The relaxation method of finding the common point of convex sets and its application to the solution of problems in convex programming. USSR Comput. Math. Phys. **7**, 200–217 (1967)

[21] K.P. Burnham, D.R. Anderson, *Model Selection and Multidimensional Inference: A practical Information-Theoretic Approach*, 2nd edn. (Springer, New York, Berlin, Heidelberg, 2002)

[22] O. Calin, D.C. Chang, *Geometric Mechanics on Riemannian Manifolds, Applications to Partial Differential Equations* (Applied and Numerical Analysis, Birkhäuser, 2004)

[23] O. Calin, D.C. Chang, P. Greiner, *Sub-Riemannian Geometry on Heisenberg Group and Its Generalization* (AMS/IP, Boston, 2007)

[24] O. Calin, D.C. Chang, K. Furutani, C. Iwasaki, *Heat Kernels for Elliptic and Sub-elliptic Operators, Methods and Techniques* (Applied and Numerical Harmonic Analysis, Birkhäuser, 2010)

[25] O. Calin, H. Matsuzoe, J. Zhang, *Generalizations of Conjugate Connections.* Trends in Differential Geometry, Complex Analysis and Mathematical Physics (World Scientific, Singapore, 2009), pp. 24–34

[26] N.N. Chentsov, Statistical Decision Rules and Optimal Inference (AMS, Rhode Island, 1982) (originally published in Russian, Nauka, Moscow, 1972)

[27] H. Chernoff, A measure of asymptotic efficiency for tests of a hypothesis based on a sum of observations. Ann. Math. Stat. **23**, 493–507 (1952)

[28] B.S. Clarke, A.R. Barron, Information theoretic asumptotics of Bayes methods. IEEE Trans. Inform. Theory **36**, 453–471 (1990)

[29] C. Corcodel, Hessian structures, special functions and evolution metrics. Ph.D. Thesis, University Polytehnica of Bucharest, 2010

[30] J.M. Corcuera, F. Giummolé, *A Characterization of Monotone and Regular Divergences*, Annals of the Institute of Statistical Mathematics **50**(3), 433–450 (1998)

[31] I. Csiszár, Information type measures of difference of probability distributions and indirect observations. Stud. Sci. Math. Hung. **2**, 299–318 (1967)

[32] I. Csiszár, On topological properties of $f$-divergence. Stud. Sci. Math. Hung. **2**, 329–339 (1967)

[33] A.P. Dawid, Discussion to Efron's paper. Ann. Stat. **3**, 1231–1234 (1975)

[34] M.P. do Carmo, *Differential Geometry of Curves and Surfaces* (Prentice-Hall, Englewoods Cliffs, 1976)

[35] M.P. do Carmo, *Riemannian Geometry* (Birkhäuser, Basel, 1992)

[36] M.P. do Carmo, *Differential Forms and Applications* (Universitext, Springer, Berlin, 1994)

[37] B. Efron, Defining the curvature of a statistical problem (with application to second order efficiency). Ann. Stat. **3**, 1189–1242 (1975)

[38] S. Eguchi, Second order efficiency of minimum contrast estimators in a curved exponential family. Ann. Stat. **11**, 793–803 (1983)

[39] S. Eguchi, A characterization of second order efficiency in a curved exponential family. Ann. Inst. Stat. Math. **36**, 199–206 (1984)

[40] S. Eguchi, A differential geometric approach to statistical inference on the bias of contrast functionals. Hiroshima Math. J. **15**, 341–391 (1985)

[41] S. Eguchi, Geometry of minimum contrast. Hiroshima Math. J. **22**, 631–647 (1992)

[42] A. Erdélyi et al., *Higher Transcendental Functions*, vol. I, IV. Bateman Manuscript project (McGraw-Hill, New York, 1955)

[43] S. Hawking, *A Brief History of Time* (Bantam Books, New York, 1988)

[44] S. Helgason, *Differential Geometry, Lie Groups, and Symmetric Spaces* (Academic Press, New York, 1978)

[45] A.T. James, in *The Variance Information Manifold and the Function on It.* ed. by P.K. Krishnaiah, *Multivariate Analysis* (Academic Press, New York, 1973), pp. 157–169

[46] H. Jeffreys, An invariant form for the prior probability in estimation problems. Proc. Roy. Soc. Lond. Ser. A. **196**, 453–461 (1946)

[47] H. Jeffreys, *Theory of Probability Theory*, 2nd edn. (Oxford University Press, Oxford, 1948)

[48] A.M. Kagan, On the theory of Fisher's amount of information. Dokl. Akad. Nauk SSSR **151**, 277–278 (1963)

[49] R.E. Kass, P.W. Vos, *Geometrical Foundations of Asymptotic Inference, Wiley Series in Probability and Statistics* (Wiley, New York, 1997)

[50] S. Kobayashi, K. Nomizu, *Foundations of Differential Geometry*, vol. I-II (Wiley Interscience, New York, 1996)

[51] S. Kullback, R.A. Leibler, On information and sufficiency. Ann. Math. Stat. **22**, 79 (1951)

[52] S. Kullback, R.A. Leibler, *Information Theory and Statistics* (Wiley, New York, 1959)

[53] S. Kullback, R.A. Leibler, *Letter to the editor: The KullbackLeibler distance*. Am. Stat. **41**(4), 340341 (1987). JSTOR 2684769

[54] S. Lauritzen, in *Statistical Manifolds*, ed. by S. Amari, O. Barndorff-Nielsen, R. Kass, S. Lauritzen, C.R. Rao, *Differential Geometry in Statistical Inference*. IMS Lecture Notes, vol. 10 (Institute of Mathematical Statistics, Hayward, 1987), pp. 163–216

[55] H. Matsuzoe, S. Amari, J. Takeuchi, Equiaffine structures on statistical manifolds and Bayesian statistics. Differ. Geom. Appl. **24**(6), 567–578 (2006)

[56] T. Matumoto, Any statistical manifold has a contrast function - On the $C^3$- functions taking the minimum at the diagonal of the product manifold. Hiroshima Math. J. **23**, 327–332 (1993)

[57] L.R. Mead, N. Papanicolaou, Maximum entropy in the problem of moments. J. Math. Phys. **25**(8), 2404–2417 (1984)

[58] R.S. Millman, G.D. Parker, *Elements of Differential Geometry* (Prentice-Hall, Englewoods Cliffs, 1977)

[59] C.R. Min, W.H. Ri, H. Chol, Equiaffine structure and conjugate Ricci-symmetry of a statistical manifold. arXiv:1302.3167v1[math.DS]13 Feb.2013

[60] M.K. Murrey, J.W. Rice, *Differential Geometry and Statistics* Chapman & Hall, Monographs on Statistics and Applied Probability, **48** (Chapman, London, 1993)

[61] H. Nagaoka, S. Amari, Differential geometry of smooth families of probability distributions. Technical Report (METR) 82–7, Dept. of Math. Eng. and Instr. Phys., Univ. of Tokyo, 1982

[62] K. Nomizu, T. Sasaki, *Affine Differential Geometry - Geometry of Affine Immersions* (Cambridge University Press, Cambridge, 1994)

[63] J.M. Oller, Information metric for extreme value and logistic probability distributions. Sankhya Indian J. Stat. Ser. A **49**, 17–23 (1987)

[64] J.M. Oller, J.M. Corcuera, Intrinsic analysis of statistical estimation. Ann. Stat. **23**, 1562–1581 (1995)

[65] J.M. Oller, C.M. Cuadras, Rao's distance for negative multinomial distributions. Sankhya Indian J. Stat. Ser. A **47**, 75–83 (1985)

[66] B. O'Neill, *Elementary Differential Geometry*, 2nd edn. (Academic Press, New York, 1997)

[67] O. Onicescu, Energie informationala. St. Cerc. Math. **18**, 1419–1430 (1966)

[68] O. Onicescu, Théorie de L'information. Energie Informalionnelle. C. R. Acad. Sci. Paris Ser. A **26**(263), 841–842 (1966)

[69] J. Pfanzagl, Asymptotic expansions related to minimum contrast estimators. Ann. Stat. **1**, 993–1026 (1973)

[70] C.R. Rao, Information and accuracy attainable in the estimation of statistical parameters. Bull. Calcutta Math. Soc. **37**, 81–91 (1945)

[71] C.R. Rao, *Linear Statistical Inference and Its Applications*. Wiley Series in Probability and Mathematical Statistics (Wiley, New York, London, Sydney, 1965)

[72] C.R. Rao, in *Differential Metrics in Probability Spaces*, ed. by S. Amari, O. Barndorff-Nielsen, R. Kass, S. Lauritzen, C.R. Rao, Differential Geometry in Statistical Inference, IMS Lecture Notes, vol. 10 (Hayward, CA, 1987), pp. 217–240

[73] C. Shannon, A mathematical theory of communication. Bell. Syst. Tech. J. **27**, 379–423, 623–656 (1948)

[74] H. Shima, *Hessian Geometry* (in Japanese) (Shokabo, Tokyo, 2001)

[75] H. Shima, K. Yagi, Geometry of Hessian manifolds. Differ. Geom. Appl. **7**, 277–290 (1997)

[76] U. Simon, in *Affine Differential Geometry*, ed. by F. Dillen, L. Verstraelen *Handbook of Differential Geometry*, vol. I (Elsevier Science, Amsterdam, 2000), pp. 905–961

[77] M. Spivak, *Calculus on Manifolds* (Addison-Wesley, Reading, 1965)

[78] M. Spivak, *A Comprehensive Introduction to Differential Geometry*, vol. I–V. (Publish or Perish Inc., Boston, 2005)

[79] J. Takeuchi, S. Amari, $\alpha$-parallel prior and its properties. Number 51. IEEE Trans. Inform. Theory **51**(3), 1011–1023 (2005)

[80] Gr. Tsagas, C. Udriste, *Vector Fields and Their Applications.* Monographs and Textbooks, vol. 5 (Geometry Balkan Press, Bucharest, 2002)

[81] C. Udriste, *Convex Functions and Optimization Methods on Riemannian Manifolds.* Mathematics and Its Applications, vol. 297 (Kluwer Academic, Dordrecht, Boston, London, 1994)

[82] C. Udriste, *Geometric Dynamics.* Mathematics and Its Applications, vol. 513 (Kluwer Academic, Dordrecht, Boston, London, 2000)

[83] C. Udriste, M. Ferrara, D. Opris, *Economic Geometric Dynamics.* Monographs and Textbooks, vol. 6 (Geometry Balkan Press, Bucharest, 2004)

[84] K. Uohashi, On $\alpha$-conformal equivalence of statistical manifolds. J. Geom. **75**, 179–184 (2002)

[85] D.D. Wackerly, W.M. Mendenhall, R.L. Scheaffer, *Mathematical Statistics with Applications*, 7th edn. (Brooks/Cole Cengage Learning, Belmont, CA, 2008)

[86] J. Zhang, Divergence function, duality, and convex analysis. Neural Comput. **16**, 159–195 (2004)

[87] J. Zhang, *A Note on Curvature of* $\alpha$*-Connections of a Statistical Manifold*, vol. 59, Annals of the Institute of Statistical Mathematics, Springer Netherlands, 161–170 (2007)

# Index



O. Calin and C. Udrişte, *Geometric Modeling in Probability and Statistics*,
DOI 10.1007/978-3-319-07779-6,